Undergraduate Texts in Mathematics

Undergraduate Texts in Mathematics

Readings in Mathematics

Undergraduate Texts in Mathematics are generally aimed at third- and fourth-year undergraduate mathematics students at North American universities. These texts strive to provide students and teachers with new perspectives and novel approaches. The books include motivation that guides the reader to an appreciation of interrelations among different aspects of the subject. They feature examples that illustrate key concepts as well as exercises that strengthen understanding.

For further volumes:
http://www.springer.com/series/666
and
http://www.springer.com/series/4672

Alexander Ostermann · Gerhard Wanner

Geometry by Its History

 Springer

Alexander Ostermann
Department of Mathematics
University of Innsbruck
Innsbruck
Austria

Gerhard Wanner
Section de Mathématiques
Université de Genève
Genève
Switzerland

ISSN 0172-6056
ISBN 978-3-642-44469-2
DOI 10.1007/978-3-642-29163-0
Springer Heidelberg New York Dordrecht London

ISBN 978-3-642-29163-0 (eBook)

Mathematics Subject Classification (2010): 01-01, 51-01, 51-03

Springer is part of Springer Science+Business Media (www.springer.com)

In Memoriam

ALFRED FRÖLICHER

(1927–2010)

Preface

ἀγεωμέτρητος μηδεὶς εἰσίτω

(literally: "non-geometers don't enter." Inscription at the entrance
to Plato's Academy 387 B.C. (perhaps only a legend), and on the
frontispiece of Copernicus' *De revolutionibus* 1543 A.D.)

Geometry, so named at least since Plato's times, is the oldest branch of mathematics. It contains many beautiful results, elegant ideas and surprising connections, to which many great thinkers have contributed through the centuries. Among these are Thales, Pythagoras, Euclid, Apollonius, Archimedes, Ptolemy, Pappus, the Arabs, Regiomontanus, Copernicus, Viète, Kepler, Descartes, Newton, the Bernoullis, Euler, Monge, Poncelet and Steiner. In this book, we study geometry from the texts of these masters as closely as we judge useful and roughly in historic order, in accordance with the famous maxim of Niels Henrik Abel, *"I learned from the masters and not from the pupils"*. This explains "by Its History" in the title, without aiming at a complete history of the subject.[1]

Geometry arose in the dawn of science and later became one of the *septem artes liberales* as part of the *quadrivium*. We start from the beginning of geometry, motivated by practical problems of measurement (μετρέω), then follow its development into a rigorous abstract science by the Greek philosophers, until the rich period with more and more sophisticated problems and methods in the later Greek and Arab period. The second part of the book then describes the victory of the methods of algebra and linear algebra, the growing audacity in dealing with infinite processes and we see how all other branches of science grew out of geometry: algebra, calculus, mechanics (in particular celestial mechanics). So we hope that the book not only constitutes a good introduction to the study of higher geometry, but also to the study of other branches of science, especially for those students who intend to become teachers.

However, due to the rapid success of algebraic methods, synthetic geometry slowly lost its place in university education, a development already deplored by Newton, more than three centuries ago (first sentence of his *Treatise of the Methods of Series and Fluxions*, 1671, p. 33) ...

"Observing that the majority of geometers, with an almost complete neglect of the ancients' synthetical method, now for the most part apply

---

[1]Michel Chasles' *Aperçu historique* has 800 pages and weighs five pounds.

themselves to the cultivation of analysis [...] I found it not amiss, for the satisfaction of learners, to draw up the following short tract ..."

... a development which accelerated during the twentieth century (Coxeter, *Introduction to Geometry*, 1961, p. iv):

"For the last thirty or forty years, most Americans have somehow lost interest in geometry. The present book constitutes an attempt to revitalize this sadly neglected subject."

What is not honoured at the university also disappears, a generation later, from the high-school. We quote from A. Connes (*Newsletter of the EMS*, March 2008, p. 32):

"We must absolutely train very young people to do mathematical exercises, in particular geometry exercises — this is very good training. I find it awful when I see that, in school, kids are taught recipes, just recipes, and aren't encouraged to think. When I was at school, I remember that we were given problems of [...] geometry. We went to a lot of trouble to solve them. It wasn't baby geometry. [...] It's a shame we don't do it anymore."

We have made all efforts to produce an interesting and enjoyable book, intended mainly for students of science (at the beginning) and teachers (throughout their career), by including many illustrations, figures, exercises and references to the literature, so that we suggest with Copernicus' *De revolutionibus*

> *Igitur eme, lege, fruere* [Therefore buy, read and enjoy].

Acknowledgements. It is our duty — and pleasure — to thank many people for their help, beginning with Ernst and Martin Hairer, the assistants P. Henry, W. Pietsch, A. Musitelli, M. Baillif, C. Extermann and the librarians A.-S. Crippa, B. Dudez and T. Dubois at the University of Geneva. A long discussion with J.-C. Pont, expert in the history of mathematics and science, helped to reshape the first two chapters. Our colleagues and friends J. Cash, B. Gisin, H. Herdlinger and C. Lubich read the entire book and gave numerous suggestions. Further we thank our wives Barbara and Myriam.

Our special thanks, however, go to J. Steinig, who read the entire book *four times (!!)* and suggested thousands of corrections and grammatical improvements. He pointed out errors and sloppy arguments, and supplied us with many references and better proofs. And if ever "eme, lege, fruere" is really justified, it is also his merit.

Innsbruck and Geneva,
December 2011

Alexander Ostermann
Gerhard Wanner

Contents

Part I

Classical Geometry

γῆ,   earth (including land and sea) ...

μετρέω,   measure out ...

(Liddel and Scott, *Greek-English Lexicon*, Oxford)

"In the tomb of Khaemhet at Thebes we see a number of men equipped with ropes and writing material measuring a field, ..."

(T.E. Peet, *Rhind mathematical papyrus*, 1923, p. 32)

"The Mathematick Lecturer to read first some easy & usefull practical things, then Euclid, Sphericks, the Projections of the Sphere, the Construction of Mapps, Trigonometry, Astronomy, Opticks, Musick, Algebra, &c."   (I. Newton, *Of Educating Youth in the Universities*, MS. Add. 4005, fol. 14–15, Cambridge 1690)

"Development of Western science is based on two great achievements: the invention of the formal logical system (in Euclidean geometry) by the Greek philosophers, and the discovery of the possibility to find out causal relationships by systematic experiment (during the Renaissance)."

(A. Einstein in a letter to J.S. Switzer, 23 Apr. 1953)

"Quoique la Géométrie soit par elle-même abstraite, il faut avoüer cependant que les difficultés qu'éprouvent ceux qui commencent à s'y appliquer, viennent le plus souvent de la maniére dont elle est enseignée dans les Elémens ordinaires. On y débute toûjours par un grand nombre de définitions, de demandes, d'axiomes, & de principes préliminaires, qui semblent ne promettre rien que de sec au lecteur".   (A.-C. Clairaut, *Elémens de Géométrie*, 1741)

We see in the chronology below that Euclid, who lived around 300 B.C., was not the first great geometer, despite the fact that his famous *Elements* "with all its definitions, postulates, axioms & preliminary principles, which seem to

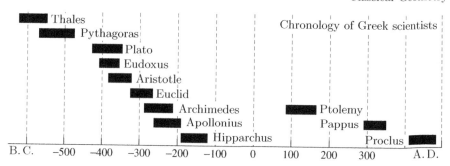

promise nothing but arid reading" (see the above quotation from Clairaut) usually serve as a model for the beginning of a course on geometry. But mathematical results had already been obtained in the preceding centuries, in order to measure (μετρέω) land (γῆ), to survey fields after the regular floods of the Nile, to compute the quantity of corn in a cylindrical container, and to construct spectacular temples and pyramids. We therefore start in Chap. 1 with "some easy & usefull practical things", the theorems of Thales and Pythagoras, which are the oldest theorems of humanity and fundamental tools for geometry. They allow one to deal with most practical applications.

A first flaw in this paradise was revealed by the discovery of irrational numbers, which showed that the concept and the proof of Thales' theorem were not as simple as had been thought. In parallel with this were the efforts, influenced by the Greek philosophers, from the Pythagoreans to Plato, to separate geometry from its practical applications, to raise it to an abstract science studying unchangeable objects and to lift the soul towards eternal truth. The nails, ropes and walls used by the temple builders were replaced by mathematical points, lines, rectangles etc., objects of pure reasoning, which require a list of definitions, axioms and postulates (see Chap. 2). This is the origin of the style of nearly all mathematical thought and exposition since then.

In Chaps. 3 and 4 we describe the achievements of the post-Euclidean period, the new curves and theorems invented by Apollonius, Nicomedes, Archimedes and Pappus, often in order to solve one of the three great problems of Greek geometry: squaring the circle, trisecting any angle or duplicating the cube. Chap. 4 also contains many more recent beautiful results, which the Greeks *could* have found with their methods.

Chap. 5 is devoted to the last great creation of the Greek period, plane and spherical trigonometry by Hipparchus and Ptolemy and their application to one of the dreams of mankind, understanding the movements of the heavenly bodies. This gave rise to modern astronomy and the physical sciences.

1

Thales and Pythagoras

"... la théorie des lignes proportionnelles et la proposition de Pythagore, qui sont les bases de la Géométrie ... [the theory of proportional lines and the theorem of Pythagoras, which form the basis of geometry]"   (J.-V. Poncelet, 1822, p. xxix)

"... the original works of the forerunners of Euclid, Archimedes and Apollonius are lost, having probably been discarded and forgotten almost immediately after the appearance of the masterpieces of that great trio."   (T.L. Heath, 1926, vol. I, p. 29)

The most beautiful discoveries of this period concern relations between *lengths* (Thales' intercept theorem), *angles* (the central angle theorem or Eucl. III.20) and *areas* (the Pythagorean theorem). A quick look at the index shows that these three theorems are by far the most basic and frequently used results of geometry.

The only original documents which have survived from the pre-Euclidean period are some cuneiform Babylonian tablets (from approximately 1900 B.C.), the Egyptian *Rhind papyrus* and the *Moscow papyrus* from approximately the same period. The achievements of Thales, Pythagoras and his pupils the Pythagoreans are only documented in commentaries, often contradictory, written many centuries later.

1.1 Thales' Theorem

"I tried (unsuccessfully) to get each high school in which my children were enrolled to go outside during geometry and find out how tall the oak in the yard really is."
(D. Mumford, President IMU; Preface in H. M. Enzensberger, *Zugbrücke außer Betrieb [Drawbridge Up]*, 1999)

Thales was born in Miletus (Asia Minor, nowadays Turkey). He travelled to Babylon and to Egypt, calculated the height of the pyramids by measuring

A. Ostermann and G. Wanner, *Geometry by Its History*,
Undergraduate Texts in Mathematics, DOI: 10.1007/978-3-642-29163-0_1,
© Springer-Verlag Berlin Heidelberg 2012

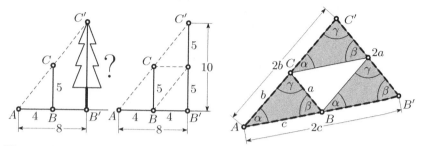

Fig. 1.1. The "oak" in Mumford's school yard and Thales' theorem for ratio 2

the length of their shadow, calculated the distance of ships from the shore, and predicted a solar eclipse in 585 B.C.

Thales is certainly the man to tell us how to measure the height of a tree $B'C'$, without having to climb it (see Fig. 1.1, left). Let AB' be the shadow of the tree; we erect a vertical stick BC in such a manner that AB is the shadow of the stick.[1] We then measure the distance AB, say 4 metres, the distance AB', say 8 metres, and the stick BC, say 5 metres. By parallel displacements of the triangle ABC we see that, since AB' measures twice AB, the height $B'C'$ will measure twice BC (see the middle picture), hence $B'C' = 2 \cdot 5 = 10$

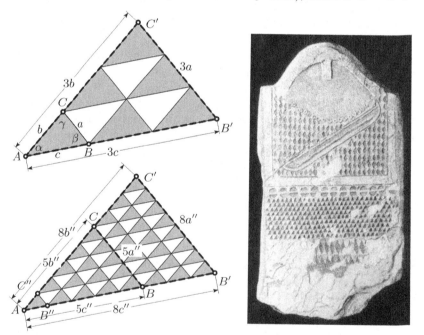

Fig. 1.2. The proof of Thales' theorem; right: Neolithic stele, Sion 2500 B.C. (courtesy Prof. A. Gallay)

---

[1]As recorded by Plutarch; see Heath (1921, p. 129)

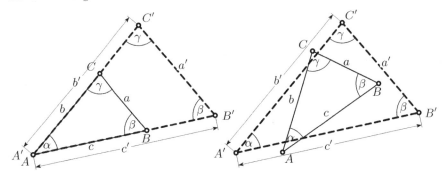

Fig. 1.3. Thales' intercept theorem

metres. The same argument can be applied to translations of any triangle ABC (see Fig. 1.1, right). We see that if a side of a triangle is doubled and the angles are preserved, then the other two sides are also doubled.

If our tree were still taller, we might have to displace our triangle *three* times and would arrive at the situation of Fig. 1.2 (upper left) where the sides of $AB'C'$ are three times as long as the sides of ABC.

By using still finer subdivisions, we arrive at the lower left picture of Fig. 1.2 where the ratios of these lengths are 8:5. We have thus discovered that the following theorem is valid for any rational fraction. We call this proof, which could have been inspired by the Neolithic stele from 2500 B.C., and which will be severely criticised later, the *Stone Age proof*.

Theorem 1.1 (Thales' intercept theorem). *Consider an arbitrary triangle ABC (see Fig. 1.3, left) and let AC be extended to C' and AB to B', so that $B'C'$ is parallel to BC. Then the lengths of the sides satisfy the relations*

$$\frac{a'}{a} = \frac{b'}{b} = \frac{c'}{c} \qquad \text{and hence} \qquad \frac{a'}{c'} = \frac{a}{c}, \qquad \frac{c'}{b'} = \frac{c}{b}, \qquad \frac{b'}{a'} = \frac{b}{a}.$$

These proportions are also preserved when the triangle is displaced and rotated, see Fig. 1.3 (right). As a consequence we get the following result. *If corresponding angles of two triangles are equal, then corresponding sides are proportional.* Triangles having these properties are called *similar*.

1.2 Similar Figures

A more general view of Thales' theorem appeared in the works of Clavius, Viète and others: figures are said to be *similar with similarity centre O* when corresponding points A_i, B_i, C_i lie on lines through O, and the corresponding lines A_iA_j, B_iB_j, C_iC_j are parallel (see Fig. 1.4). Applying Thales' theorem to selected pairs of triangles with a vertex in O shows that all corresponding lengths of similar figures are proportional. Such similar figures were an important source of inspiration for many of the great masters (see Fig. 1.5).

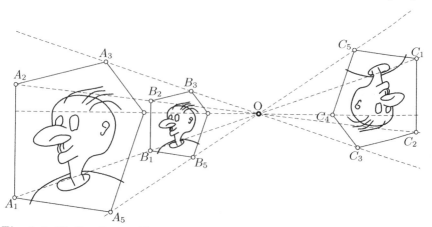

Fig. 1.4. Similar figures: illustration inspired by Clavius and Viète, improved by modern computer technology

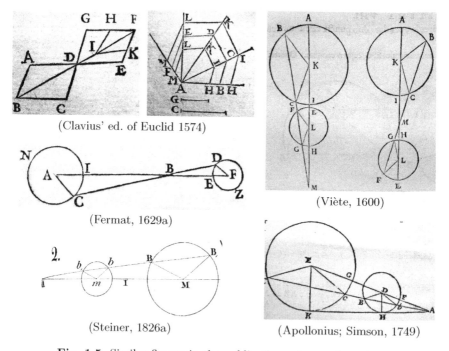

(Clavius' ed. of Euclid 1574)

(Viète, 1600)

(Fermat, 1629a)

(Steiner, 1826a)

(Apollonius; Simson, 1749)

Fig. 1.5. Similar figures in the publications of several masters

Constructing rational lengths. Consider two distinct points 0 and 1 on a line. We call the length of the segment joining these two points the *unit length*. By carrying this unit forward on the line, we easily construct the integer points 2, 3, etc. But how can we construct points corresponding to rational values? For this we draw an arbitrary ray, not parallel to the line, through the point 0. We then carry forward several times (five times, say)

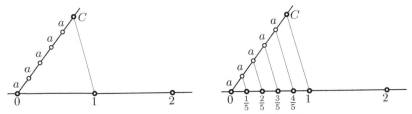

Fig. 1.6. Constructing rational lengths

an arbitrary length a. This construction yields a point C, see Fig. 1.6 (left). If we now draw the corresponding parallels to the line joining C with 1 (see Fig. 1.6, right), we obtain by Thales' theorem the required points $\frac{1}{5}$, $\frac{2}{5}$, etc. (this procedure will later be called Eucl. VI.9).

1.3 Properties of Angles

Emil Artin (1898–1962) was famous for the extremely clear and extraordinarily well presented lectures that he always gave without any notes. One day, midway in a proof, he suddenly hesitated and said: "this conclusion is trivial". After a few seconds, he repeated: "it is trivial, but I no longer know why". He then thought about the question for another minute and said: "I *know* that it is trivial, but I no longer understand it". He reflected on it a few moments more and finally said: "excuse me, I have to look at my lecture notes". He then left the room and came back ten minutes later saying: "it *really* is trivial".

(Witnessed by Prof. Josef Schmid, Fribourg)

"I still remember a guy sitting on the couch, thinking very hard, and another guy standing in front of him, saying, 'And therefore such-and-such is true.' 'Why is that?' the guy on the couch asks. 'It's trivial! It's trivial!' the standing guy says ..."

(R.P. Feynman,

souvenir from the math-physics common lounge at Princeton; quoted from *Surely You're Joking, Mr. Feynman*, 1985, p. 69)

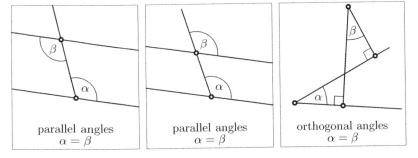

Fig. 1.7. Parallel and orthogonal angles

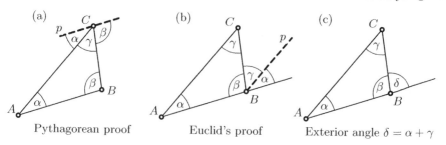

Fig. 1.8. Angles in a triangle (Eucl. I.32)

Angles in a triangle. Some basic equality properties of angles (parallel and orthogonal angles) are shown in Fig. 1.7. As in pre-Euclidean times, we consider (for the moment) these properties to be "trivial". A more thorough treatment will follow in Chap. 2.

> "Die Winkelsumme im Dreieck kann nicht nach den Bedürfnissen der Kurie abgeändert werden. [The sum of the angles of a triangle can not be modified according to the requirements of the Curia.]"
> (B. Brecht, *Leben des Galilei*, 1939, scene 8)

Theorem 1.2 (Eucl. I.32). *The sum of the three angles of an arbitrary triangle ABC is equal to two right angles:*[2]

$$\alpha + \beta + \gamma = 2\,\llcorner = 180°. \tag{1.1}$$

For its *proof*, the Pythagoreans draw a line p through C parallel to the opposite side AB, see Fig. 1.8 (a). Euclid extends the side AB, draws a parallel to AC through B (Fig. 1.8 (b)) and uses the parallel angles α and γ.

Euclid's method yields the following corollary.

Corollary 1.3. *Each exterior angle is the sum of the non-adjacent interior angles, see Fig. 1.8 (c):*

$$\delta = \alpha + \gamma. \tag{1.2}$$

Angles in a circle. On a circle with centre O and diameter AB, we choose an arbitrary point C (other than A or B) and join it to A and to O, see Fig. 1.9 (a). Since the triangle AOC is isosceles, we have the same angle β at A as at C (see Eucl. I.5 in Sect. 2.1). Hence, by (1.2), the angle BOC is twice the angle BAC. We shall call BOC the *central angle on the arc BC*, and BAC an *inscribed angle* on this arc. More generally, in Fig. 1.9 (b), we call CAD an inscribed angle on the arc CD and COD the central angle on this arc. We next choose an arbitrary point D on the circle, such that C and D are on opposite sides of the diameter AB, see Fig. 1.9 (b). Deleting this diameter, we obtain in Fig. 1.9 (c) an important relation for $\alpha = \beta + \gamma$:

---

[2]Inspired by Euclid (cf. Euclid's Postulate 4 in Sect. 2.1), we use a specific symbol $\llcorner$ for a right angle; similarly, Steiner (1826c) used the symbol R, and Miquel (1838a) the symbol d (*angle droit*), so we are in good company.

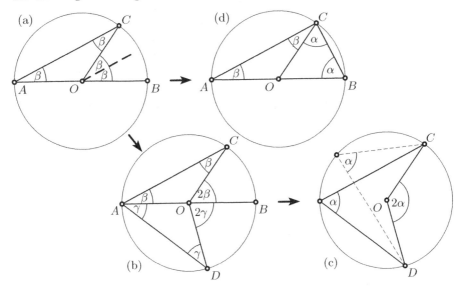

Fig. 1.9. Central angle and inscribed angle

Theorem 1.4 (Eucl. III.20). *A central angle of a circle is twice any inscribed angle on the same arc, see Fig. 1.9 (c).*

Theorem 1.5 (Eucl. III.31). *If AB is a diameter and C a point (other than A or B) on the circle, then ACB is a right angle, see Fig. 1.9 (d).*

Proof. This follows from the equality of the two angles denoted by α and β and Eucl. I.32, because $2\alpha + 2\beta = 2\llcorner$ implies $\alpha + \beta = \llcorner$. It can also be considered as a special case of Eucl. III.20 by taking $2\alpha = 2\llcorner$ in Fig. 1.9 (c). □

The Thales circle. The circle with a given segment AB as diameter is called the *Thales circle* of the segment, see also Fig. 2.1, Def. 21. Any triangle ABC with C on this circle is right-angled. For the converse to Theorem 1.5, see Exercise 4 of Chap. 2, page 54.

1.4 The Regular Pentagon

Regular polygons have fascinated geometers since the dawn of science. The Babylonians had understood the equilateral triangle and the square (see Sect. 1.6 below), therefore the Greeks directed their attention to the regular pentagon, a polygon with five vertices.

Length of the diagonal. By drawing all the diagonals of a regular pentagon, we obtain a star as shown in Fig. 1.10 (b).

This beautiful star was for the Pythagoreans a symbol of recognition between members, a tradition which has survived until today in revolutionary movements and luxury hotels.

We will determine the length, say Φ, of the diagonal of a regular pentagon of side length 1, see Fig. 1.10 (a). Since the central angles on the arcs AB, BC, etc. are $72°$ by construction, the inscribed angles on these arcs are $\alpha = 36°$ (Eucl. III.20). We consider the triangle ACD, see Fig. 1.10 (c). It contains the smaller triangle CDF which is similar to ACD. Hence, we get

$$\text{Thales:} \quad s = \frac{1}{\Phi} \qquad \text{isosceles:} \quad \Phi = 1 + s \tag{1.3}$$

which leads to

$$\Phi^2 = \Phi + 1 \qquad \text{and} \qquad s^2 + s = 1 \,. \tag{1.4}$$

A geometrical construction for these values, showing that $\Phi = \frac{\sqrt{5}}{2} + \frac{1}{2}$ and $s = \frac{\sqrt{5}}{2} - \frac{1}{2}$, was probably known to the Pythagoreans, and is numbered II.11 in Euclid's *Elements* (see Exercise 15 on page 57).

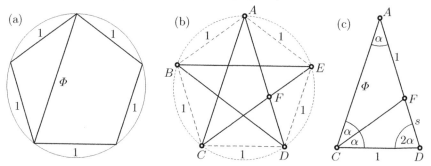

Fig. 1.10. The regular pentagon

The number Φ is called *the golden ratio*. The fact that many beautiful ancient buildings, in particular the *Parthenon* on the Acropolis, fit so perfectly into a "golden rectangle" (a rectangle with sides length 1 and Φ) led to the notation Φ in honour of Φειδίας, its architect.

The discovery of irrational numbers. *All is number*, claimed Pythagoras, who apparently had only *rational numbers* in mind. However, it was soon discovered that $\sqrt{2}$ and Φ are not rational.

To give a proof for Φ, we assume that the rational number $\frac{m}{n}$ is a solution of (1.3). We further assume that this fraction is reduced, i.e. that m and n are relatively prime. Hence, by (1.3), we have

$$\frac{m}{n} = 1 + \frac{1}{\frac{m}{n}} = 1 + \frac{n}{m} = \frac{m+n}{m} \,. \tag{1.5}$$

But if m and n are relatively prime, so are m and $m + n$ (for more details, see Eucl. VII.2 in Sect. 2.4, in particular Fig. 2.19). Hence, the fractions m/n and $(m + n)/m$ cannot be equal and Φ *cannot be a rational number*.

The fact that the regular pentagon, considered holy by the Pythagoreans, has a non-measurable diagonal was a real shock. A legend says that Hippasus, having discovered this fact and talked too much, was drowned at sea.

This discovery was also a major upset to the theory: the proof given above for Thales' theorem is not valid for irrational proportions. This considerably complicated Euclid's *Elements*, see Chap. 2.

1.5 The Computation of Areas

> A study for the Department of Education ... found nearly one in three adults (29%) in England could not calculate the floor area of a room in feet or metres — with or without calculators or paper and pens.
>
> (BBC News Online [Education], Sunday, May 5, 2002)

The calculation of areas will lead us to the Pythagorean theorem, the third pillar of this chapter, after Thales' theorem and Eucl. III.20. We start with the *area of a rectangle*, which is $a \cdot b$. This is the number of wine bottles (28) that can be stored in a bin holding 4 layers, each of 7 bottles, see Fig. 1.11, left.

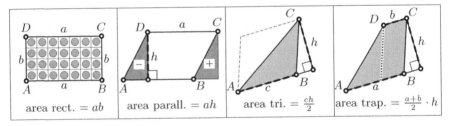

Fig. 1.11. Areas of rectangle, parallelogram, triangle and trapezium

The area of a parallelogram is $a \cdot h$, where h is the *altitude* of the parallelogram (Eucl. I.35). There are two ways to see this: (a) We cut off the triangle on the left and add it on the right to obtain a rectangle (Euclid's proof, see the second figure in Fig. 1.11); (b) We cut the parallelogram parallel to AB into a large number of very slim rectangles ("method of exhaustion" of Eudoxus and Archimedes, in this form in the commentaries of Legendre (1794); see also Fig. 2.34, right).

The *area of a triangle* is half the area of the parallelogram,

$$\mathcal{A} = \text{ area of triangle } = \text{ base} \times \text{altitude divided by } 2 = \frac{c \cdot h}{2} \qquad (1.6)$$

(Eucl. I.41), see third picture of Fig. 1.11.

Finally the *area of a trapezium* (see Fig. 1.11, right) is found by cutting the trapezium into a parallelogram and a triangle, which gives by combining the two previous results $\mathcal{A} = bh + \frac{a-b}{2} \cdot h = \frac{a+b}{2} \cdot h$.

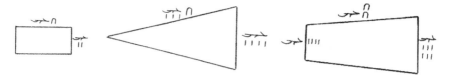

Fig. 1.12. Area calculations in Rhind papyrus; rectangle $10 \cdot 2 = 20$ (No. 49, left); triangle $\frac{4 \cdot 10}{2} = 20$ (No. 51, middle); trapezium $\frac{4+6}{2} \cdot 20 = 100$ (No. 52, right); reproductions of transcriptions by Peet (1923)

The Rhind papyrus. In 1858, the Scottish egyptologist A.H. Rhind bought in a market place at Luxor two pieces of a papyrus roll (now papyri 10057 and 10058 of the British Museum) which was written around 1650 B.C. and claimed to be a copy of a still older document from the 19th century B.C. The first extensive analysis and translation was made by A. Eisenlohr (1877), who numbered the examples of the papyrus from 1 to 84. In 1898 a facsimile was published by the British Museum. A very careful treatment with transcriptions directly from the papyrus was given by T.E. Peet (1923). The Egyptians noted numbers in a decimal system, using the symbols $1 = |$, $10 = \cap$, $100 = \mathcal{C}$, $1000 = $
, so that, for example, the number

$$4678 = \text{IIII} \,\mathcal{CCC}\,\mathcal{CCC}\,\cap\cap\cap\cap \; \text{IIII}$$

requires a great deal of writing. In the Rhind papyrus, the area of a rectangle is treated in No. 49 (see Fig. 1.12, left). The result for the rectangle of sides $||$ and $\cap$ *khet*, which should be $\cap\cap$ *setat*, is unfortunately buried under "scribe's errors of the worst description" (Peet 1923, p. 90). However, in No. 51 (Fig. 1.12, middle) the area of a triangle of base $||||$ and altitude $\cap$ is correctly computed as $\cap\cap$, but the discussion of whether the Egyptian scribe correctly understood the meaning of the altitude, fills four pages in Peet (1923), pp. 91–94). In No. 52 we find the correct computation of the area $\mathcal{C}$ of a trapezium with sides $|||||$ and $||||$ and altitude $\cap\cap$. Again the meaning of the altitude is not completely clear (Fig. 1.12, right).

Areas of similar triangles. Take the triangle ABC of Fig. 1.2 (Stone Age proof of Thales' theorem) with sides that are 5 times longer than those of the triangle $AB''C''$. It is composed of

$$1 + 3 + 5 + 7 + 9 = \;\; \cdots \;\; = 5^2 \tag{1.7}$$

copies of the small triangle (this was one of the favourite arguments of Pythagoras). In the same way, the triangle $AB'C'$ contains 8^2 copies. Hence, the area of $AB'C'$ is $\frac{8^2}{5^2}$ times that of ABC.

We thus obtain the following result.[3]

Theorem 1.6 (Eucl. VI.19). *A similar triangle with q times longer sides has q^2 times larger area.*

1.6 A Remarkable Babylonian Document

Figure 1.13 displays a Babylonian tablet dating from 1900 B.C., hence much older than Nebuchadnezzar or Tutankhamun. This tablet shows a square with sides of length 30. On its diagonal the sexagesimal digits 1, 24 51 10 and 42, 25 35 are engraved (in Babylonian notation '𒁹' stands for 1, '𒌋' stands for 10).

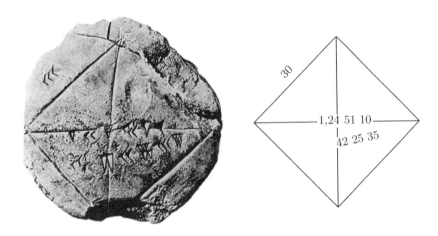

Fig. 1.13. Babylonian cuneiform tablet YBC7289 from 1900 B.C. (image enhanced by S. Cirilli)

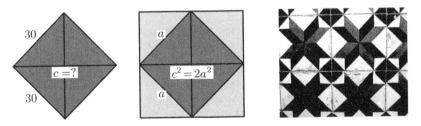

Fig. 1.14. Length of the diagonal of a square (left); ornamental tessellation seen in an old chapel in Crete (right)

---

[3]Another way of obtaining this result is based on (1.6).

Explanation. If we have a square of side length $a = 30$ and diagonal c (Fig. 1.14, left), the square on its diagonal is twice as large (composed of 8 triangles instead of 4; Fig. 1.14, middle). Thus, $c^2 = 2a^2$ and $c = a\sqrt{2}$. Another way of obtaining this result would be to meditate on one of the ornamental tessellations (Fig. 1.14, right) which were so frequent in antiquity. The above numbers, written in base 60, are

$$\sqrt{2} = 1, 24\ 51\ 10\ 7\ 46\ 6\ 4 \dots , \qquad 30 \cdot \sqrt{2} = 42, 25\ 35\ 3\ 53\ 3\ 2 \dots$$

and we see that the digits shown on the tablet are all correct (see Exercise 7 below for the computation).

The tablet thus gives evidence that Pythagoras' theorem (for the case of an isosceles triangle) was already known to the Babylonians, as were the rules of proportions. This knowledge was combined with an admirable ability for calculation.

1.7 The Pythagorean Theorem

> "This great theorem is universally associated with the name of Pythagoras. Proclus says 'If we listen to those who wish to recount ancient history, we find some of them referring this theorem to Pythagoras and saying that he sacrificed an ox in celebration of his discovery.'"   (T.L. Heath, *Euclid in Greek*, 1920, p. 219)

Millions of pupils around the world have had to learn the formula

$$a^2 + b^2 = c^2 \qquad\qquad (1.8)$$

relating the three sides of a right-angled triangle; fewer by far know a proof, or even its precise meaning. This theorem, often considered the *first great theorem* of mankind, is attributed to Pythagoras (see the quotation), but it is not known how the original discovery was achieved.

Classical proofs. Figure 1.15 spans three civilisations: Chinese, Indian and Arabic. We start with the square of area c^2, slightly tilted as in Fig. 1.15 (a).

The Chinese proof. Adding four right-angled triangles with sides a and b, we arrive at Fig. 1.15 (b) and get the large square of area $(a+b)^2 = a^2 + 2ab + b^2$. Since the areas of the four triangles add up to $2ab$, the square of area c^2 also has area $a^2 + b^2$. This is the proof of Chou-pei Suan-ching (China, 250 B.C.; see van der Waerden, 1983, p. 27). In the pictures of Fig. 1.15 (right), this transformation is obtained by translating the three triangles 2, 3 and 4. The fact that the lower picture is precisely the picture of Eucl. II.4 on page 38 gives strong evidence that this was also Pythagoras' original proof.

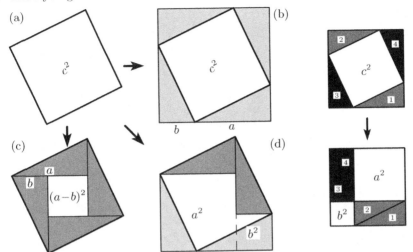

Fig. 1.15. Left: three classical proofs (Chou-pei Suan-ching (b), Bhāskara (c), Thābit ibn Qurra (d)); right: transforming Chou-pei's figure by translating triangles into Eucl. II.4

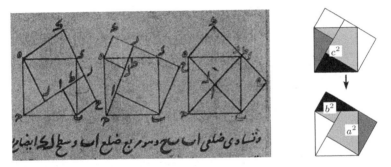

Fig. 1.16. Manuscript by Naṣīr al-Dīn al-Ṭūsī 1201–1274 with Thābit ibn Qurra's proof of Pythagoras' theorem (left); explanation (right)

The Indian proof. Bhāskara (born in 1114 A.D. in India) *removes* these four triangles to get $(a - b)^2$ and concludes the proof by saying simply "Look!", see Fig. 1.15 (c).

The Arabic proof. But why not remove *two* triangles and add them on the opposite sides, see Fig. 1.15 (d) and Fig. 1.16? By this construction, the square of area c^2 is transformed directly, without any additional triangle calculation, into two squares of total area $a^2 + b^2$. This elegant proof is attributed to Thābit ibn Qurra (826–901).

Proofs using tessellations. A legend says that Pythagoras discovered his theorem by observing a tiled floor in the palace of Polycrates, the tyrant of Samos. Since the legend does not describe the floor he considered, we have to rely on conjectures. Some possible patterns are displayed in Fig. 1.17. The

Fig. 1.17. Various patterns which could have tesselated Polycrates' palace

first picture shows a tessellation of a hypothetic hall in this palace by squares of two different areas, say a^2 and b^2, the two kinds in equal number. Looking at the dotted lines, we can also imagine this floor tiled by the same number of squares of area, say, c^2. It is thus intuitively clear that $a^2 + b^2$ should be equal to c^2 (see also Penrose, 2005, pp. 26–27). In order to convert this intuition into a more convincing proof, we isolate *one* square of area c^2 and transform it by parallel translations of the quadrilaterals 2, 3 and 4 as in Fig. 1.18 into two squares of areas a^2 and b^2. The truth of Pythagoras' theorem is now immediately obvious. If we place the stars at one of the *vertices* of the squares c^2 (and not at their centres), we obtain in a similar way the Arabic proof (see also Exercise 11 below).

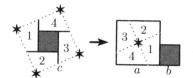

Fig. 1.18. Displacing tiles in Polycrates' tessellations for Pythagoras' theorem

The second pattern in Fig. 1.17 (proposed by Antje Kessler) might give the idea for the Chinese proof. Finally, the third pattern, with the Swiss crosses of area 5, indicates the truth of Pythagoras' theorem for a particular triangle, with sides 1, 2 and $\sqrt{5}$.

Euclid's proof. This brilliant proof was much admired by Proclus (see Heath, 1926, vol. I, p. 349). The idea is to attach the three squares of areas a^2, b^2 and c^2 to the right-angled triangle $AB\Gamma$ as in Fig. 1.19. The two grey triangles $BA\Delta$ and $BZ\Gamma$ are identical and just rotated by 90°. The triangle $BZ\Gamma$ has the same base and altitude as the square $BAHZ$; the triangle $BA\Delta$ has the same base and altitude as the rectangle $B\Delta\Lambda$. These two quadrilaterals thus have the same area. The same proof applies to the quadrilaterals on the right. The Pythagorean theorem now follows by adding the two results.

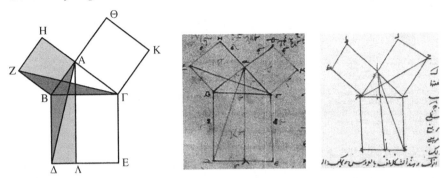

Fig. 1.19. Left: Euclid's proof; middle: Greek manuscript; right: Arabic manuscript (Thābit ibn Qurra, Baghdad 870)

Leonardo Pisano's proof. Leonardo Pisano (Fibonacci) proved Pythagoras' theorem in his *Practica Geometriae* (1220) by using Thales' theorem ("ut Euclides in sexto libro demonstravit") as follows (see Fig. 1.20).

Drawing the altitude through C gives two pairs of similar triangles: DBC, CBA and DAC, CAB; see Fig. 1.20. We thus get

$$\left.\begin{array}{l} \dfrac{a}{p} = \dfrac{c}{a} \quad \Longrightarrow \quad a^2 = pc \\[2mm] \dfrac{b}{q} = \dfrac{c}{b} \quad \Longrightarrow \quad b^2 = qc \end{array}\right\} \quad \Longrightarrow \quad a^2 + b^2 = (p+q)c = c^2. \qquad (1.9)$$

Note for later use that we also have

$$\frac{p}{h} = \frac{h}{q} \quad \Longrightarrow \quad h^2 = pq \qquad \textit{(the altitude theorem).} \qquad (1.10)$$

Naber's proof. B.L. van der Waerden (1983, p. 30) attributes this proof to H.A. Naber (Haarlem 1908); Heath (1921, p. 148) presents it as one of the most probable original proofs of Pythagoras.

Without doubt, this proof is the most elegant of all. The four triangles in Fig. 1.21 are similar. If the area of the first, with hypotenuse 1, is denoted by k, the areas of the others are, by Theorem 1.6, equal to ka^2, kb^2, and kc^2,

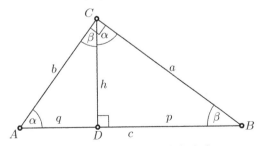

Fig. 1.20. A proof using Thales' theorem

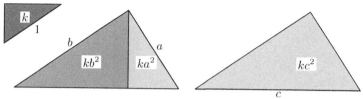

Fig. 1.21. Naber's proof

respectively. By comparing the figures, we see that, obviously, $ka^2 + kb^2 = kc^2$, and it only remains to divide this formula by k.

For more proofs of the Pythagorean theorem, we recommend some exercises below and the book by Loomis (1940), which enumerates 370 proofs. One of these proofs is even due to a president of the United States (James Garfield), from those beautiful times when mathematics was more fascinating than oil.

Application to regular polygons. The above pre-Euclidean results allow us to demystify many regular polygons and to compute the radii ρ of their incircle and R of their circumcircle. The results are collected in Table 1.1.

Table 1.1. Radius of incircle (ρ) and radius of circumcircle (R) for regular polygons with side length 1

n		R	ρ
3	△	$R = \dfrac{\sqrt{3}}{3}$	$\rho = \dfrac{\sqrt{3}}{6}$
4	◇	$R = \dfrac{\sqrt{2}}{2}$	$\rho = \dfrac{1}{2}$
5	⬠	$R = \dfrac{1}{\sqrt{3-\Phi}} = \dfrac{\sqrt{2+\Phi}}{\sqrt{5}}$	$\rho = \dfrac{\sqrt{3+4\Phi}}{2\sqrt{5}}$
6	⬡	$R = 1$	$\rho = \dfrac{\sqrt{3}}{2}$
10	⬟	$R = \Phi$	$\rho = \dfrac{\sqrt{3+4\Phi}}{2}$

Proofs. One always has $\rho = \sqrt{R^2 - \frac{1}{4}}$ by Pythagoras. For $n = 3$ and 5, the quantities h and ℓ (defined in Fig. 1.22) are calculated with Pythagoras; ℓ simplifies by using $\Phi^2 = \Phi + 1$. This gives

$$h = \sqrt{1 - \frac{1}{4}} = \frac{\sqrt{3}}{2}, \qquad \ell = \sqrt{1 - \frac{\Phi^2}{4}} = \frac{\sqrt{3-\Phi}}{2}.$$

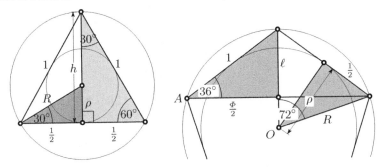

Fig. 1.22. Equilateral triangle and pentagon

The values for R are obtained by applying Thales' theorem to the grey triangles in Fig. 1.22. For $n = 10$, see Fig. 1.10 (c); since $\alpha = 36°$, ten of these triangles arranged as in a cake form a regular decagon. □

1.8 Three Famous Problems of Greek Geometry

The following three problems appeared during the pre-Euclidean period and occupied the Greek geometers for at least three centuries. The new curves and algebraic tools which were needed to solve them contributed for another two millennia to the development of geometry, algebra and analysis.

Squaring the circle. Finding a square whose area is equal to that of a given rectangle was an easy exercise after the altitude theorem (1.10) was discovered. The next challenge was then to find areas of certain regions bounded by *curves*. In particular, the *squaring of a given circle* exercised great fascination throughout the centuries. The earliest known result is given in the examples No. 48 and 50 of the Rhind papyrus, see the pictures of Fig. 1.23 (left): a circle in a square of $9 \times 9 = 81$ units is squared by cutting off corners with two sides of length 3 units. This creates a surface of $81 - 18 = 63$ units. Since 63 is close to $64 = 8^2 = (9 - 1)^2$, we obtain the "Egyptian algorithm"

subtract one ninth of the diameter, then square.

This is demonstrated in No. 50, where (reproductions from Peet, 1923)

the area of a circle of diameter 9 is 64 ꓵꓵꓵ|| ꓵꓵꓵ|| .

In Rhind No. 42, while computing the volume of a cylindrical container, the area for diameter 10 is given as $79\frac{1}{108} + \frac{1}{324}$ or ꓵꓵꓵꓵ|| ꫀ||| ꫀꫀꫀꓵ||| , which is $79\frac{1}{81}$, the correct value. In modern notation these values correspond to the approximation $\pi \approx \frac{256}{81} = 3.1605$. Only during the Greek period were rigorous results obtained. Archimedes showed is his *celebrissimo* work (*Measurement of a circle*, Heath, 1897, p. 91), with virtuoso estimates from above and below that

$$3\,\frac{10}{71} < \pi < 3\,\frac{1}{7}. \tag{1.11}$$

The details are given in Exercise 22 on page 58. In Chap. 8 we will see why all the efforts of the Greek geometers to obtain an exact solution were doomed to failure.

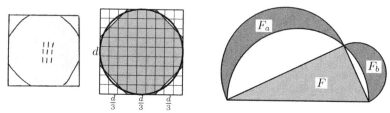

Fig. 1.23. Squaring the circle in Rhind papyrus (left pictures, reproduced from Rhind No. 48 in Peet, 1923); the quadrature of the lunes of Hippocrates (right)

The lunes of Hippocrates. However, *one* precise result in this direction was found during the Greek period, the squaring of the lunes by Hippocrates of Chios.[4] Let two lunes be cut out by three semicircles drawn on the sides of a right-angled triangle (see Fig. 1.23, right). Then their areas satisfy the relation

$$F_a + F_b = F \quad \text{(area of the triangle)}. \tag{1.12}$$

To see this, let F'_a, F'_b and F'_c be the areas of the semicircles with diameters a, b and c. Then we see from the figure that $F'_a + F'_b + F = F_a + F_b + F'_c$. We have to know that Theorem 1.6, i.e. the fact that the areas of the semicircles are proportional to the squares of the diameters, remains valid here (this result will later be Eucl. XII.2). Then the terms $F'_a + F'_b$ and F'_c cancel by Pythagoras' theorem.

Doubling the cube. The problem is: *find a cube whose volume is twice that of a given cube* (see Fig. 1.24, left). Ancient sources give two different versions for the origin of the problem: according to one source, King Minos of Crete wanted Glaucus' tomb to be doubled (see Heath, 1921, p. 245); according to the other source, the *oracle of Delos* ordered the altar to be doubled in order to stop a plague epidemic. When the people went to Plato asking for help with the solution, he replied that the oracle did not mean that the actual doubling of the altar would heal the people, but that the advances in mathematics required for this construction would do so. For the geometers, who already knew how to *double a square* (see Section 1.6), this problem, which consists in

---

[4]who lived in the 5th century B.C., not to be mistaken with his contemporary Hippocrates of Kos, the famous physician.

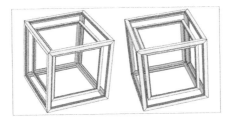
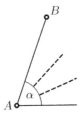

Fig. 1.24. Doubling the cube (left; the picture is a stereogram, if you stare at it by merging the two images with the left and right eye, you'll see it in 3d); trisecting an angle (right)

constructing $\sqrt[3]{2}$, was an interesting challenge. We will see how this problem led to the discovery of the first conic sections (Chap. 3) and many other new curves, one of which is the conchoid (see Chap. 4). Today's science would not be the same without the theory of conics (Chap. 5).

Trisecting an angle. The regular polygons with their divine beauty have fascinated geometers since time immemorial. The square and the equilateral triangle were known to the Babylonians, the regular pentagon was demystified by the Greeks (see above). Since it is easy to *bisect* an angle (e.g. with Eucl. III.20), we have no difficulty in constructing a hexagon, octagon, decagon, dodecagon or any 2^k-gon. The next challenges are thus the regular heptagon (7-gon) and the regular enneagon (nonagon, 9-gon). This last problem would require one to *trisect* the angle of 120°. From this question arose (probably) the challenge of trisecting *any* given angle (see Fig. 1.24, right). The solution of these problems contributed considerably to the development of algebra (see Chap. 6).

1.9 Exercises

1. Ptolemy gives the approximation

$$\sqrt{3} \approx 1, 43\ 55\ 23$$

   in base 60 (see Heath, 1926, vol. II, p. 119). Check whether he is accurate.

2. Modify the proof of Theorem 1.4 for the case in which the points C and D are *not* on opposite sides of AB, see Fig. 1.25 (left). This time, α will be the *difference* of two angles β and γ.

3. Let ABC be a triangle inscribed in a circle, as in Fig. 1.25 (right). Show that the size of α is independent of the position of A on the circle (Eucl. III.21).

4. In order to approximate the golden ratio we consider the sequence of rational numbers given recursively by

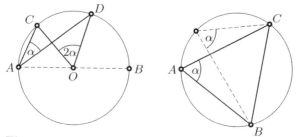

Fig. 1.25. Eucl. III.20 modified (left); Eucl. III.21 (right)

$$r_{k+1} = 1 + \frac{1}{r_k}, \qquad k \ge 0$$

with $r_0 = 1$. Find a relation to (1.3) and discover, by considering the denominators of the fractions r_k, an interesting sequence, the *Fibonacci numbers*.

5. Let a "golden" rectangle with sides Φ and 1 be given. Show that cutting off a square from this golden rectangle produces another golden rectangle with sides smaller by the factor $1/\Phi$ (see Fig. 1.26). The procedure can be repeated and produces an embedded sequence of golden rectangles. If we draw a quarter of a circle in each of these squares, we obtain a beautiful spiral which is said to possess great mystical power ...

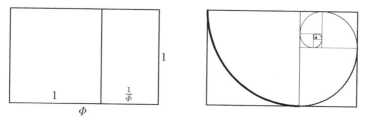

Fig. 1.26. A golden rectangle and its subdivisions

6. Find the error in the "proof" presented in Fig. 1.27, where different arrangements of identical pieces suggest that $273 = 272$.

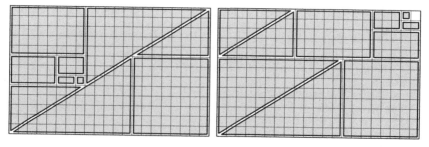

Fig. 1.27. A curious proof that $273 = 272$

7. The Ancients were skilled at extracting square roots as this was necessary for applying Pythagoras' theorem (see also Exercise 22 of Chap. 2). Rediscover the method which (probably) allowed the Babylonians, nearly 4000 years ago, to find the excellent value $1, 24\ 51\ 10$ in base 60 for $\sqrt{2}$.

   Hint. On another Babylonian tablet, which lists squares of numbers, you will discover that a square of sides $1, 25$ has an area very close to 2, because $(1, 25)^2 = 2, 00\ 25$ in base 60. Cut two strips of width δ from this square in order to reduce the area to 2.

8. Triangular arrangements of dots of the form

   were sacred figures for the Pythagoreans, especially the *holy tetractys* with 10 dots, by which the Pythagoreans used to swear. Find a general expression for t_n, the number of dots of the n-th figure.

9. Find a general formula for the *pentagonal numbers*

   $\bullet = 1, \quad = 5, \quad = 12, \quad = 22, \ \ldots \ .$

10. (Inspired by a picture of Eugen Jost, 2010.) Guess a formula for the number of dots forming an equilateral triangle on a hexagonal grid

    $\bullet = 1, \quad = 4, \quad = 9, \quad = 16, \quad = 25, \ \ldots$

    and explain the result.

11. Glue the drawings of Fig. 1.28 onto some cardboard (or make a Xerox copy if you want to preserve this beautiful book undamaged). Carefully cut out the pieces to obtain two jig-saw puzzles that allow one to *grasp* (literally) a 2500-year-old theorem. Which theorem is this?

12. Explain another version of Euclid's proof of Pythagoras' theorem (see Fig. 1.29, left and middle pictures): Produce ZH and $K\Theta$ to find a point Π such that Π, A, Λ are collinear and $\Pi A = B\Delta$ (why?). Move the area a^2 first upwards parallel to ZH and then downwards parallel to ΠA.

13. (A discovery of Heron.) Show that in Euclid's figure for the proof of the Pythagorean theorem the lines ΓZ, BK and $A\Lambda$ are concurrent, see Fig. 1.29 (right).

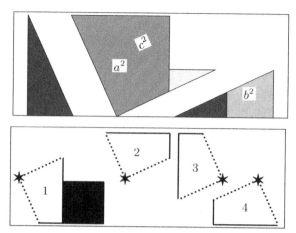

Fig. 1.28. Two jig-saw puzzles of high educational value

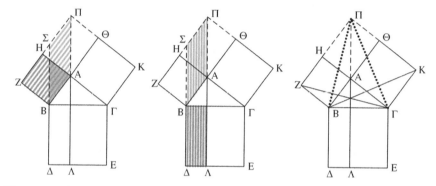

Fig. 1.29. Another proof of Pythagoras' theorem (left); Heron's discovery about Euclid's figure (right)

14. Given an angle AOB with vertex O and a point P inside the angle, construct perpendiculars PA, PB, and OC, PD, see Fig. 1.30 (left). Then show that $AC = BD$ (Hartshorne, 2000, p. 62).

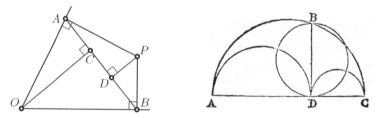

Fig. 1.30. Diagonal in a particular quadrilateral (left); Archimedes' Lemma (right; copied from Peyrard's edition of Archimedes' *Opera*, vol. 2, Paris 1808)

15. Prove one of Archimedes' Lemmata (see Fig. 1.30, right): *The area of the moon-like region bounded by the semicircles AC, CD and DA is equal to the area of the circle with diameter BD.*

16. A young couple, to celebrate their golden wedding, set up a tent whose base is a square of side length the golden ratio Φ (what else), held up by 5 tubes of length 1, see Fig. 1.31. Show that the polygons AEB, $BEFC$, CFD and $DFEA$ are parts of a regular pentagon. Further, show that the angles of the faces AEB and $BCEF$ with the base add up to a right angle ⌐. With these two results we at once understand the construction of the dodecahedron (Eucl. XIII.17) by attaching six of these tents to a *golden* cube, see Fig. 2.37 in Sect. 2.6.

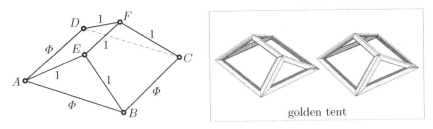

golden tent

Fig. 1.31. The golden tent

17. (Pythagorean triples.) Find (all) right-angled triangles with all sides of integer length.

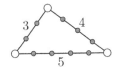

18. Show that

$$x = \frac{1 - u^2}{1 + u^2}, \quad y = \frac{2u}{1 + u^2}, \quad u \in \mathbb{Q} \tag{1.13}$$

represent all points with rational coordinates on the unit circle, except $(-1, 0)$ (which corresponds to $u = \infty$).

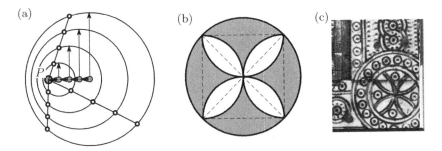

Fig. 1.32. Geneva duck theorem (a); ornamental figure (b); ornament from a reliquary casket, 8th century, Abbey church of Saint Ludger, Essen-Werden (c)

19. Prove the famous "Geneva duck theorem": A duck moves on the Lake of Geneva at constant speed towards a point P and creates circles at a constant rate (see Fig. 1.32 (a)). Prove (with Thales) that any half-line through P is cut by the circles into intervals of the same length (the situation is slightly more complicated if the movement is "supersonic").

20. An ancient ornamental figure (see Figs. 1.32 (b) and (c)) consists of a circle (which we take of radius 1) from which a cross is cut out. The cross is bordered by eight circular arcs which are either tangent to each other or cut orthogonally at the centre. Find the area of the part shaded in grey.

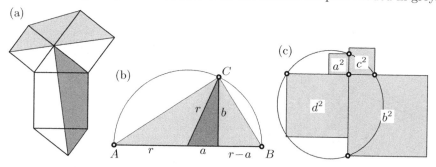

Fig. 1.33. Leonardo's proof (a); altitude theorem (b); four squares (c)

21. Explain Leonardo da Vinci's proof (see Fig. 1.33 (a)) of Pythagoras' theorem, which is striking by — its beauty!

22. Use Fig. 1.33 (b) to deduce the altitude theorem (1.10) for triangle ABC from Pythagoras' theorem for the small dark triangle, and conversely. (This will be Eucl. II.14 in the next chapter.)

23. Solve a "beau problème de géométrie", inspired by a serigraph of Max Bill (1908–1994) and communicated to the authors by P. Zabey, Geneva: Let $ABCD$ be a square whose side length is taken as 1. Let E be the midpoint of BC. Construct a square $EFGH$ such that D is the midpoint of FG. This creates six triangles whose angles and areas are requested.

24. The oldest theorems of humanity in this chapter provide nice discoveries even now in the 21st century. Prove the following result, due to Nelsen (2004): If two chords of a circle intersect at right angles forming four segments a, b, c, d, then $a^2+b^2+c^2+d^2 = D^2$, where D is the diameter of the circle (see Fig. 1.33 (c)).

25. Analyse Dürer's *circling of the square* (Underweysung, book 2) and its error for π.

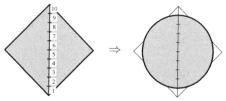

2

The Elements of Euclid

"At age eleven, I began Euclid, with my brother as my tutor. This was one of the greatest events of my life, as dazzling as first love. I had not imagined that there was anything as delicious in the world."   (B. Russell, quoted from K. Hoechsmann, *Editorial, π in the Sky*, Issue 9, Dec. 2005. A few paragraphs later K. H. added: An innocent look at a page of contemporary theorems is no doubt less likely to evoke feelings of "first love".)

"At the age of 16, Abel's genius suddenly became apparent. Mr. Holmboë, then professor in his school, gave him private lessons. Having quickly absorbed the *Elements*, he went through the *Introductio* and the *Institutiones calculi differentialis* and *integralis* of Euler. From here on, he progressed alone."

(Obituary for Abel by Crelle,
J. Reine Angew. Math. 4 (1829) p. 402; transl. from the French)

"The year 1868 must be characterised as [Sophus Lie's] breakthrough year. ... as early as January, he borrowed [from the University Library] Euclid's major work, *The Elements* ..."   (*The Mathematician Sophus Lie* by A. Stubhaug, Springer 2002, p. 102)

"There never has been, and till we see it we never shall believe that there can be, a system of geometry worthy of the name, which has any material departures ... from the plan laid down by Euclid."

(A. De Morgan 1848; copied from the *Preface* of Heath, 1926)

"Die Lehrart, die man schon in dem ältesten auf unsere Zeit gekommenen Lehrbuche der Mathematik (den Elementen des Euklides) antrifft, hat einen so hohen Grad der Vollkommenheit, dass sie von jeher ein Gegenstand der Bewunderung [war] ... [The style of teaching, which we already encounter in the oldest mathematical textbook that has survived (the *Elements* of Euclid), has such a high degree of perfection that it has always been the object of great admiration ...]"   (B. Bolzano, *Grössenlehre*, p. 18r, 1848)

A. Ostermann and G. Wanner, *Geometry by Its History,*
Undergraduate Texts in Mathematics, DOI: 10.1007/978-3-642-29163-0_2,
© Springer-Verlag Berlin Heidelberg 2012

Euclid's *Elements* are considered by far the most famous mathematical *oeuvre*. Comprising about 500 pages organised in 13 books, they were written around 300 B.C. All the mathematical knowledge of the period is collected there and presented with a rigour which remained unequalled for the following two thousand years.

Over the years, the *Elements* have been copied, recopied, modified, commented upon and interpreted unceasingly. Only the painstaking comparison of all available sources allowed Heiberg in 1888 to essentially reconstruct the original version. The most important source (M.S. 190; this manuscript dates from the 10th century) was discovered in the treasury[1] of the Vatican, when Napoleon's troops invaded Rome in 1809. Heiberg's text has been translated into all scientific languages. The English translation by *Sir Thomas L. Heath* in 1908 (second enlarged edition 1926) is completed by copious comments.

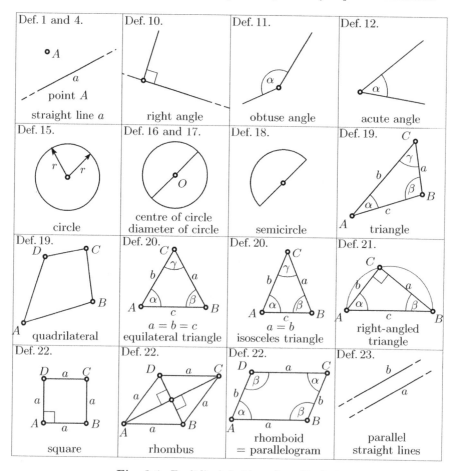

Fig. 2.1. Euclid's definitions from Book I

---

[1]That's where invading troops go first ...

2.1 Book I

The definitions. The *Elements* start with a long list of 23 *definitions*, which begins with

Σημεῖόν ἐστιν, οὗ μέρος οὐθέν   (A *point* is that which has no part)

and goes on until the definition of parallel lines (see the quotation on p. 36).

Euclid's definitions avoid figures; in Fig. 2.1 we give an overview of the most interesting definitions in the form of pictures. Euclid does not distinguish between straight lines and segments. For him, two segments are apparently "equal to one another" if their *lengths* are the same. So, for example, a circle is defined to be a plane figure for which all radius lines are "equal to one another".

The postulates.[2] Let the following be postulated:

1. To draw a straight line from any point to any point.

2. To produce a finite straight line continuously in a straight line.

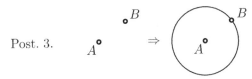

3. To describe a circle with any centre and distance.

4. That all right angles are equal to one another.

5. That, if a straight line falling on two straight lines make the interior angles on the same side less than two right angles, the two straight lines, if produced indefinitely, meet on that side on which are the angles less than two right angles.

---

[2]English translation from Heath (1926).

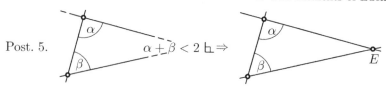

Post. 5. $\alpha + \beta < 2 \llcorner \Rightarrow$

Remark. The first three postulates raise the usual constructions with *ruler*[3] (Post. 1 and 2) and *compass* (Post. 3) to an intellectual level. The fourth postulate expresses the homogeneity of space in all directions by using the right angle as a universal measure for angles; the fifth postulate, finally, is the celebrated *parallel postulate.* Over the centuries, it gave rise to many discussions.

The postulates are followed by *common notions* (also called *axioms* in some translations) which comprise the usual rules for equations and inequalities.

The propositions. Then starts the sequence of *propositions* which develops the entire geometry from the definitions, the five postulates, the axioms and from propositions already proved. Among others, the *trivialities* of Chap. 1 now become real propositions. A characteristic of Euclid's approach is that the alphabetic order of the points indicates the order in which they are constructed during the proof.

In order to give the flavour of the old text, we present the first two propositions in full and with the original Greek letters; but we will soon abandon this cumbersome style[4] and turn to a more concise form with lower case letters for side lengths (Latin alphabet) and angles (Greek alphabet), as has become standard, for good reason, in the meantime.

Eucl. I.1.   *On a given finite straight line AB to construct an equilateral triangle.*

The construction is performed by describing a circle Δ centred at A and passing through B (Post. 3) and another circle E centred at B and passing through A (Post. 3). Their point of intersection Γ is then joined to A and to B (Post. 1). The distance $A\Gamma$ is equal to $B\Gamma$ and to AB, which makes the triangle equilateral.

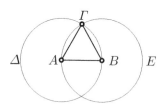

Remark. The fact that Euclid assumes without hesitating the existence of the intersection point Γ of two circles has repeatedly been criticised (Zeno, Proclus, ...). Obviously, a *postulate of continuity* is required. For a detailed discussion we refer the reader to Heath (1926, vol. I, p. 242).

---

[3]In order to emphasise that this ruler has no markings on it, some authors prefer to use the expression straightedge instead.

[4]"... statt der grässlichen Euklidischen Art, nur die Ecken mit Buchstaben zu markieren; [... instead of the horrible Euclidean manner of denoting only the vertices by letters;]" (F. Klein, *Elementarmathematik, Teil II*, 1908, p. 507; in the third ed., 1925, p. 259 the adjective *horrible* is omitted).

Eucl. I.2. *To place at a given point A a straight line AE equal to a given straight line BΓ.*

For the construction, one erects an equilateral triangle $AB\Delta$ on the segment AB (Eucl. I.1), produces
the lines ΔB and ΔA (Post. 2) and describes the circle with centre B passing through Γ (Post. 3) to find
the point H on the line ΔB. Then one draws the circle with centre Δ passing through H (Post. 3). The
intersection point E of this circle with the line ΔA
has the required property. Indeed, the distance $B\Gamma$
equals the distance BH, and the distance ΔH equals
the distance ΔE. Hence, the distance AE equals the
distance BH, since the distance ΔB equals ΔA.

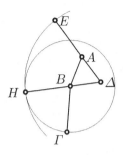

Remark. Post. 3 only allows one to draw a circle with given centre A and passing through a given point B. The aim of this proposition is to show that one
is now allowed to draw a circle with a *compass-carried* radius. This proof also
was criticised by Proclus. Depending on different positions of the points A, B
and Γ, various cases must be distinguished, with a slightly different argument
in each case. To prove all particular cases separately already here becomes
cumbersome. Therefore, Euclid's method will henceforth be our model: as
soon as *one* case is understood, the others are left to the intelligent reader.

Eucl. I.4. *Given two triangles with $a = a'$, $b = b'$, $\gamma = \gamma'$, then all sides and
angles are equal.*

This result is a cornerstone for all
that follows. In its proof, Euclid speaks
vaguely of *applying* the triangle ACB
onto the triangle DFE, of *placing* the
point C on the point F, of placing the
line a on the line a', etc. Of course, this
lack of precision attracted much criti-

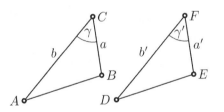

cism.[5] Note that in Hilbert's axiomatic formulation of geometry, see Sect. 2.7,
this *proposition* becomes an *axiom*.

Eucl. I.5 (commonly known as *Pons Asinorum*, i.e. asses' bridge). *If in a
triangle $a = b$, then $\alpha = \beta$.*

One of the *trivialities* of the previous section thus becomes a real theorem.
Let us see how Euclid proved this proposition. One produces (see Fig. 2.2,

---

[5] "Betrachten wir aber andererseits - das scheint noch die einzig mögliche Lösung
in diesem Wirrwarr - diese Nr. 4 als ein späteres Einschiebsel ... [If we consider on
the other hand — and this seems to be the only possible solution in this chaos — this
No. 4 as a later insertion ...]" (F. Klein, *Elementarmathematik, Teil II*, 1908, p. 416;
third ed., 1925, p. 217 with a modified wording).

left) CA and CB (Post. 2) to the points F and G with $AF = BG$ (Eucl. I.2), and joins F to B and A to G (Post. 1). Thus the triangles FCB and GCA are equal by Eucl. I.4, i.e. $\alpha + \delta = \beta + \varepsilon$, $\eta = \zeta$ and $FB = GA$. Now, by Eucl. I.4, the triangles AFB and BGA are equal and thus $\delta = \varepsilon$. Using the above identity, one has $\alpha = \beta$. This seems to be a brilliant proof, but is in fact needlessly complicated. Pappus remarked 600 years later that it would be sufficient to apply Eucl. I.4 to the triangles ACB and BCA with A and B interchanged, see Fig. 2.2, centre and right.

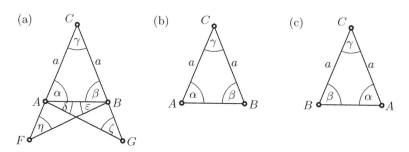

Fig. 2.2. Angles in an isosceles triangle

This proposition is immediately followed by Eucl. I.6, where the converse implication is proved: $\alpha = \beta$ implies $a = b$.

The next two propositions treat the problem of uniquely determining a triangle by prescribing the length of the three sides.

Eucl. I.7. *Consider the two triangles of Fig. 2.3 (a), erected on the same base AB and on the same side of it. If $a = a'$ and $b = b'$, then $C = D$.*

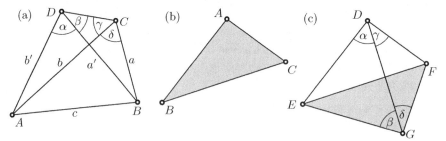

Fig. 2.3. Triangles with equal sides

Proof by Euclid. Suppose that $C \neq D$. Since DAC is isosceles by hypothesis, $\alpha + \beta = \gamma$ (Eucl. I.5). Since DBC is isosceles, $\beta = \gamma + \delta$ (Eucl. I.5). Thus we have on the one hand $\gamma > \beta$, and on the other hand $\gamma < \beta$, which is impossible.   □

This is our first *indirect proof*. More than two thousand years later, a school of mathematics rejected this kind of reasoning, because "one can not prove something true with the help of something false" (L.E.J. Brouwer, 1881–1966).

Eucl. I.8. *If two triangles ABC and DEF have the same sides, they also have the same angles.*

The proof of Philo of Byzantium, which is given here, is more elegant than Euclid's. We apply the triangle ABC (see Fig. 2.3 (b)) onto the triangle DEF in such a manner that the line BC is placed on EF and the point A which becomes G lies on the opposite side of EF to D (see Fig. 2.3 (c)). By hypothesis, DEG is isosceles and thus $\alpha = \beta$ (Eucl. I.5). But DFG is also isosceles and hence $\gamma = \delta$ (Eucl. I.5). Thus the angle at A ($= \beta + \delta$) is equal to the angle at D ($= \alpha + \gamma$). For the other angles, one repeats the same reasoning, placing first AC on DF, then AB on DE.

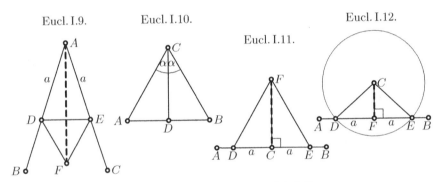

Fig. 2.4. Propositions I.9–I.12

Eucl. I.9–I.12. These propositions treat the bisection of an angle BAC (see Fig. 2.4.I.9), the bisection of a line AB (see Fig. 2.4.I.10) and the erection of the perpendicular to a line AB at a point C on it (see Fig. 2.4.I.11). The common tool for solving these three problems is the equilateral triangle (Eucl. I.1). Finally, the construction of a perpendicular to a line AB *from* a point C outside of it (see Fig. 2.4.I.12) is achieved with the help of a circle (Post. 3) and the midpoint of DE (Eucl. I.10).

The entrance of Postulate 4.

> "When a straight line set up on a straight line makes the adjacent angles equal to one another, each of the equal angles is *right*, and the straight line standing on the other is called a *perpendicular* to that on which it stands".
> (Def. 10 of Euclid's first book in the transl. of Heath, 1926).

The fourth postulate expresses the homogeneity of the plane, the absence of any privileged direction, and allows one to compare, add and subtract the angles around a point. It does this by defining *the right angle* as a universal unit. We denote this angle (90°) by the symbol $\llcorner$.

Eucl. I.13. *Let the line AB cut the line CD (Fig. 2.5). Then $\alpha + \beta = 2\llcorner$.*

Proof. Draw the perpendicular BE, which divides the angle β into $\llcorner + \eta$. Thus

$$\left.\begin{array}{c} \beta = \llcorner + \eta \\ \alpha + \eta = \llcorner \end{array}\right\} \quad \Rightarrow \quad \alpha + \beta + \eta = 2\llcorner + \eta$$

which proves the assertion. □

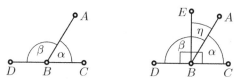

Fig. 2.5. Eucl. I.13 (left) and its proof (right)

Eucl. I.14. *In the situation of Fig. 2.6 (left), let $\alpha + \beta = 2\llcorner$. Then C lies on the line DB.*

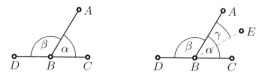

Fig. 2.6. Eucl. I.14 (left) and its proof (right)

Proof. Let E lie on the line DB, i.e. by Eucl. I.13, let $\gamma + \beta = 2\llcorner$. By hypothesis, $\alpha + \beta = 2\llcorner$. These angles are equal by the fourth postulate, hence $\gamma = \alpha$. Therefore, E and C lie on the same line. □

Eucl. I.15. *If two straight lines cut one another, they make the opposite angles equal to one another, i.e. $\alpha = \beta$ in Fig. 2.7 (left).*

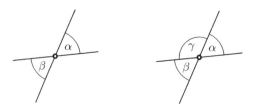

Fig. 2.7. Eucl. I.15 (left) and its proof (right)

Proof. By Eucl. I.13, we have $\alpha + \gamma = 2\llcorner$ and also $\gamma + \beta = 2\llcorner$. By Post. 4, $\alpha + \gamma = \gamma + \beta$. The result then follows from subtracting γ from each side. □

Eucl. I.16. *If one side of a triangle is produced at C (see Fig. 2.8), the exterior angle δ satisfies δ > α and δ > β.*

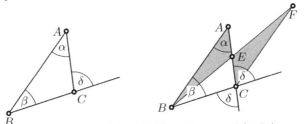

Fig. 2.8. Eucl. I.16 (left) and its proof (right)

Proof. Let E be the midpoint of AC (Eucl. I.10). We produce BE (Post. 2) and cut off the distance EF such that $EF = BE$ (Post. 3). The grey angles at E are equal (Eucl. I.15), hence the two grey triangles are identical (Eucl. I.4). Thus the grey angle at C is α, which is obviously smaller than δ. For the second inequality, one proceeds similarly with the angle on the other side of C (which is equal to δ by Eucl. I.15).   □

Remark. In the geometry on the *sphere*, which we will discuss in more detail in Section 5.6, Eucl. I.16 is the first of Euclid's propositions which does not remain valid. Suppose, for example, that B is at the North Pole and A, E and C lie on the Equator. Then $\alpha = ∟$ and $\delta = ∟$, hence the inequality $\delta > \alpha$ is false. The reason is that the point F, which in our example becomes the South Pole, is no longer certain to remain in the open sector between the produced lines CA and BC.

Eucl. I.17–I.26. Various theorems of Euclid on the congruence of triangles determined by certain side lengths or angles (see Fig. 2.9). The ambiguous case ASS (last picture) is not mentioned by Euclid. For an inequality involving the angles and sides of a triangle (Eucl. I.18), see Exercise 11 below.

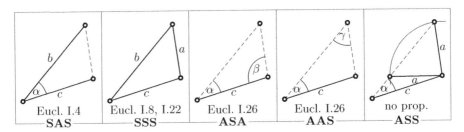

Fig. 2.9. Congruence theorems for triangles

Eucl. I.20 states the famous *triangle inequality*

$$a < b + c, \qquad b < c + a, \qquad c < a + b \tag{2.1}$$

(see Exercise 12 below). This result has been ridiculed as being evident even to an ass. For if one puts the ass at one vertex of the triangle and hay at another, the ass will follow the side that joins the two vertices and will not make the detour through the third vertex (*digni ipsi, qui cum Asino foenum essent,* Heath, 1926, vol. I, p. 287). Proclus gave a long logical-philosophical answer. Instead, he could have said briefly: "The *Elements* were not written for asses".

> "Parallel straight lines are straight lines which, being in the same plane and being produced indefinitely in both directions, do not meet one another in either direction".
>
> (Def. 23 of Euclid's first book in the transl. of Heath, 1926).

Fig. 2.10. Eucl. I.27 (left) and its proof (right)

Eucl. I.27. *If some line cuts two lines a and b under angles α and β (see Fig. 2.10), then $\alpha = \beta$ implies that the lines are parallel. In this case, we write $a \parallel b$ for short.*

Proof. If a and b were not parallel, they would meet in a point G, see Fig. 2.10. Then EGF would be a triangle having α as exterior angle. Therefore, α would be greater than β (Eucl. I.16), which contradicts the assumption. □

The entrance of Postulate 5. Eucl. I.27, which ensures the *existence* of parallels (simply take $\alpha = \beta$ and you have a parallel), is the last of the propositions, carefully collected by Euclid at the beginning of his treatise, which do not require the fifth postulate for its proof. This part of geometry is called *absolute geometry*. For all that follows we need the *uniqueness* of parallels, which requires the fifth postulate.

Eucl. I.29. *If $a \parallel b$ (see Fig. 2.11), then $\alpha = \beta$.*

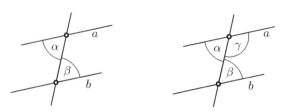

Fig. 2.11. Eucl. I.29 (left) and its proof (right)

Proof. Suppose $\alpha > \beta$. By Eucl. I.13, $\alpha + \gamma = 2\llcorner$, hence $\beta + \gamma < 2\llcorner$. By the fifth postulate, these lines have to meet, which is a contradiction. A similar reasoning shows that $\alpha < \beta$ is also impossible.   □

Remark. Combined with Eucl. I.15, the propositions Eucl. I.27 and Eucl. I.29 give variants, one of which formulates the *trivial properties of parallel angles* of Fig. 1.7 (Eucl. I.28).

Remark. For more than 2000 years, geometers conjectured that Eucl. I.29 could be established without appealing to the fifth postulate. Many attempts were made to prove this conjecture, without success. We shall return to this question in Section 2.7.

Eucl. I.30. *For any three lines a, b, c with a ∥ b and b ∥ c, we have a ∥ c.*

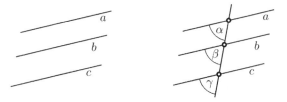

Fig. 2.12. Eucl. I.30 (left) and its proof (right)

Proof. By Eucl. I.27 and Eucl. I.29, the lines a and b are parallel if and only if the angles α and β are equal.   □

Eucl. I.31. *Drawing a parallel to a given line through a given point A.*

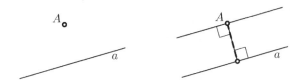

Fig. 2.13. Eucl. I.31 (left) and the proposed construction (right)

Proof. Euclid's proof makes use of Eucl. I.23 which is itself a consequence of Eucl. I.22. One can also use two orthogonal lines (Eucl. I.12 followed by Eucl. I.11).   □

Remark. Proclus made the following statement in his commentary: *There exists* at most one *line through a given point A which is parallel to a given line.* This statement turns out to be equivalent to the fifth postulate. In the form just given, it is called *Playfair's axiom* (1795).

Eucl. I.32 gives the formula $\alpha + \beta + \gamma = 2\llcorner$ for the three angles of an arbitrary triangle, see (1.1) and the proof in Fig. 1.8. This is a very old theorem, certainly known to Thales. It comes quite late in Euclid's list, since its proof requires the fifth postulate.

The remainder of Book I. Eucl. I.33–34 treat parallelograms; Eucl. I.35–41 the areas of parallelograms and triangles; Eucl. I.42–45 the construction of parallelograms with a prescribed area; Eucl. I.46 treats the construction of a square. The highlight of the first book, however, is Pythagoras' theorem (Eucl. I.47, see the proof on page 16 and Fig. 1.19) and its converse: *if a, b, c are the sides of a triangle and $a^2 + b^2 = c^2$, then the triangle is right-angled.*

Book II. This book contains *geometrical algebra*, i.e. algebra expressed in geometric terms. For instance, the product of two numbers a, b is represented geometrically by the area of a rectangle with sides a and b. We have for example the following relations, Eucl. II.1 and Eucl. II.4:

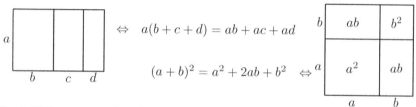

$$\Leftrightarrow \quad a(b + c + d) = ab + ac + ad$$

$$(a + b)^2 = a^2 + 2ab + b^2 \quad \Leftrightarrow$$

Eucl. II.5 concerns the identity

$$a^2 - b^2 = (a + b)(a - b)$$

(see Fig. 2.14 left). The two light grey rectangles are the same. If one adds the dark rectangle to each, one obtains on the left the rectangle $(a + b) \times (a - b)$, and on the right an L-shaped "gnomon", which is the difference of a^2 and b^2.

Eucl. II.8. The identity $(a + b)^2 - (a - b)^2 = 4ab$ (see Exercise 14 below).

Eucl. II.13. The identity[6]

$$2uc = b^2 + c^2 - a^2 \tag{2.2}$$

for the segment u cut off from the side of a triangle by the altitude (see Fig. 2.14, middle). Euclid obtains this result from $c^2 + u^2 = 2cu + (c - u)^2$ (which is Eucl. II.7, a variant of Eucl. II.4), by adding h^2 on both sides and applying Eucl. I.47 twice.

---

[6]The original text, in Heath's translation, is as follows: "In acute-angled triangles the square of the side subtending the acute angle is less than the squares on the sides containing the acute angle by twice the rectangle contained by one of the sides about the acute angle, namely that on which the perpendicular falls, and the straight line cut off within by the perpendicular towards the acute angle." We see how complicated life was before the invention of good algebraic notation; and the case of an obtuse angle, where u becomes negative, required another proposition (Eucl. II.12).

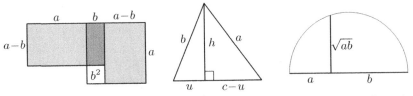

Fig. 2.14. Eucl. II.5 (left), Eucl. II.13 (middle), and Eucl. II.14 (right)

Remark. For a *direct* proof of (2.2), without using Pythagoras' theorem, see Exercise 18 below. With the advance of algebra, the above propositions can all be obtained from Eucl. II.1 by simple calculations. However, Euclid's figures remain beautiful illustrations for these algebraic identities and, moreover, pictures such as that in Fig. 2.14 (left) appeared at the very beginning of this algebra (see Fig. II.1 below).

Eucl. II.14 proves the *altitude theorem* (1.10), by using Eucl. II.8 in the same way[7] as in Exercise 22 of Chap. 1. It allows the *quadrature of a rectangle*, i.e. the construction of a square with an area equal to that of a given rectangle (see Fig. 2.14 right).

2.2 Book III. Properties of Circles and Angles

The third book is devoted to circles and angles. For instance, Eucl. III.20 is the *central angle theorem*, see Theorem 1.4 and Fig. 1.9; Eucl. III.21 is a variant of this theorem, see Exercise 3 of Chap. 1.

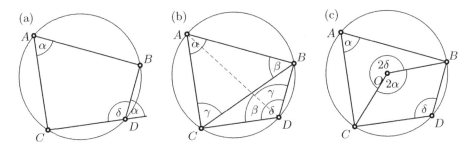

Fig. 2.15. Angles of a quadrilateral inscribed in a circle (Eucl. III.22)

Eucl. III.22. *Let $ABDC$ be a quadrilateral inscribed in a circle, as shown in Fig. 2.15 (a). Then the sum of two opposite angles equals two right angles:*

$$\alpha + \delta = 2\,\llcorner\,. \tag{2.3}$$

---

[7]It also follows from Eucl. III.35 below, for the particular case where AB is a diameter and CD is orthogonal to AB.

Proof by Euclid. We consider the triangle ABC in Fig. 2.15 (b). By Eucl. III.21, we have the two angles β and γ at the point D. This shows that $\delta = \beta + \gamma$. The result is thus a consequence of Eucl. I.32.   □

Another proof of Eucl. III.22. It is clear from Fig. 2.15 (c) that the central angles cover the four right angles around O, i.e., by applying Eucl. III.20, we have $2\alpha + 2\delta = 4\llcorner$. (Euclid did not consider angles greater than $2\llcorner$; hence he would not have presented such a proof.)   □

Eucl. III.35. *If two chords AB and CD of a circle intersect in a point E inside the circle (see Fig. 2.16 (a)), then*

$$AE \cdot EB = CE \cdot ED. \tag{2.4}$$

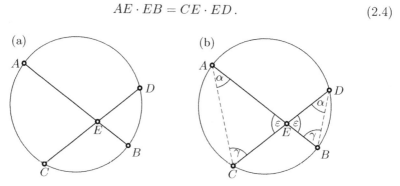

Fig. 2.16. Eucl. III.35 (a) and its proof by Thales' theorem

Proof. Concerned by rigour, Euclid persistently refuses to use Thales' theorem. Hence his proof, repeatedly using Pythagoras' theorem (Eucl. I.47), requires $1\frac{1}{2}$ pages. Being less scrupulous, we see by Eucl. III.21 that the triangles AEC and DEB are similar, see Fig. 2.16 (b). Hence (2.4) follows from Thales' theorem.   □

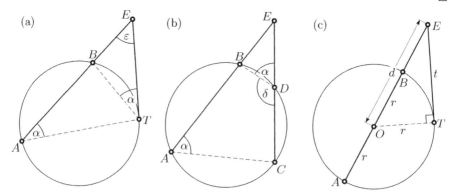

Fig. 2.17. Eucl. III.36 (a); Clavius' corollary (b); relation with Pythagoras' theorem and Steiner's power of a point with respect to a circle (c).

Eucl. III.36. *Let E be a point outside a circle and consider a line through E that cuts the circle in two points A and B. Further let T be the point of tangency of a tangent through E (see Fig. 2.17 (a)). Then*

$$AE \cdot BE = (TE)^2 . \tag{2.5}$$

Proof. The two angles marked α in Fig. 2.17 (a) are equal by Eucl. III.21, because they are inscribed angles on the arc BT (the second one is a limiting case as in Eucl. III.32, cf. Exercise 17 on page 57). Hence ATE is similar to TBE and the result follows from Thales' theorem. This, again, is not Euclid's original proof.   □

Corollary (Clavius 1574). *Let A, B, C and D denote four points on a circle. If the line AB meets the line CD in a point E outside the circle (see Fig. 2.17 (b)), then*

$$AE \cdot BE = CE \cdot DE . \tag{2.6}$$

Proof. This is clear from Eucl. III.36, because $AE \cdot BE$ and $CE \cdot DE$ are both equal to $(TE)^2$.

We can also prove this corollary directly by Eucl. III.22, because the triangles AEC and DEB are similar. Then Eucl. III.36, as well as the picture Fig. 2.17 (a), would be limiting cases where C and D coincide.   □

Remark. The particular case of Eucl. III.36, in which AB is a diameter of the circle (see Fig. 2.17 (c)), leads to $t^2 = (d + r)(d - r) = d^2 - r^2$. This is in accordance with Pythagoras' theorem since the angle at T is right by Eucl. III.18 (see Exercise 16). The quantity $d^2 - r^2$ is called the *power of the point E with respect to the circle*, an important concept introduced by Steiner (1826a, §9).

Book IV. This book treats circles, inscribed in or circumscribed to triangles, squares, regular pentagons (Eucl. IV.11), hexagons (Eucl. IV.15). Without Thales' theorem, the treatment of the pentagon is still unwieldy. The more elegant proof that we gave in Chap. 1 appears much later in the *Elements* (Eucl. XIII.9). The book ends with the construction of the regular 15-sided polygon (Eucl. IV.16, see Fig. 2.18).

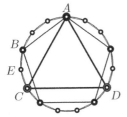

Fig. 2.18. Eucl. IV.16 (left); application to modern car technology (right).

2.3 Books V and VI. Real Numbers and Thales' Theorem

> "There is nothing in the whole body of the *Elements* of a more subtile invention, nothing more solidly established, and more accurately handled than the doctrine of proportionals."
>
> (I. Barrow; see Heath, 1926, vol. II, p. 186)

Book V. The theory of proportions. This theory is due to Eudoxus and has been greatly admired. It concerns ratios of irrational quantities and their properties. One constantly works with inequalities that are multiplied by integers. One thereby *squeezes* irrational quantities between rational ones, somewhat in the style of *Dedekind cuts* 2200 years later.

Book VI. Thales-like theorems. Once the theory of proportions is established, one can finally give a rigorous proof of Thales' theorem.

Eucl. VI.2. *If BC is parallel to DE, then* $\dfrac{a}{c} = \dfrac{b}{d}$ *(see the figure on the left).*

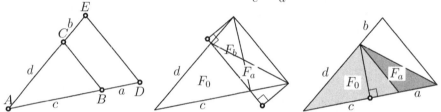

Proof. One joins B to E and C to D. This gives two triangles with the same base CB and the same altitude, hence with the same area $F_a = F_b$, see the second figure. Thus, if F_0 denotes the area of ABC,

$$F_a = F_b \quad \Rightarrow \quad \frac{F_a}{F_0} = \frac{F_b}{F_0} \quad \Rightarrow \quad \frac{a}{c} = \frac{b}{d}$$

since $\dfrac{F_a}{F_0} = \dfrac{a}{c}$. (We use here the fact that both triangles have the same altitude on AD, see the figure on the right.) □

Eucl. VI.3 (Theorem of the angle bisector). *Let CD be the bisector of the angle γ. Then* $\dfrac{a}{b} = \dfrac{p}{q}$ *(see the figure on the left).*

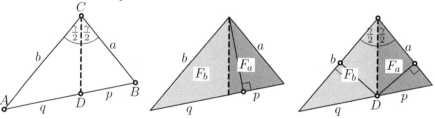

Proof. Euclid proves this theorem as an application of Eucl. VI.2. We, however, use the spirit of the above proof and consider the areas F_a and F_b of the triangles DBC and ADC, respectively. These triangles have the same altitude on AB (see second figure). As the points on the angle bisector have the same distance from both sides (a consequence of Eucl. I.26), the triangles have the same altitude on AC and BC, respectively, see the figure on the right. Thus we have on the one hand

$$\frac{F_a}{F_b} = \frac{p}{q}, \quad \text{and on the other hand} \quad \frac{F_a}{F_b} = \frac{a}{b}. \qquad \square$$

The subsequent propositions are variants of Thales' theorem and their converses; Eucl. VI.9 explains how to cut off a rational length from a line, see Fig. 1.6; Eucl. VI.19 proves Theorem 1.6 on the areas of similar triangles. It is only now that Euclid is fully prepared for Naber's proof of the Pythagorean theorem, see Fig. 1.21.

2.4 Books VII and IX. Number Theory

These books introduce a completely different subject, the theory of numbers (divisibility, prime numbers, composite numbers, even and odd numbers, square numbers, perfect numbers). The later development of this theory, now called *number theory*, with results that are simple to enunciate, but whose proofs require the deepest thought and the most difficult considerations, became the favourite subject of the greatest among the mathematicians (Fermat, Euler, Gauss[8]) and is still full of mysteries and open problems.

The results are not geometrical, but the way of thinking is, at least for Euclid.

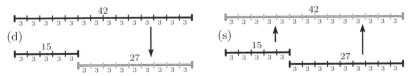

Fig. 2.19. Measure of difference (d) and sum (s) of two numbers

The book starts with propositions about the divisibility of numbers. The main tool is the observation, already known from Book V (in particular Eucl. V.1 and V.5), that if a number divides (Euclid says "measures") *two* quantities, it also divides their *difference* (see Fig. 2.19, (d)), and their *sum* (Fig. 2.19, (s)). This leads to Eucl. VII.2, better known as the Euclidean algorithm.

---

[8] "Die schönsten Lehrsätze der höheren Arithmetik ... haben das Eigne, dass ... ihre Beweise ... äusserst versteckt liegen, und nur durch sehr tief eindringende Untersuchungen aufgespürt werden können. Gerade diess ist es, was der höheren Arithmetik jenen zauberischen Reiz gibt, der sie zur Lieblingswissenschaft der ersten Geometer gemacht hat." (Gauss, 1809; *Werke*, vol. 2, p. 152)

Eucl. VII.2. *Given two numbers not relatively prime, to find their greatest common measure.*

The Euclidean algorithm.[9] Given a pair of distinct positive integers, say a, b with $a > b$, subtract the smaller from the larger. Then repeat this with the new pair $a - b, b$. Any common divisor of a and b also divides $a - b$ and b, and conversely. Therefore, the last non-zero difference is divisible by the greatest common divisor of a and b, and divides it. Hence it is their *greatest common divisor*.

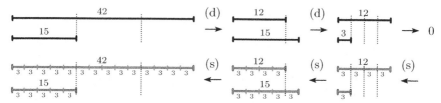

Fig. 2.20. Euclidean algorithm for the greatest common measure of two numbers

Other highlights of these books are Eucl. VII.34 on the least common multiple of two numbers and Eucl. IX.20 on the fact that the number of primes is infinite.

Book X. A classification of irrational numbers

This book is the culmination of the mathematical theory of the *Elements*, using the tools from analysis (Books V and VI) and number theory (Books VII–IX) in order to set up an immense classification of irrationals (with 115 propositions in all).

Eucl. X.1. This is the first convergence result in history, telling us that for n sufficiently large, $a \cdot 2^{-n}$ becomes smaller than any number $\varepsilon > 0$.[10] The main advantage of this proposition is to terminate proofs which otherwise would go on indefinitely (see e.g. Eucl. X.2 and Eucl. XII.2 below).

Eucl. X.2 applies the algorithm of Eucl. VII.2 to *real* numbers. If the algorithm never terminates, the ratio of the two initial numbers $a > b$ is *irrational*.[11] Two thousand years later, this led to the theory of *continued fractions* (see e.g. Hairer and Wanner, 1997, p. 67).

Example. In Fig. 2.21 we see the Euclidean algorithm applied to $a = \Phi$ (resp. $a = \sqrt{2}$) and $b = 1$. We see that we obtain an infinite sequence of similar triangles (resp. squares) and an unending sequence of remainders $c = a - b$,

---

[9]The Arabic word "algorithm" only appeared some thousand years later.

[10]The ε, though a Greek letter, came into use for this purpose only with Weierstrass many many centuries later. If you want to know, Euclid used a capital Γ at this place.

[11]In Euclid's words: a and b are *incommensurable*.

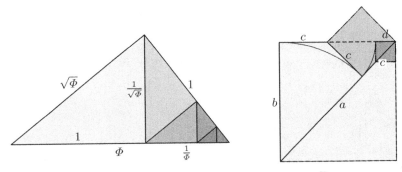

Fig. 2.21. Euclidean algorithm for Φ and $\sqrt{2}$

$d = b - c$, $e = c - d$ (resp. $c = a - b$, $d = b - 2c$, $e = c - 2d$), etc. Hence, both Φ and $\sqrt{2}$ must be irrational. The second picture is inspired by a drawing in Chrystal (1886, vol. I, p. 270), the first by a result of Viète (1600), who discovered that Φ, $\sqrt{\Phi}$ and 1 form a Pythagorean triple.

Other highlights of this book are Eucl. X.9, which shows that numbers like $\sqrt{2}$, $\sqrt{3}$, $\sqrt{5}$, $\sqrt{6}$, etc. are irrational, and Eucl. X.28, which contains the construction of Pythagorean triples.

2.5 Book XI. Spatial Geometry and Solids

Book XI introduces solids (στερεός). Euclid gives the definition of a *pyramid* (πῠρᾰμίς; a solid formed by a polygon, an apex and triangles; see Fig. 2.22),

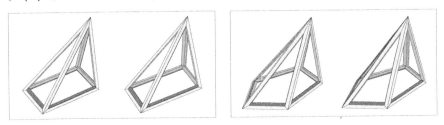

Fig. 2.22. Pyramids over a rectangle and over a pentagon, respectively

a *prism* (πρῖσμα; a solid formed by a polygon, a second identical polygon parallel to the first one, and parallelograms; see Fig. 2.23, left),

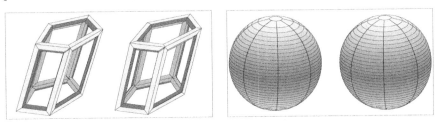

Fig. 2.23. Prism over a pentagon (left) and sphere (right)

a *sphere* (σφαῖρᾰ; a solid obtained by rotating a semicircle around the diameter; see Fig. 2.23, right), a *cone* (κῶνός; a solid formed by rotating a right-angled triangle around a leg; see Fig. 2.24, left), a *cylinder* (κύλινδρος, rotation of a rectangle around a side; see Fig. 2.24, right),

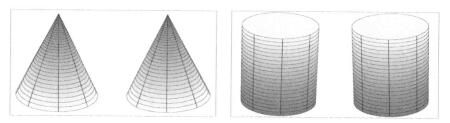

Fig. 2.24. Cone and cylinder

a *cube* (κῦβος; see Fig. 2.25, left), an *octahedron* (ὀκτάεδρον from ὀκτάεδρος – eight-sided; see Fig. 2.25, right)

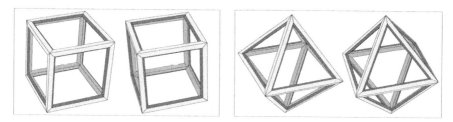

Fig. 2.25. Cube and octahedron

an *icosahedron* (εἰκοσάεδρον; see Fig. 2.26, left), and finally a *dodecahedron* (δωδεκάεδρον; see Fig. 2.26, right).

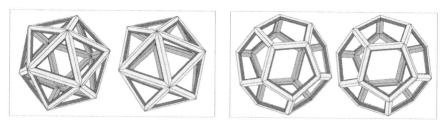

Fig. 2.26. Icosahedron and dodecahedron

The four last ones, together with the *tetrahedron* (τετράεδρον, with four faces) which Euclid does not define, form the class of regular polyhedra. This class is identical to that of the *Platonic solids* or *cosmic figures*; Plato described them in his *Timæus* and associated them to the *five elements* (cube ↔ earth, icosahedron ↔ water, octahedron ↔ air, tetrahedron ↔ fire, dodecahedron ↔ ether). An illustration by Kepler is reproduced in Fig. 2.27.

Fig. 2.27. Platonic solids (drawings by Kepler, *Harmonices mundi*, p. 79, 1619)

We further note the interesting fact that tetrahedron $\leftrightarrow$ tetrahedron, octahedron $\leftrightarrow$ cube, and dodecahedron $\leftrightarrow$ icosahedron are seen to be dual by joining the *centres* of the faces of the regular polyhedra, see Figs. 2.28–2.30.

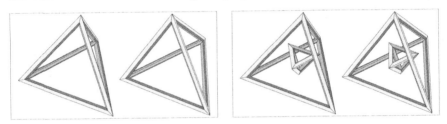

Fig. 2.28. Self-duality of tetrahedron

Fig. 2.29. Duality between cube and octahedron

Fig. 2.30. Duality between icosahedron and dodecahedron

Euclid omitted the definition of the *parallelepiped* (παραλληλεπίπεδον, a solid with parallel surfaces) and of the *right-angled parallelepiped* (where all angles are right), see Fig. 2.31.

Fig. 2.31. Parallelepiped and right-angled parallelepiped

Eucl. XI.1–XI.26. Properties of planes, lines and angles in space. We postpone these questions to Part II where we will discuss them using tools from linear algebra.

Eucl. XI.27 ff. Volume of prisms and parallelepipeds. We have

$$\mathcal{V} = \mathcal{A} \cdot h \qquad \text{where} \quad \mathcal{A} = \text{area of the base;} \quad h = \text{altitude.} \qquad (2.7)$$

The proofs are in the style of the second figure of Fig. 1.11 (cut off a piece and add it onto the other side). An alternative proof—in the spirit of Archimedes—can be given by cutting the solid into thin slices (*exhaustion method*); for an illustration, see Fig. 2.32).

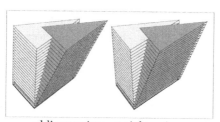

oblique prism → right prism

Fig. 2.32. Transformation of an oblique prism into a right prism

2.6 Book XII. Areas and Volumes of Circles, Pyramids, Cones and Spheres

Areas and volumes of more complicated figures are the topic of Book XII. Euclid starts with circles.

Eucl. XII.2. *The areas $\mathcal{A}_1$ and $\mathcal{A}_2$ of two circles C_1 and C_2 of radii r_1 and r_2, respectively, satisfy*

$$\frac{r_2}{r_1} = q \qquad \Rightarrow \qquad \frac{\mathcal{A}_2}{\mathcal{A}_1} = q^2. \qquad (2.8)$$

Proof. The proof is based upon Eucl. VI.19, see Theorem 1.6. Its rigour is impressive.

Suppose that $\frac{A_2}{A_1} > q^2$, i.e.

$$q^2 A_1 < A_2 . \qquad (2.9)$$

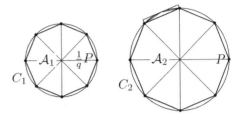

Fig. 2.33. Proof of Eucl. XII.2

We now apply an idea, called the *method of exhaustion* and attributed by Archimedes to Eudoxus: we inscribe in the circle C_2 a polygon P whose area fits in the gap given by (2.9). In order to see that this is possible, one shows that by doubling the number of points of P, the difference of the areas diminishes by at least the factor $\frac{1}{2}$ (see the small rectangle in Fig. 2.33, right). One then applies Eucl. X.1 and obtains for the area of P

$$q^2 A_1 < P < A_2 . \qquad (2.10)$$

The polygon P is then divided by q and transferred into C_1. Then, by Eucl. VI.19, and because $\frac{1}{q}P$ is contained in C_1,

$$\frac{1}{q^2} P < A_1 .$$

If this inequality is multiplied by q^2, we obtain a contradiction with (2.10).

For the assumption $\frac{A_2}{A_1} < q^2$ one exchanges the roles of C_1 and C_2 and arrives at a similar contradiction. Thus, the only possibility is $\frac{A_2}{A_1} = q^2$. $\square$

Euclid, with his disdain for all practical applications, says not a word about the actual value of the similarity factor, which is today denoted by π. With the famous estimate (1.11) we obtain

$$A = r^2 \pi \qquad \text{where } \pi \text{ is a number satisfying} \quad 3\,\frac{10}{71} < \pi < 3\,\frac{1}{7} \qquad (2.11)$$

(see Exercise 22 below).

Eucl. XII.3–XII.9. Volumes of pyramids. The result is

$$V = \frac{A \cdot h}{3} \qquad \text{where} \quad A = \text{area of the base,} \quad h = \text{altitude.} \qquad (2.12)$$

We again prefer to give a proof by using thin slices, see Fig. 2.34. To make the factor 1/3 convincing, Euclid decomposes a triangular prism into *three* pyramids which have — two by two — the same base and altitude. Thus, all three have the same volume (see upper picture of Fig. 2.35). A simpler proof (Clairaut, 1741) is obtained by cutting a cube into *six* pyramids of altitude $\frac{h}{2}$ (see lower left picture of Fig. 2.35). Cavalieri (1647, Exercitatio Prima,

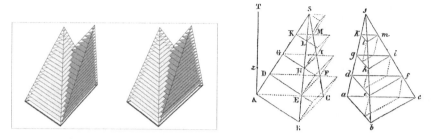

Fig. 2.34. Volume of a pyramid; on the right: drawing by Legendre (1794), p. 203

Prop. 24) shows by calculus that, in modern notation, $\int_0^1 x^2\, dx = \frac{1}{3}$. This is illustrated by a skew quadratic pyramid which, when assembled as in the lower right picture of Fig. 2.35, shows once again that the volumes of the solids "erunt in ratione tripla".

Eucl. XII.10–XII.15. (Volumes of cylinders and cones.) We have:

$$\mathcal{V}_{\text{cylinder}} = r^2\pi h, \qquad \mathcal{V}_{\text{cone}} = \frac{r^2\pi h}{3}. \tag{2.13}$$

Eucl. XII.17. *The volumes $\mathcal{V}_1$ and $\mathcal{V}_2$ of two spheres with radius r_1 and r_2, respectively, satisfy*

$$\frac{r_2}{r_1} = q \qquad \Rightarrow \qquad \frac{\mathcal{V}_2}{\mathcal{V}_1} = q^3. \tag{2.14}$$

The proof is similar to that of Eucl. XII.2, but more involved.

Later, Archimedes (see *On conoids and spheroids*, Prop. XXVII) found that

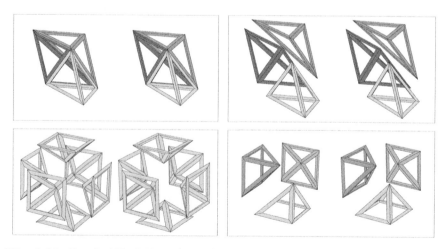

Fig. 2.35. Proof of Eucl. XII.7 (above); proof by Clairaut (below left), Cavalieri (below right)

$$\mathcal{V}_{\text{sphere}} = \frac{4\pi r^3}{3}$$

and the beautiful relation

$$\boxed{\mathcal{V}_{\text{cone}} : \mathcal{V}_{\text{sphere}} : \mathcal{V}_{\text{cylinder}} = 1 : 2 : 3} \tag{2.15}$$

for a cylinder circumscribing the sphere, and a double-cone with the same radius and altitude as the cylinder.

Archimedes' proof uses *slim slices* by observing that, slice by slice, the area $\mathcal{A}$ of the cross-section of the sphere

$$\mathcal{A}_{\text{sphere}} = \rho^2\pi = r^2\pi - x^2\pi = \mathcal{A}_{\text{cylinder}} - \mathcal{A}_{\text{cone}}$$

equals that of the cylinder minus that of the cone. This is obvious from Fig. 2.36, which shows that $\rho = \sqrt{r^2 - x^2}$.

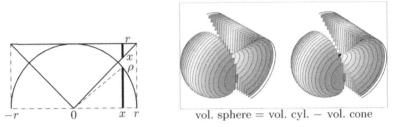

vol. sphere = vol. cyl. − vol. cone

Fig. 2.36. Volume of sphere, cylinder and cone

Book XIII. Construction and properties of the Platonic solids

Eucl. XIII.1–12 are concerned with the golden ratio, the regular pentagon and isosceles triangles, see Chap. 1.

Eucl. XIII.13–18. Euclid constructs the tetrahedron, octahedron, cube, icosahedron and dodecahedron. For the dodecahedron, he starts from a cube by adding *hipped roofs* on each face, as shown in Fig. 2.37, see also Exercise 16 in Sect. 1.9.

Fig. 2.37. The dodecahedron built on a cube

2.7 Epilogue

"Some time ago in Berlin, a brilliant young man from a respected
family was dining with an elderly man, to whom he explained
enthusiastically all the research he was carrying out in geometry,
which is so easy at the beginning and becomes difficult only later.
'For me', said the elderly man, 'the first principles are very difficult
and contain complications which I cannot resolve'. The young man
smiled sarcastically, until someone whispered in his ear: 'Do you
know to whom you are talking? To Euler!' "

(Testimony of L. Hoffmann 1786; quoted from Pont, 1986, p. 467)

"Die vorliegende Untersuchung ist ein neuer Versuch, für die Geometrie ein *vollständiges* und *möglichst einfaches* System von Ax-
iomen aufzustellen und aus denselben die wichtigsten geometri-
schen Sätze ... abzuleiten, ... [The following investigation is a new
attempt to choose for geometry a *simple* and *complete* set of independent axioms and to deduce from them the most important
geometrical theorems ...]"

(D. Hilbert, 1899, p. 1; Engl. trans. by E.J. Townsend, 1902)

"Studying the foundations is not an easy task. If the reader en-
counters difficulties when reading the first chapter ... he may skip
the proofs ... "

(M. Troyanov, 2009, p. 3; transl. from the French)

"Ich habe noch einen kurzen Schlusssatz hinzugefügt – für ungläu-
bige und formale Gemüther. [I have also added a short closing
sentence — for unbelieving and formal minds.]"

(D. Hilbert, letter to F. Klein, 4. 3. 1891)

For more than 2000 years, the *Elements* of Euclid have served as a basic
text in geometry. Their austere beauty has fascinated readers throughout the
ages. However, the *Elements* have also received much critical attention from
the very beginning, examples of which we have already seen in our discussions
following Eucl. I.1 and Eucl. I.4. Authors have repeatedly tried to improve on
Euclid's axioms. A particularly thorough contribution was Legendre's book
(1794), which was reprinted in many editions during more than a century. But
only during the 19th century were final breakthroughs made in two directions:
(a) in relaxing one of Euclid's postulates, creating *non-Euclidean geometry*;
(b) in laying firmer foundations for classical geometry by a complete reorgan-
isation and strengthening of the axioms (Hilbert).

Non-Euclidean geometry. During all these 2000 years, Euclid's Postulate 5
on parallel lines was suspected of being superfluous; this caused an enduring
discussion with innumerable attempts to deduce it from the other postulates.
The continued failure of all these efforts finally aroused the suspicion that
such a proof is impossible. Gauss expressed in several letters to his friends,
but not in print, the idea that one could create an entirely new geometry

which does *not* satisfy Postulate 5. The construction of this so-called *hyperbolic geometry* was carried out and published independently by Bolyai (1832) and Lobachevsky (1829/30) and was the origin of non-Euclidean geometry. The originally very complicated theory was later simplified by the models of Beltrami (see Fig. 7.25 on page 213), Klein and Poincaré. For more details we refer to the textbooks by Gray (2007, Chaps. 9, 10, 11), Hartshorne (2000) and the article Milnor (1982). Many interesting details are given in Klein (1926, pp. 151–155). Very careful historical notes accompany the advanced text Ratcliff (1994) and a complete epistemological account of all the actors of this long development is given in Pont (1986).

Hilbert's axioms. The ongoing formalisation of mathematics in the second half of the 19th century also called for firmer foundations of classical geometry. In 1899, Hilbert came up with a new and "simple" system of 21 axioms, later reduced to 20, because the axiom II.4 was seen to be redundant. This system of axioms characterises plane and solid Euclidean geometry. Many of Euclid's vague definitions for the principal objects of Euclidean geometry, namely *points*, *straight lines* and *planes*, are simply omitted[12] and Hilbert characterises them by their mutual relations, such as *situated*, *between*, *parallel*, and *congruent*. The actual calculations are based on a so-called segment arithmetic, leading first to Pappus' theorem (see Thm. 11.3 on page 325), and then to Thales' theorem as a consequence.

During the 20th century, attempts were made to reduce the large number of Hilbert's axioms. The main idea for this was to assume the real numbers to be known, which allowed, for example in Birkhoff (1932), the introduction of a set of four postulates to axiomatically describe plane Euclidean geometry. His postulates are based on the use of a (scaled) ruler and a protractor; this is made possible by accepting the fundamental properties of the real numbers. In this approach, Thales' theorem is simply postulated.

Despite the great importance of axiomatic systems, their austere character often discourages beginners (see the quotation above). We will therefore abandon at this point the axiomatic bones and turn our attention to a meatier fare. It is interesting to note that Hilbert himself, in his later book written with Cohn-Vossen, *Geometry and the Imagination* (1932), did not mention his own system of axioms at all.

---

[12]In Hilbert's own words, such basic objects may be replaced by *tables*, *chairs* and *beer mugs*, as long as they meet the required relations.

2.8 Exercises

1. Prove the extension by Proclus of Eucl. I.32 (cf. Heath, 1926, vol. I, p. 322): *for any polygon with n vertices the sum of the interior angles satisfies*

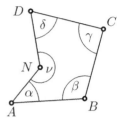

$$\alpha + \beta + \gamma + \ldots + \nu = 2 \llcorner (n - 2). \qquad (2.16)$$

2. The assertion of the first two pictures of Fig. 1.7 (see Chap. 1) for *parallel* angles are Eucl. I.29 together with I.15. Prove the last assertion, for *orthogonal* angles.

3. (Golden ratio with ruler and rusty compass; Hofstetter, 2005.) Extend the construction of Eucl. I.1 and Eucl. I.10, by adding another circle of the same radius centred at the midpoint M (see figure at right), to obtain the point F which divides the segment AB in the golden ratio.

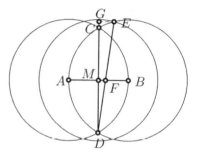

4. Let ABC be a triangle with right angle at C. Show that the vertex C lies on the Thales circle of the hypotenuse AB.

5. Close a gap in the "Stone Age proof" of Thales's theorem in Chap. 1 (see Fig. 1.2): It is *not* evident that the points D and E, after the parallel translations of the triangle ABC, must *really* coincide.

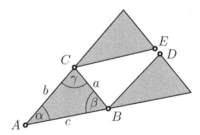

"Figures don't lie, but liars figure."
(Mark Twain [from an e-mail by Jerry Becker])

6. Criticise the "proof" by W.W. Rouse Ball (see Hartshorne, 2000, p. 36) of a wrong variant of Eucl. I.5: *Every triangle is isosceles*, which goes as follows: Let E be the intersection of the angle bisector at A and the perpendicular bisector of BC, see Fig. 2.38, left. Drop the perpendiculars EF and EG. Then use all the valid propositions of Euclid to show that $AF = AG$ and $FB = GC$. From this the "result" follows.

A clever student might object that the intersection point E could be *outside* the triangle. However, this situation is not much better, see Fig. 2.38, right.

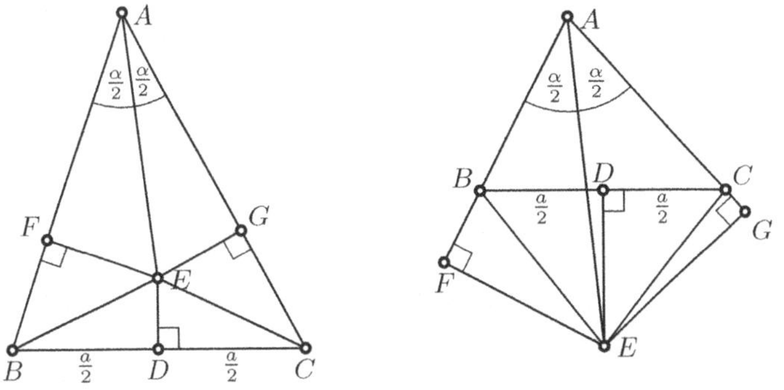

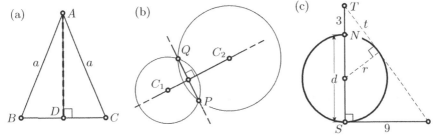

Fig. 2.38. The proof that every triangle is isosceles

7. Let ABC be an isosceles triangle and D the midpoint between B and C (see Fig. 2.39 (a)). Use judiciously chosen propositions of Euclid to prove that the line AD is perpendicular to BC. In the language of Chap. 4 below, we say that the median through A, the bisector of the angle BAC, the perpendicular bisector of BC and the altitude through A coincide.

Fig. 2.39. Median of an isosceles triangle (a); radical axis of two circles (b); the problem of Qin Jiushao (c)

8. Use the result of the previous exercise to show that the *radical axis QP* of two circles (see Fig. 2.39 (b)) is perpendicular to the line joining the two centres.

9. Solve a problem by Qin Jiushao, China 1247:[13] Given a circular walled city of unknown diameter with four gates, one at each of the four cardinal points. A tree T lies 3 li[14] north of the northern gate N. If one turns and walks eastwards for 9 li immediately on leaving the southern gate S, the tree just comes into view. Find the diameter of the city wall (see Fig. 2.39 (c) and Dörrie, 1943, §262).

[13] English wording by J.J. O'Connor and E.F. Robertson, The MacTutor History of Mathematics Archive, http://www-history.mcs.st-andrews.ac.uk/index.html
[14] A *li* is a traditional Chinese unit of length, nowadays 500 m.

10. Prove that the diagonals of a parallelogram
    bisect each other and that, in addition, the
    diagonals of a rhombus are perpendicular
    to each other (see the figure to the right,
    and Def. 22 of Fig. 2.1).

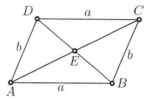

11. Reconstruct Euclid's proof for Eucl. I.18: *In any triangle the greater side
    subtends the greater angle*, i.e. show that if in a triangle AC is greater
    than AB, then β is greater than γ.
    Hint. Insert a point D such that $AB = AD$; see Fig. 2.40 (a).

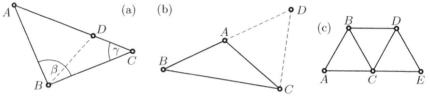

Fig. 2.40. Eucl. I.18; Eucl. I.20 and Eucl. IV.15

12. Give Euclid's proof of the triangle inequality (Eucl. I.20) with the help of
    Fig. 2.40 (b); i.e. show that $AB + AC$ is greater than BC. The auxiliary
    point D is found by producing line AB so that $AD = AC$.

13. The following exercise is the basis for understanding the regular hexagon
    (Eucl. IV.15): if three equal equilateral triangles are as in Fig. 2.40 (c),
    then ACE is a straight line.

14. Find a geometric proof for Eucl. II.8, which expresses the algebraic identity

$$(a + b)^2 - (a - b)^2 = 4ab$$

and was a key relation in the search for Pythagorean triples. (*Hint.* A look
at Fig. 12.1 might help.)

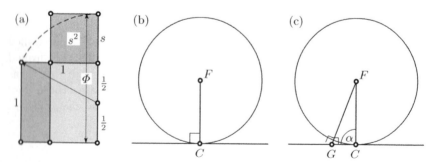

Fig. 2.41. Proof of Eucl. II.11 (a); property of the tangent to a circle (b); Euclid's
proof of Eucl. III.18 (c)

15. Explain the solution of Eucl. II.11 in Fig. 2.41 (a) for the computation of the golden ratio Φ determined by equation (1.4).

16. Discover Euclid's proof for Eucl. III.18: *If a straight line touches a circle with centre F at a point C, then FC is perpendicular to this line* (see Fig. 2.41 (b)). (*Hint.* A look at Fig. 2.41 (c) might help.)

17. Find a proof of Eucl. III.32, which states that *if a line EF touches a circle at B, and if C and D are points on this circle, then the angle DCB is equal to the angle DBE* (see Fig. 2.42, left).

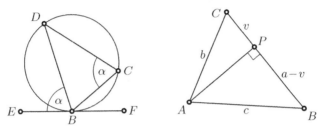

Fig. 2.42. Eucl. III.32 (left); Eucl. II.13 (right)

18. Eucl. II.13, i.e. formula (2.2), written for the situation of Fig. 2.42 (right), reads as
$$a^2 + b^2 - 2av = c^2, \tag{2.17}$$
and is a direct extension of Pythagoras' theorem (1.8). *Question:* can you, inspired by Euclid's proof of Fig. 1.19, find a *direct* proof of (2.17)?

19. Let two circles intersect in two points P and Q (see Fig. 2.43 (a)). From a point T on one of the circles, produce TP and TQ to cut the other circle at A and B. Show that the tangent at T is parallel to AB.

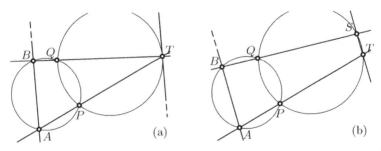

Fig. 2.43. Property of the tangent to a circle (left); two secants to two circles (right)

20. Prove a beautiful result, generally attributed to Jacob Steiner, the *four-circles theorem*: Suppose that four circles intersect in points A, A', B, B', C, C' and D, D' as shown in Fig. 2.44 (a). Show then that A, B, C, D are concyclic (i.e. lie on a circle) if and only if A', B', C', D' are.

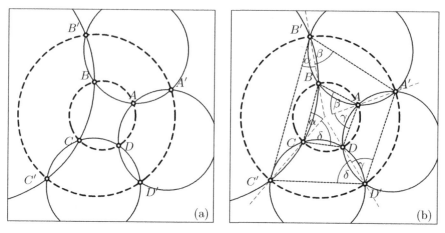

Fig. 2.44. The four-circles theorem (left); its proof (right)

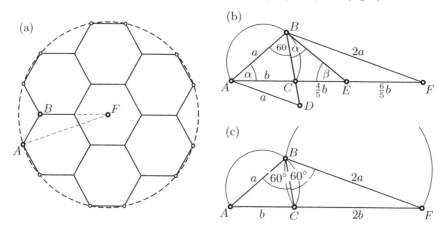

Fig. 2.45. Pappus' hexagon problem

21. Solve "Pappus' last mathematical problem" (from *Collection*, Book VIII, Prop. 16, see Fig. 2.45 (a)): Inscribe in a given circle with radius AF seven identical regular hexagons of maximal size. The problem reduces to the question: Given a segment AF, find a point B such that $BF = 2 \cdot AB$ and the angle ABF is $120°$.

    (a) Verify Pappus' construction (Fig. 2.45 (b)): Insert on the segment AF points C and E such that $AC = \frac{1}{3} \cdot AF$ and $CE = \frac{4}{5} \cdot AC$. Draw on AC a circle containing an angle of $60°$ (by Eucl. III.21), and draw EB, tangent to the circle at B. Then B is the required point.

    (b) Is there an easier solution?

22. (Archimedes' calculation of π.) Compute the perimeters of the regular inscribed and circumscribed 96-gons of a circle of radius 1 to show that

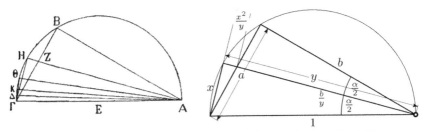

Fig. 2.46. Archimedes' computation of the regular inscribed 96-gon

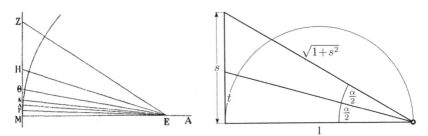

Fig. 2.47. Archimedes' computation of the regular circumscribed 96-gon

$$3\frac{10}{71} < \pi < 3\frac{1}{7} \, .$$

(a) Apply Pythagoras, Thales and Eucl. III.20 to find $x = H\Gamma$ if $a = B\Gamma$ is known (see Fig. 2.46) and H is the midpoint of the arc $B\Gamma$. This allows one to compute successively, starting from the hexagon, the perimeters of the regular dodecagon, 24-gon, 48-gon and 96-gon.

Hint. The triangles ABZ, $AH\Gamma$ and ΓHZ are similar.

(b) Apply Eucl. VI.3 to find $t = H\Gamma$ if $s = Z\Gamma$ is known (see Fig. 2.47). This will lead similarly to the perimeters of the circumscribed regular n-gons.

23. (Another of the divine discoveries of Euler.) Count, for each of the polyhedra from Euclid's Book XI drawn above,

$$s_0 \quad \ldots \quad \text{the number of vertices,}$$
$$s_1 \quad \ldots \quad \text{the number of edges,}$$
$$s_2 \quad \ldots \quad \text{the number of faces.}$$

Make a list of these values and discover Euler's famous relation (Euler, 1758).

3

Conic Sections

"The cream of the classical period's contributions are Euclid's
Elements and Apollonius' *Conica*."   (M. Kline, 1972, p. 27)

"Quotusquisque Mathematicorum est, qui tolerat laborem per-
legendi Appollonii Pergaei Conica? [How few mathematicians
would endure the effort of reading the entire Conics of Apollonius of Perga?]"   (J. Kepler, 1609, from the introduction)

"... i libri di Apollonio, ... delle quali sole siamo bisogni nel presente
trattato [the books of Apollonius, the only tools which we require
in the present treatise]"   (Galilei, *Discorsi* 1638, fourth day)

"A peine la Géométrie sortoit-elle de l'enfance, qu'elle s'occupa
des Sections coniques, ... [Barely out of infancy, geometry devoted
itself to conic sections ...]"   (G. Cramer, 1750, p. vi)

We now turn our attention to another of the great treatises of the classical period, the *Conics* of Apollonius of Perga. Apollonius wrote eight books
on conic sections; the first four have survived in the original Greek text (a
critical edition was published by Heiberg 1893), books V, VI and VII were
reconstructed from arabic texts by E. Halley (1710), the last volume is lost.
We base our quotations on the French translation by Ver Eecke (1923). An au-
thoritative English edition, slightly arranged and adapted to modern notation,
was published by Heath (1896).

The theory of conics was taken up again by Kepler (1604), who included a
short section on conics in his book on astronomy and optics. He emphasised
the two particularly important points of a conic and called them *foci* ("Nos
lucis causâ, et oculis in Mechanicam intensis ea puncta Focos appellabimus").
Many of Apollonius' proofs were later simplified with the use of analytic methods, see Chap. 7. Even the most elegant *geometric idea* in this field, *Dandelin
spheres*, had to wait another 2000 years before being discovered by a Belgian
army engineer (G.P. Dandelin, 1794–1847). This discovery turned the presentation of conics upside down.

A. Ostermann and G. Wanner, *Geometry by Its History*,
Undergraduate Texts in Mathematics, DOI: 10.1007/978-3-642-29163-0_3,
© Springer-Verlag Berlin Heidelberg 2012

Origin of the conics. Hippocrates of Chios solved the problem of doubling the cube, i.e. of finding $x = \sqrt[3]{2}$ (see Section 1.8) or solving $x \cdot x \cdot x = 2$, in the following way: separate *two* of these factors $x \cdot x = y$ to obtain the equations

$$x \cdot y = 2, \qquad y = x^2. \qquad (3.1)$$

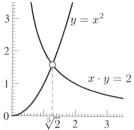

Menaechmus, a pupil of Eudoxus and Plato, discovered that the two curves defined by these equations are generated by the intersection of a plane with a cone. This is how the theory of conics was born (see Viète, 1593b, Caput II, *Historia duplicationis cubi*, for details). It was further developed in (lost) works of Euclid and the famous treatise of Apollonius.

3.1 The Parabola

"And Jesus answered and spoke unto them again by parables, ..."
(The Holy Bible, Matthew 22.1)

παραβολή, comparison, illustration, juxtaposition, analogy ...
(Liddel and Scott, *Greek-English Lexicon*, Oxford)

Definition of a parabola (Pappus, *Collection*, Book VII, Prop. 238). Let d be a line, called the *directrix*, and F be a point, called the *focus*, at distance p from the directrix. The locus of all points P that have the same given distance ℓ from F as from d (see first picture in Fig. 3.1) is called a *parabola*. A parabola is symmetric with respect to the normal to d through the focus. This line of symmetry is called the *axis* of the parabola. It intersects the parabola at its *vertex*.

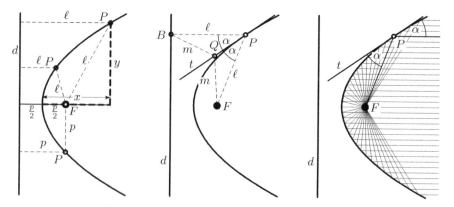

Fig. 3.1. Definition and tangent of a parabola

Theorem 3.1 (Apoll. I.11). *If a cone is cut by a plane that has the same slope as the generators of the cone, then the intersection is a parabola.*

Proof (Dandelin 1822). We use the fact that the tangents from a point P to a sphere all have the same length. They form a cone and touch the sphere along a circle, see Fig. 3.2 (left).

The crucial idea is now to choose a sphere (*Dandelin's*) that touches both the cone (along a circle $AA\ldots$) and the plane π (at a point F that will turn out to be the focus; see the two pictures on the right of Fig. 3.2). Let P be an arbitrary point on the intersection of the cone with the plane, and let A be the intersection point of the circle $AA\ldots$ with the generator of the cone through P. The intersection of the plane containing the circle $AA\ldots$ with π defines a line, the directrix. Let B denote the point on the directrix above P.

We have $PF = PA$, since both segments are tangent to the sphere. But since the plane has the same slope as the cone, we also have $PA = PB$. This concludes the proof. $\qquad\square$

Tangents to a parabola. Let P be an arbitrary point on the parabola, and t the bisector of the angle BPF; see the second picture of Fig. 3.1. For any other point Q of t, we have $BQ = QF$, since the triangles BPQ and FPQ are congruent.[1] But QF is *longer* than the distance of Q from the directrix d, as BQ is *not* orthogonal to d. Thus all points of the line t other than P lie outside the parabola and t is the *tangent* at P.

One of the consequences (use Eucl. I.15, if you like) is that each ray which is parallel to the axis is reflected by a parabolic mirror through the focus of the parabola, see the third picture of Fig. 3.1. The parabolic antennas on our balconies, the parabolic mirrors used in headlights and for astronomical telescopes are all based on this principle.

Defining equation. Denote by x and y the coordinates of the point P on a parabola with respect to the vertex, see the first picture of Fig. 3.1.

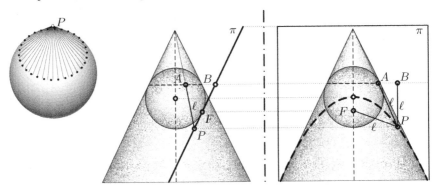

Fig. 3.2. A parabola as the intersection of a cone with a plane

---

[1]Two figures in geometry are called congruent if they can be transformed into another by a combination of translations, reflections and rotations.

Pythagoras' theorem shows that

$$\left(x - \frac{p}{2}\right)^2 + y^2 = \left(x + \frac{p}{2}\right)^2 \qquad \text{whence} \qquad y^2 = 2xp\,, \qquad (3.2)$$

i.e. the area of the square $y \cdot y$ equals the area of the rectangle $x \cdot 2p$. This "comparison, analogy ..." is the origin of the name *parabola* (given by Apollonius, see the quotation). The value $2p$ is called the *latus rectum*, i.e. the length of the vertical segment through the focus.

3.2 The Ellipse

ἐλλῐπής ,   leaving out, omitting, lack ...
(Liddel and Scott, *Greek-English Lexicon*, Oxford)

We perform a construction similar to the preceding one, but suppose now that the intersecting plane π is *less steep* than the generators of the cone, see Fig. 3.3. Consequently, the segment PB is *longer* than PA by a factor that we denote by $\frac{1}{e}$. Here e is a number satisfying $0 \leq e < 1$. It is called the *eccentricity*.

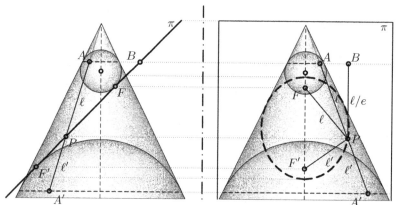

Fig. 3.3. An ellipse as the intersection of a cone with a plane

First definition of an ellipse (Pappus). Consider the focus F at distance p/e from a line d, with $0 \leq e < 1$. The locus of all points P for which the ratio of the distances to the point F and to the line d equals e is called an *ellipse*, see Fig. 3.4 (left). The line d is called the *directrix*.

Defining equation. In the same way as for the parabola, Pythagoras' theorem gives us (we now denote the coordinate from the *vertex* V by u, see Fig. 3.4, left),

$$\left(u - \frac{p}{1+e}\right)^2 + y^2 = e^2\left(u + \frac{p/e}{1+e}\right)^2 = \left(eu + \frac{p}{1+e}\right)^2,$$

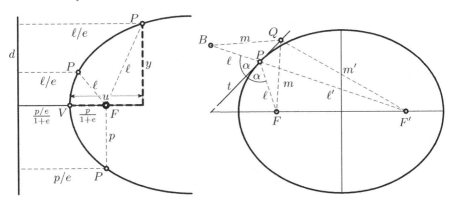

Fig. 3.4. Definition of an ellipse and construction of a tangent

whence

$$y^2 = 2up - (1 - e^2)u^2 \qquad \text{or} \qquad (1 - e^2)u^2 - 2up + y^2 = 0, \qquad (3.3)$$

i.e. the area of the square $y \cdot y$ is *smaller* than that of the rectangle $u \cdot 2p$. This "omission" or "lack" motivated Apollonius (Apoll. I.13) to call such a curve an *ellipse*, see the above quotation.

Second definition of an ellipse. We now place a *second* Dandelin sphere on the other side of the plane π (see Fig. 3.3), which touches the plane at a second focus F'. The two spheres touch the cone along two parallel circles. Consequently, the *sum* of the two distances PA and PA' is a constant. By the same reasoning as before we get that

$$\text{the sum of the distances } PF \text{ and } PF' \text{ is constant.} \qquad (3.4)$$

For Apollonius, this is Prop. 52 of Book III (Apoll. III.52).

Tangents to an ellipse (Apoll. III.48). We find the tangents to an ellipse by an idea very similar to that used for the parabola: let P be a point on the ellipse (see Fig. 3.4, right), join P to F and F' at distance ℓ and ℓ', respectively. Then produce $F'P$ by the distance ℓ to obtain B, so that the distance $F'B$, by (3.4), is the same for all points of the ellipse. We draw the bisector t of the angle BPF. Consequently, we have $BQ = QF = m$ for any other point Q on t since the triangle BQF is isosceles. But $m + m'$ is *longer* than $F'B$, because the line BQF' is not straight (Eucl. I.20). Thus all points of the line t other than P lie outside the ellipse and t must be the *tangent* at P.

The mirror property of parabolas is thus modified as follows: *all rays emitted by one focus are reflected into the other focus* (see the figure on the right).

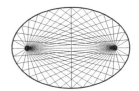

Second defining equation. We choose several particular locations for the point P on the ellipse (see the pictures on the left of Fig. 3.5):

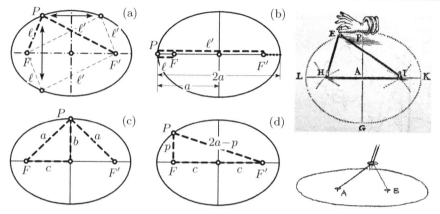

Fig. 3.5. An ellipse and its foci; right: drawing from van Schooten (1657, above), drawing by R. Feynman (conference on Newton's law of gravitation, 1965, below)

(a) The ellipse is *symmetric*, firstly with respect to the symmetry axis FF', secondly with respect to the symmetry axis which is the perpendicular bisector of FF'.

(b) If we place P at the left vertex on the first symmetry axis, the line FPF' extends from F to P and back to F'. By symmetry, this distance (i.e. the constant of (3.4)) is the longest diameter of the ellipse. We denote this constant by $2a$, and call it the *major axis*, so that

$$\ell + \ell' = 2a. \tag{3.5}$$

The constant a is the distance of a vertex from the centre and is called the *semi-major axis*.

(c) By placing P on the second symmetry axis, we see by Pythagoras that

$$b^2 = a^2 - c^2, \quad \text{or} \quad c^2 = a^2 - b^2, \tag{3.6}$$

where b is the *semi-minor axis* and c is the distance of a focus from the centre.

(d) We place P vertically above F (see Fig. 3.5 (d)). The quantity $p = PF$ is called the *semi-latus rectum*. We have by Pythagoras $(2a-p)^2 = p^2+(2c)^2$, which simplifies to

$$a^2 - ap = c^2, \quad \text{or} \quad b^2 = ap, \quad p = \frac{b^2}{a}. \tag{3.7}$$

(e) Finally, we place P at the right vertex, where $u = 2a$. This is the point where y^2 in (3.3) vanishes for the second time, i.e. where $2p-(1-e^2)u = 0$. This leads to the relations

$$1 - e^2 = \frac{b^2}{a^2} \quad \text{and} \quad c = ea. \tag{3.8}$$

The last formula, which is obtained from (3.6), motivated the name *eccentricity* for e.

By inserting the relations (3.7) and (3.8) into (3.3) we get $\frac{b^2}{a^2} u^2 - 2u \frac{b^2}{a} + y^2 = 0$. For more symmetry between u and y we divide this equation by b^2 and obtain $\frac{u^2}{a^2} - 2\frac{u}{a} + \frac{y^2}{b^2} = 0$. Adding 1 on both sides in order to transform the first two terms into a "complete square" we obtain $(\frac{u}{a} - 1)^2 + \frac{y^2}{b^2} = 1$. If we now set $u - a = x$ or $u = x + a$, which means that the coordinate x measures the horizontal distance from the centre of the ellipse (see Fig. 3.6, left), we finally obtain the equation

$$\frac{x^2}{a^2} + \frac{y^2}{b^2} = 1, \tag{3.9}$$

which is as wonderfully symmetric as the curve itself.

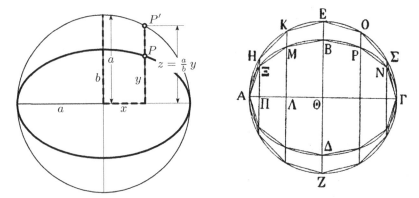

Fig. 3.6. Construction of an ellipse from a circle by a similarity transformation; right: drawing by Archimedes (*On conoids and spheroids*)

Relation with a circle. Those readers still not satisfied by the simplicity of formula (3.9) can set $z = \frac{a}{b} y$, so that this equation becomes

$$\frac{x^2}{a^2} + \frac{z^2}{a^2} = 1, \qquad \text{or} \qquad x^2 + z^2 = a^2,$$

the equation of a circle with radius a. Thus, each *slim slice* of the ellipse is shorter by a factor $\frac{b}{a}$ than the corresponding one of the circle. This property was used by Archimedes in one of his first theorems in *On conoids and spheroids* to conclude that (see the pictures in Fig. 3.6)

$$\mathcal{A}_{\text{ellipse}} = \frac{b}{a} \cdot \mathcal{A}_{\text{circle}} = \frac{b}{a} \cdot a^2\pi = ab\pi. \tag{3.10}$$

Proclus' construction of an ellipse. We carry out the similarity transformation $y \mapsto \frac{a}{b} y$ with the help of Thales' theorem. The ellipse is thus generated by two circles of radius a and b, respectively. Rotating the ray OBA around

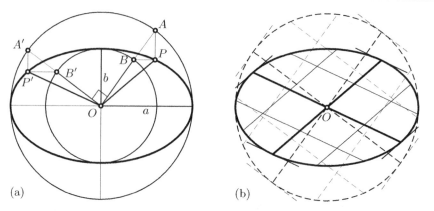

Fig. 3.7. Construction of an ellipse by Proclus (a) and conjugate diameters (b)

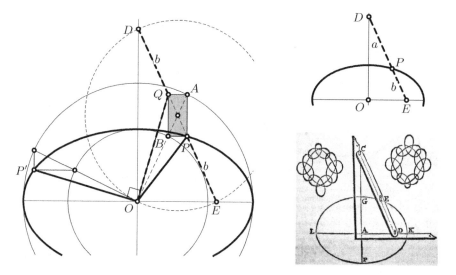

Fig. 3.8. Rytz' construction of an ellipse from two conjugate diameters (left); construction with the help of a gliding stick (right, together with illustration from van Schooten 1657)

the centre, one obtains the point P of the ellipse by a horizontal projection of B and a vertical projection of A, see Fig. 3.7 (a).

Conjugate diameters. The two diameters in Proclus' construction, coming from the two *orthogonal* rays OA and OA', are called *conjugate diameters*, see Fig. 3.7 (a). Each diameter is parallel to the tangents at the endpoints of its conjugate diameter and cuts its conjugate diameter in the midpoint (Apoll. II.6; see Fig. 3.7 (b)). These properties follow from the corresponding properties of orthogonal diameters of a circle by using again and once more Thales' theorem.

Rytz' construction of the semi-axes of an ellipse from two conjugate diameters. We now treat the problem of finding the semi-axes of an ellipse from a given pair of conjugate diameters. A first construction is due to Pappus (*Collection*, Book VIII, §XVII, see Exercise 19 below). Step by step, starting with Frézier (1737, p. 132)[2] and Euler (1753), simpler and simpler constructions were found. The crucial idea, independently found by Frézier and Euler (cf. his last construction in E192), was to rotate one of the diameters by $\llcorner$. Consequently, we rotate by $\llcorner$ in Fig. 3.7 (left) the semi-diameters OP' together with the triangle attached at P'. We so obtain the segment OQ and the triangle QBA, see Fig. 3.8 (left). This triangle joins the supporting triangle of P to form a rectangle which is parallel to the axes. Since the distances $AO = QE = PD = a$ and $BO = PE = QD = b$, the midpoint M between P and Q has the same distance $\frac{a+b}{2}$ from O, from E and from D. This leads to the following construction:

Let OP and OP' be a given pair of conjugate semi-diameters. Rotate OP' by $\llcorner$ towards OP to obtain the segment OQ. Let M denote the midpoint between P and Q. Draw the circle with centre M passing through O. This circle will cut the line PQ at the points D and E. Then the lines OE and OD point in directions of the axes; the distances $DP = a$ and $PE = b$ are the semi-axes of the ellipse.

Since 1845 this construction has been attributed in books on descriptive geometry to Daniel Rytz, "Professor der Mathematik an der Gewerbeschule zu Aarau [professor of mathematics at the vocational school at Aarau]".

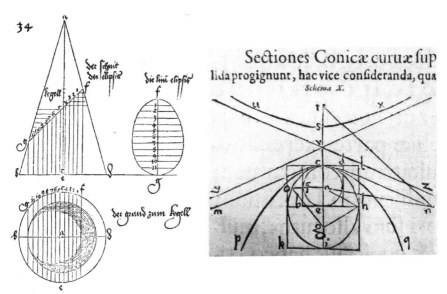

Fig. 3.9. Woodcuts by A. Dürer, Underweysung der messung, 1525 (left); Kepler, Harmonices mundi, 1619 (right)

---

[2]see also Baier (1967)

Proclus' construction with a stick. Figure 3.8 (right) indicates the following method for constructing an ellipse. Imagine that the *stick DE* of length $a+b$ glides with its extremities on the axes, see Fig. 3.8, right). Then the point P on the stick, at distance a from D and b from E, will describe an arc of the ellipse with semi-axes a and b, respectively.

Remark. We illustrate in Fig. 3.9 the growing importance that the conic sections gained for art and science during the Renaissance with two woodcuts, one by Dürer (1525) and one by Kepler (1619).

3.3 The Hyperbola

> ὑπερβολή,   a throwing beyond others, overshooting, excess ...
> (Liddel and Scott, *Greek-English Lexicon*, Oxford)

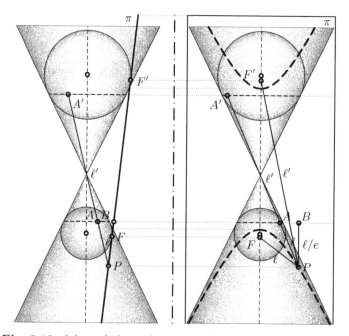

Fig. 3.10. A hyperbola as the intersection of a cone with a plane

This time, we take the plane π *steeper* than the generators of the cone, see Fig. 3.10, i.e. the eccentricity satisfies $e > 1$. Thus, (3.3) becomes

$$y^2 = 2up + (e^2 - 1)u^2, \tag{3.11}$$

and we have *excess* in the area of the square. As the prefix *hyper* is present in many words like hypersensitive, hypertension (students are much too young

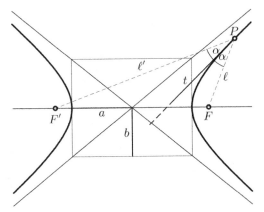

Fig. 3.11. A hyperbola and its tangent

for this), hypermarket, hyperactive, etc., we easily understand why this curve
is called a *hyperbola* (Apoll. I.12).

The theory of the hyperbola is very similar to that of the ellipse. We have,
see Fig. 3.10,

$$\ell - \ell' = \pm 2a \tag{3.12}$$

instead of (3.5), and

$$\frac{x^2}{a^2} - \frac{y^2}{b^2} = 1 \tag{3.13}$$

for the coordinates measured from the centre, instead of (3.9). The foci are
at distance c from the centre with

$$c^2 = a^2 + b^2 \, . \tag{3.14}$$

The tangent is the bisector of the angle FPF', see Fig. 3.11.

Asymptotes of the hyperbola.

> σύμπτωσις ,   falling together, collapsing, meeting ...
> (Liddel and Scott, *Greek-English Lexicon*, Oxford)

With the hyperbola we encounter a new object — the asymptotes (*symptosis*
means *a meeting*, see the quotation, the prefix a- is the negation as in atom,
atypical, asocial, anonymous). An *asymptote* is thus a line that, although
approaching the curve, *never meets* it (Apoll. II.1). We write (3.13) in the
form

$$\left(\frac{x}{a} + \frac{y}{b} \right) \cdot \left(\frac{x}{a} - \frac{y}{b} \right) = 1. \tag{3.15}$$

If the values of x and y become large, the 1 on the right-hand side becomes
negligible and the equation factors to

$$y = \frac{b}{a} \cdot x \quad \text{and} \quad y = -\frac{b}{a} \cdot x \, . \tag{3.16}$$

These are the two lines that the hyperbola approaches as x and y tend to infinity. One further sees that the equation of the hyperbola becomes very simple by taking the asymptotes as axes for the coordinates. Thus, the first curve of (3.1) is a hyperbola.

3.4 The Area of a Parabola

"Sed illum (Archimedem) plures laudant quam legant; admirantur plures quam intelligant [more people praise him (Archimedes) than read him; and more people admire him than understand him]"

(A. Taquet, Antwerpen 1672; copied from Ver Eecke, 1923)

"Qui Archimedem et Apollonium intelligit, recentiorum summorum virorum inventa parcius mirabitur. [Those who perceive the works of Archimedes and Apollonius will marvel less at the discoveries of the greatest modern scholars.]"

(G.W. Leibniz; copied from Ver Eecke 1923)

Another very famous result of Archimedes concerns the area of the parabola:

$$\mathcal{P} = \frac{4}{3} \cdot \mathcal{T} \qquad \text{where} \quad \begin{cases} \mathcal{P} = \text{area of the parabola,} \\ \mathcal{T} = \text{area of the inscribed triangle.} \end{cases} \tag{3.17}$$

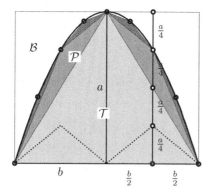 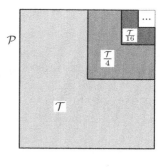

Fig. 3.12. The quadrature of the parabola

Proof. One fills the parabola by first inserting the light-grey triangle $\mathcal{T}$, then by adding the two medium-grey triangles of area $\frac{\mathcal{T}}{8}$ each (same area as the triangles with dotted sides; use (3.2) and Eucl. I.41), then by adding four dark-grey triangles of area $\frac{\mathcal{T}}{64}$ etc., see first picture in Fig. 3.12. Thus

$$\mathcal{P} = \mathcal{T} + 2 \cdot \frac{\mathcal{T}}{8} + 4 \cdot \frac{\mathcal{T}}{64} + \ldots = \mathcal{T} + \frac{\mathcal{T}}{4} + \frac{\mathcal{T}}{16} + \frac{\mathcal{T}}{64} + \ldots .$$

The second picture in Fig. 3.12 (and Fig. 3.13) is particularly ingenious. It shows that the above sum equals $\frac{4}{3} \cdot \mathcal{T}$, since $\mathcal{T}$ (or A in Fig. 3.13) covers three quarters of the square. □

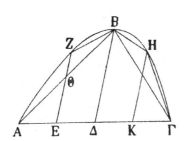 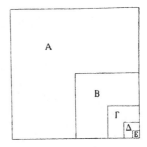

Fig. 3.13. Drawings by Archimedes (quadrature of the parabola)

Remarks. (a) Nowadays, one usually considers the reversed parabola. Thus, the area $\mathcal{B}$ of the white part is *one third of the area of the square*.

(b) The area of a *hyperbola* had to wait for another 19 centuries; its computation is related to the calculus of logarithms (see e.g. Hairer and Wanner, 1997, Sect. I.3).

3.5 Exercises

1. Prove a result of Euler (1748), vol. II, §119: If OP and OP' are two conjugate semi-diameters of an ellipse (see Fig. 3.8, left), then $OP^2 + OP'^2 = a^2 + b^2$ is a constant. This result, which Euler obtained by a long trigonometric calculation, is also one of the last propositions of Apollonius, Apoll. VII.12.

2. Prove a theorem of Newton (*Principia* 1687, Liber I, Lemma XII), saying that *Parallelogramma omnia circa datam Ellipsin descripta esse inter se æqualia*. [All parallelograms circumscribed about any conjugate diameters of a given ellipse are equal.] (See also Newton, *Math. Papers* vol. IV, p. 9, note (24) and vol. VI, p. 35). This theorem earned Newton the admiration of many of his contemporaries, because "in Mathematicks" this universal genius "could sometimes see almost by Intuition, even without Demonstration" (William Whiston, 1749).

3. Show that the locus of points P which have the same distance from a given circle and a given line is a parabola (see Fig. 3.14, left).

4. Find the locus of points P which have the *same distance* from two given circles.

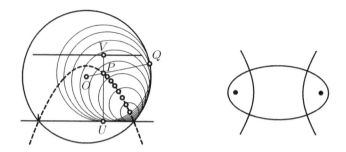

Fig. 3.14. Same distance from circle and line (left); confocal conics (right)

5. Guess a nice property of the tangents to two *confocal* conics (i.e. an ellipse and a hyperbola with the same foci) at the intersection points, and prove it (see Fig. 3.14, right).

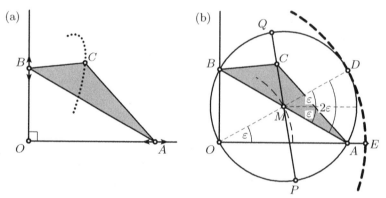

Fig. 3.15. Van Schooten's ellipse-drawing triangle machine

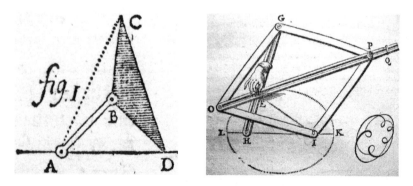

Fig. 3.16. Facsimile of van Schooten's ellipse-drawing triangle machine (left) and parallelogram machine (right)

6. Van Schooten, in his book *Exercitationum mathematicorum* (1657) dedicated an entire chapter to mechanisms for drawing ellipses, parabolas or
   hyperbolas. One of these, which we shall call his *triangle machine*, is represented in Fig. 3.15 (a), facsimile reproduction in Fig. 3.16 (left): two vertices A and B of a fixed triangle ABC glide on two fixed orthogonal lines.
   Show that the third vertex C then moves on an ellipse. Van Schooten's
   original machine of Fig. 3.16 is equivalent, because the midpoint between
   A and B moves on a circle.

7. Explain why van Schooten's *parallelogram machine* in Fig. 3.16 (right)
   produces an ellipse.

8. Prove Apoll. III.50, which states that the orthogonal projections R, R' of
   the foci F, F' onto a tangent of an ellipse lie on the circle with centre O
   and radius a (see Fig. 3.17, left).

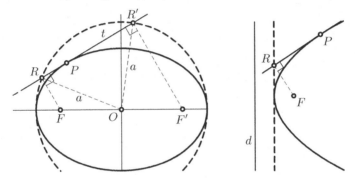

Fig. 3.17. Orthogonal projections of foci onto tangents

9. Prove a result analogous to that of the previous exercise for parabolas,
   i.e. prove that the orthogonal projection R of the focus F onto a tangent
   to a parabola lies on the tangent through the vertex of the parabola (see
   Fig. 3.17, right), so that the ordinate of R is half that of P.

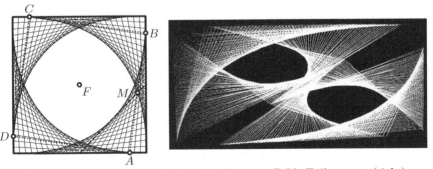

Fig. 3.18. A square rotating in a fixed square (left); Evi's carpet (right)

10. A square $ABCD$ rotates with its vertices gliding on the sides of a fixed square (see Fig. 3.18, left). Analyse the nature of the beautiful curves created by the four sides. Similar curves appear in certain masterpieces of modern art (Fig. 3.18, right).

The following six exercises retrace Apollonius' original approach to the important properties of the conics, two centuries B.C. They allow one to admire repeatedly this genius and to apply repeatedly Thales' theorem and Eucl. III.21.

11. Let AB be the major axis of an ellipse and $\Delta\Gamma$ be the tangent at Γ (see Fig. 3.19 (a)). Prove the following identities:

$$\text{Apoll. I.34:} \qquad \frac{BE}{EA} = \frac{B\Delta}{\Delta A} \qquad \text{or by Thales} \qquad \frac{u'}{u} = \frac{h'}{h},$$

$$\text{Apoll. I.36:} \qquad \frac{ZE}{ZA} = \frac{ZA}{Z\Delta},$$

$$\text{Apoll. III.42:} \qquad h \cdot h' = b^2, \qquad \text{where } b \text{ is the semi-minor axis.}$$

We will later say that the points B, A, E, Δ form a "harmonic set" (see Chap. 11; in particular, the formulas (11.10) of that chapter show the passage from Apoll. I.34 to Apoll. I.36).

Hint. Stretch the ellipse into a circle by the transformation $y \mapsto \frac{a}{b}y$.

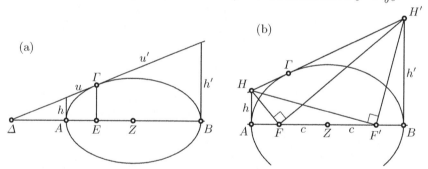

Fig. 3.19. Theorems Apoll. I.34 and I.36, Apoll. III.42 (a); Apoll. III.45 (b)

12. Apoll. III.45: Let $H\Gamma H'$ be the tangent to the ellipse at Γ. Prove, with the results of the previous exercise, that there exist two points F and F' on the major axis, such that the angles $H'FH$ and $H'F'H$ are right angles (see Fig. 3.19 (b)). These points have the same distance c from the centre of the ellipse and c is the same for all tangents (we now call these points the *foci* of the ellipse; Apollonius denoted them by Z and H).

13. Apoll. III.46: Prove, with the results of the previous exercise, that the angles called α in Fig. 3.20 (a) are all equal, as are the angles called β, and those called γ.

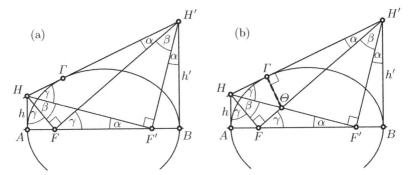

Fig. 3.20. Theorems Apoll. III.46 (a); Apoll. III.47 (b)

14. Apoll. III.47: Let Θ be the intersection of the lines $F'H$ and FH' (see Fig. 3.20 (b)). Prove, with the results of the previous exercises, that the line $\Theta\Gamma$ is perpendicular to the tangent $H\Gamma H'$.

15. We are now in a position to prove a first great result, Apoll. III.48: the angle between $F'\Gamma$ and the tangent is the same as the angle between $F\Gamma$ and the tangent. Because of Apoll. III.47 this is equivalent to: the angles $F'\Gamma\Theta$ and $\Theta\Gamma F$ are equal. Prove this.

16. Prove Apoll. III.49: Consider a situation as in Fig. 3.20 (a), and let Θ be the orthogonal projection of F' onto the tangent HH'. Then $A\Theta B$ is a right angle.

Remark. With Thales' circle for this right angle with diameter AB, we then obtain Apoll. III.50 (see Exercise 8 above) and by an argument reciprocal to the one for that exercise, we finally get Apoll. III.52 as originally proved more than 2200 years ago.

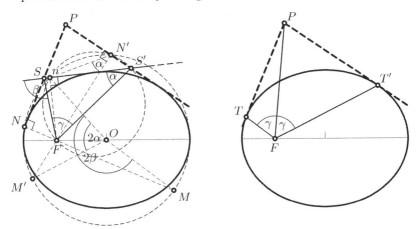

Fig. 3.21. Poncelet's first theorem (left) and its corollary (right)

17. Prove, following the preceding proofs of Apollonius, the so-called "first theorem of Poncelet" (which is "Théorème I" of Poncelet, 1817/18 together with "Théorème II"): *Let the two tangents be drawn to an ellipse from a fixed point P (see Fig. 3.21 left) and let a moving tangent cut these tangents at the points S and S'. Let F be the focus of the ellipse. Then, the angle S'FS = γ is the same for all such tangents.* As a corollary deduce that the two tangents PT and PT' are seen from a focus under the same angle γ (Fig. 3.21 right).

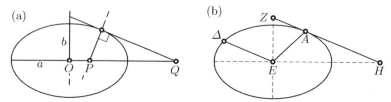

Fig. 3.22. Statement of the exercise of the Math. Assoc. of America (a); Pappus' construction of the axes of an ellipse (b)

18. (An exercise from an envelope of the Math. Assoc. of America): Prove for an ellipse that $OP \cdot OQ = a^2 - b^2$ (see Fig. 3.22 (a)).

19. Show: If EA and $E\Delta$ are conjugate and ZAH is tangent to an ellipse, where Z and H are on the axes (see Fig. 3.22 (b)), then $ZA \cdot AH = E\Delta^2$. This result (together with Eucl. III.35) is the main ingredient of Pappus' construction of the directions EH and EZ of the axes of an ellipse from two conjugate diameters (see Pappus, *Collection*, Book VIII, §XVII and Fig. 5 of Euler's E192, 1753).

20. Prove the following corollary to Apoll. III.42: if d and d' are the distances of the foci of an ellipse from a tangent, then

$$d \cdot d' = b^2 . \qquad (3.18)$$

4

Further Results in Euclidean Geometry

> "With Archimedes and Apollonius Greek geometry reached its cul-
> minating point." (T. Heath, 1921, p. 197)

Euclidean geometry is the oldest field of mathematics. The great thinkers
during all these centuries accumulated an enormous treasure of beautiful ideas
and results. We take pleasure in presenting some of them in this section.

4.1 The Conchoid of Nicomedes, the Trisection of an Angle

Let A be a fixed point, $DDD\ldots$ a fixed line at distance c from A, and b a
given positive value. The curve $CCC\ldots$ such that the distance DC equals b
for each line through A is called the *conchoid* of the line $DDD\ldots$ with respect
to A and with distance b, see Fig. 4.1. Taking b negative creates curves which
may contain a cusp or a loop. Nicomedes originally invented this curve for
doubling the cube (see Exercise 2 on page 178).

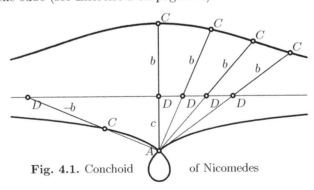

Fig. 4.1. Conchoid of Nicomedes

Trisection of an angle. Pappus discovered (see *Collection*, Prop. IV.32) that
the conchoid can also be used to trisect a given angle α at A (see Fig. 4.2,

A. Ostermann and G. Wanner, *Geometry by Its History,*
Undergraduate Texts in Mathematics, DOI: 10.1007/978-3-642-29163-0_4,
© Springer-Verlag Berlin Heidelberg 2012

left), i.e. to find an angle β satisfying $\beta = \frac{\alpha}{3}$. The construction is described in Fig. 4.2, right. Denote the length of AB by a. Draw the perpendicular BE and the parallel to AE through B. Find the points C on this parallel and D on BE by requiring that A, D, C be collinear and that the length of DC equals $2a$. In other words, C is found with the help of the conchoid of the line EB with respect to A and with distance $b = 2a$ (drawn as a dashed curve). The angle EAD is then $\frac{\alpha}{3}$.

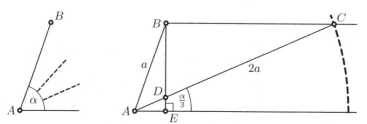

Fig. 4.2. Pappus' trisection of an angle with the help of the conchoid

Proof. Let β be the angle EAD in Fig. 4.3, and let G be the midpoint of the segment DC with $DG = GC = a$. Since G is at half height between D and B, the triangle DGB is isosceles and $GB = a$. Hence the triangles BGC and GBA are also isosceles. Consequently, we have β at C (parallel angle), β at B (isosceles triangle), 2β at G (exterior angle), and finally $2\beta + \beta = 3\beta$ at A (isosceles triangle).   □

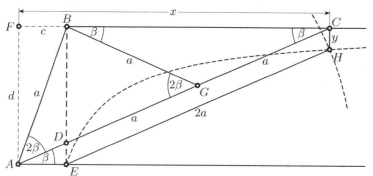

Fig. 4.3. Proof for Pappus' constructions

Pappus' variant. An elegant variant of the above construction, which avoids the conchoid and uses a hyperbola instead, was also found by Pappus (*Collection*, Book IV, Prop. 31). We draw in Fig. 4.3 the segment EH parallel to DC, whose length $2a$ is known. By applying Thales' theorem twice, we get

$$\frac{x}{c} = \frac{CA}{DA} = \frac{BE}{DE} = \frac{d}{y} \qquad \Rightarrow \qquad xy = cd. \tag{4.1}$$

This means that we can find the point H by intersecting the circle of radius $2a$ centred at E with the hyperbola passing through E with asymptotes FA and FB.

Remark. Both of the above constructions do *not* use only ruler and compass (i.e. Post. 1–3 of Euclid). Two thousand years later, it was actually proved (see Chap. 8) that the trisection of an angle with ruler and compass is in general *a mission impossible*!

4.2 The Archimedean Spiral

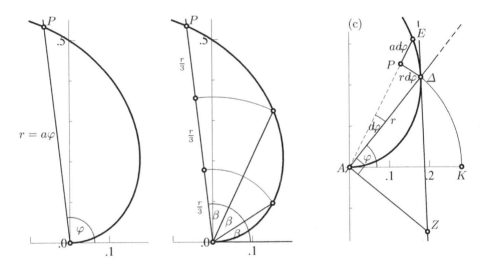

Fig. 4.4. Archimedean spiral for the trisection of an angle (middle), and its tangent (right)

Consider a ray that rotates at constant angular velocity around the origin. Let P be a point on the ray, that moves away from the origin at constant speed. Then the locus of P is a curve called an *Archimedean spiral*, see Fig. 4.4, left. This spiral is obviously a trisectrix (and even an n-sectrix, see Fig. 4.4, middle). If a denotes the quotient of the velocity of the point on the ray and the angular velocity of the ray itself, the spiral is characterised by the formula

$$r = a\varphi. \tag{4.2}$$

Archimedes (in *On spirals*, see Heath, 1897, p. 151) was mainly interested in finding its tangents and its area. It is precisely here that modern differential and integral calculus had its origin.

Prop. XX. *The tangent at a point Δ cuts the line through A orthogonal to $A\Delta$ at the point Z such that $AZ = \text{arc } \Delta K$ (see Fig. 4.4 (c)).*

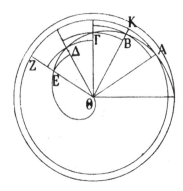

Fig. 4.5. Drawing by Archimedes for the area of the spiral (left); an Archimedean spiral on a larnax, Minoan period (right, Archaeological Museum of Rethymno, Crete)

Idea of the proof. If from a given point Δ we increase the angle φ by a small amount $d\varphi$,[1] we create a small right-angled triangle ΔPE with sides $P\Delta = r\,d\varphi$ and $PE = a\,d\varphi$ (because of (4.2)). If we choose $d\varphi$ small enough, the triangle ΔAZ is similar to the triangle $EP\Delta$. Thus by Thales $\frac{AZ}{r} = \frac{r}{a} = \varphi$, so that $AZ = r\varphi = \text{arc } \Delta K$ is the length of the arc ΔK.   □

Prop. XXIV. *The area between the spiral and the ray* $A\Delta$ *(see Fig. 4.4(c)) equals* one third *of the area of the circular sector* $A\Delta K$.

Idea of the proof. Archimedes simply uses the so-called *Riemann sums*, see Fig. 4.5, left. The factor $\frac{1}{3}$ is related to the area of the parabola (because the area of such a small slice is $\frac{1}{2}r^2 d\varphi$, see the remark at the end of Sect. 3.4), or to the volume of the pyramid.   □

4.3 The Four Classical Centres of the Triangle

> "Ànno i Trianguli rettilinei sì belle Affezioni, che meritano di esser considerate dai Geometri più di quello abbian fatto sinora. [Triangles have such beautiful properties, which deserve more consideration from geometers than they have so far received.]"
> (Conte Giulio Carlo di Fagnano, 1750, vol. II, p. 1)

> "Es ist in der That bewundernswürdig, dass eine so einfache Figur, wie das Dreieck, so unerschöpflich an Eigenschaften ist. [It is indeed remarkable that such a simple figure as a triangle has inexhaustibly many properties.]"   (A.L. Crelle, 1821/22, p. 176)

> "Down with Euclid! Death to triangles!"   (J. Dieudonné, 1959)

---

[1]Here we express Archimedes' argument from 250 B.C. using Leibniz' symbols from 1684 A.D.

Following Euclid's propositions IV.4 and IV.5, an enormous treasury of properties of the triangle was discovered through the centuries (see the first two quotations), not to everybody's delight (see the last quotation). It has become common to denote vertices, side lengths and angles of a triangle in a nicely symmetric way (see the figure).

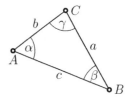

The incentre I of a triangle. Let a triangle ABC be given. The centre of any circle which touches both sides AB and AC lies on the angle bisector of α (see Fig. 4.6, left). Suppose that such a circle grows and finally touches the *third* side. Then its centre has the same distance ρ from all three sides, hence it lies on *all three* angle bisectors (Fig. 4.6, right). We have thus proved the following result.

Eucl. IV.4. *The three angle bisectors of a triangle are concurrent in a point I, called the incentre.*

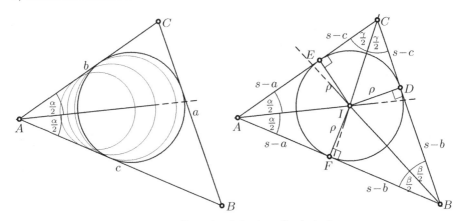

Fig. 4.6. Genesis of the inscribed circle

Remark. If we denote by D, E and F the points of contact of the incircle with the triangle, we have

$$AE = AF, \qquad BF = BD, \qquad CD = CE.$$

The sum of these three quantities is equal to the *semi-perimeter*, usually denoted by s:

$$s = \frac{a+b+c}{2}. \tag{4.3}$$

Since $AF + FB + DC = s$ and $AF + FB = c$, we see that $DC = s - c$ and we obtain similarly all the other quantities indicated in Fig. 4.6 (right).

The circumcentre O of a triangle. The centre of any circle that passes through two vertices A and B of a triangle lies on the perpendicular bisector

of AB (Fig. 4.7 (a)). If this centre moves and the circle finally passes through
the third point C, its centre has the same distance R from *all three* vertices
and thus lies on each of the perpendicular side bisectors (Fig. 4.7 (b)). This
gives the following proposition.

Eucl. IV.5. *The three perpendicular side bisectors of a triangle are concurrent
in a point O, called the circumcentre.*

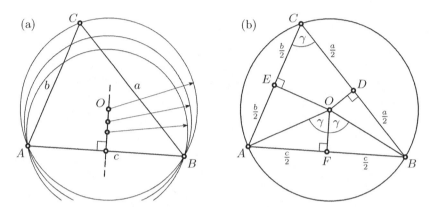

Fig. 4.7. Genesis of the circumscribed circle (a); the circumcentre O and the
surrounding angles

We have the interesting additional information that the segments AO and BO
both make the angle γ with the perpendicular side bisector OF (and similarly
for α and β). This follows from Eucl. III.20, because the central angle 2γ is
cut into halves by OF (see Fig. 4.7 (b)).

The medians and the centroid G.

> βάρος,   weight, burden, load, heaviness, ...
> (Liddel and Scott, *Greek-English Lexicon*, Oxford)

Although Euclid discovered the first two remarkable points of a triangle, it
was reserved to Archimedes, a genius not only in mathematics, but also in
mechanics, to discover the third one, the *centroid*. It was a by-product of his
efforts to find the *centre of gravity*[2] of a triangle. Archimedes proved that this
centre of gravity must lie on each of the lines (called *medians*) connecting a
vertex of the triangle to the midpoint of the opposite side (see Fig. 4.8 (a)).
This is "Proposition 13" of Archimedes' text *On the equilibrium of planes*, see
Heath (1897), p. 198. Archimedes' *first* proof involved parallel (we would say

---

[2]In mechanics, the centre of gravity of a body is usually called *barycentre* from
βάρος, still in use today, see Καθαρό βάρος **2 x 8,5 g = 17 g**. On a box with *Obesity
Management* food one can read, among languages from all rich countries of the
world, **υπερβολικού βάρους**. Geometers from all over the world understand *both* words.

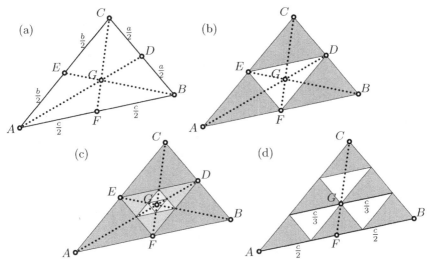

Fig. 4.8. Centroid of a triangle

today: infinitely small) strips. He then gave a *second* proof, which uses the subdivision of the triangle into four similar triangles by the so-called *medial triangle DEF*, whose vertices are the midpoints of the sides of *ABC*, as shown in Fig. 4.8 (b).

Theorem 4.1. *The three medians of a triangle are concurrent in a point G, called the centroid. The centroid divides the medians in the ratio* $2 : 1$.

Proof. By Thales, the medial triangle (see Fig. 4.8 (b)) has the same medians as the original triangle. We now apply this *medial reduction* repeatedly (see Fig. 4.8 (c)). Then the triangles shrink to a point called *G*, which must lie on all three medians.

  For a *second* proof, where the result is seen immediately, not only after an infinity of steps, we divide each side into *three* equal parts (see Fig. 4.8 (d)), as in the Stone Age proof for Thales' theorem of Chap. 1. Then we see by Thales' theorem that the median *CF* passes through *G*, and similarly for the other medians.   □

The altitudes and the orthocentre *H*.

ὄρθιος,   straight up, upright, steep, uphill, ...
(Liddel and Scott, *Greek-English Lexicon*, Oxford)

An *altitude* is a line through a vertex of a triangle, *orthogonal* to the opposite side. We again have a remarkable result:

Theorem 4.2. *The three altitudes of a triangle are concurrent in a point H, called the orthocentre.*

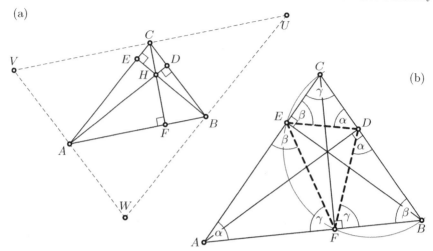

Fig. 4.9. Gauss' proof of Thm. 4.3 for the orthocentre (a); the orthic triangle (b)

Proof. This theorem is contained in a lost manuscript of Archimedes and reappears several times in the work of Pappus, Proclus, Regiomontanus, Ludolph van Ceulen, but either without a proof or with an incorrect one (see the footnotes in Newton's Mathematical Papers, ed. by D.T. Whiteside, vol. 4, p. 454). The first known proofs are from the 17th century and use Thales' theorem (see Exercise 4 on page 108). The apparently most elegant of all possible proofs was found by Gauss (*Werke*, vol. IV, p. 396): We apply the medial reduction of Fig. 4.8 (b) *backwards*, i.e. *we look for a triangle UVW whose medial triangle is ABC*. This is done by drawing through each of the points *A*, *B* and *C* a line parallel to the opposite side (Fig. 4.9 (a)). This yields the parallelograms *ACBW*, *ABCV* and *ABUC* and shows that *A*, *B* and *C* are the midpoints of *VW*, *WU* and *UV*, respectively. Hence the new triangle *UVW* has the altitudes of *ABC* as perpendicular side bisectors. By Eucl. IV.5, these lines must meet at the circumcentre of the triangle *UVW*, *which is thus the orthocentre of the triangle ABC*.   □

The orthic triangle. The feet of the altitudes of a given triangle *ABC* (i.e. the points *DEF* of Fig. 4.9 (a)) form a new triangle which is worth studying. It is called the *orthic triangle* of *ABC* (see Fig. 4.9 (b)). The discovery of this triangle (and its properties) by Giov. Fagnano (1770, 1779) was related to the minimisation problem discussed in Exercise 14 on page 236.

Theorem 4.3. *(a) The orthic triangle DEF of ABC determines three triangles AEF, DBF and DEC which are all similar to ABC, but inversely oriented.*

(b) The altitudes AD, BE and CF of ABC are the angle bisectors of the orthic triangle.

(c) The segments AO, BO and CO, connecting the vertices of ABC to its circumcentre O in Fig. 4.7 (b), are perpendicular to the sides EF, FD and DE, respectively, of the orthic triangle.

Proof. The proof is displayed in Fig.4.9 (b).

(a) We draw the circle with diameter BC, on which the points E and F are located by orthogonality. Thus $CEFB$ is a cyclic quadrilateral, i.e. a quadrilateral inscribed in a circle. By Eucl. III.22 the angle BFE is $2\llcorner - \gamma$, hence the complementary angle EFA is γ. The same argument applies to all other angles.

(b) Since the segments EF and DF make the same angle γ with AB at F, they make the same angle $\llcorner - \gamma$ with CF.

(c) The angle $\gamma = EFA$ in Fig. 4.9 (b) is an orthogonal angle to the angle AOF in Fig. 4.7 (b). □

Remark. Since by property (b) the altitudes of ABC pass through the incentre of DEF, we have, using this time Eucl. IV.4, another proof of Theorem 4.2.

4.4 The Theorems of Menelaus and Ceva

Menelaus of Alexandria lived around 100 A.D. and was, with Hipparchus and Ptolemy, one of the founders of spherical trigonometry. The following theorem was used in the third volume of his *Sphaerica* to solve spherical triangles. It had been used by Ptolemy in his *Almagest* and became famous through this work. For details see Chasles (1837, Chap. I, §22 and "Note VI"). It reappeared in Carnot (1803), where its fundamental importance for the foundation of geometry was recognised.

We are interested in answering the following question. Suppose that a given triangle ABC is cut by a line EDK (see Fig. 4.10, left). What can be said about the lengths $a_1, a_2, b_1, b_2, d_1, d_2$?

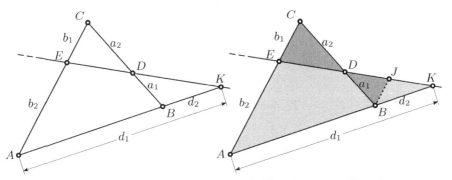

Fig. 4.10. Menelaus' theorem (left) and its proof (right)

Theorem 4.4 (Menelaus). *The points E, D, K in Fig. 4.10 are collinear if and only if*

$$\frac{a_1}{a_2} \cdot \frac{b_1}{b_2} \cdot \frac{d_1}{d_2} = 1 \, .$$

Proof. Since we have a famous theorem to prove, just looking at the picture on the left side of Fig. 4.10 doesn't help very much. In order to be able to apply Thales, we must draw a line somewhere, parallel to some other line. We choose the point B and draw BJ parallel to AEC. This creates two pairs of similar triangles (in grey in Fig. 4.10, right). The dark grey triangles give us $BJ = \frac{b_1 a_1}{a_2}$ and the light grey triangles BJK and AEK give us $\frac{d_2}{d_1} = \frac{BJ}{AE} = \frac{b_1 a_1}{b_2 a_2}$.

The converse implication follows from the uniqueness of the point D for a given ratio of $\frac{a_1}{a_2}$. □

Ceva's theorem. Select three points, one on each side of a triangle, and connect each one with the opposite vertex. Under which conditions will these three lines meet in a single point? The theorem which answers this question was for a long time attributed to Joh. Bernoulli (*Opera*, 1742, vol. 4, p. 33)[3] and was rediscovered by Crelle in 1816. Finally, Chasles (1837, "note VII") discovered that the result was already known to Giovanni Ceva (1648–1734). Ceva's insight came from mechanics by placing mass points of different weights at the vertices of the triangle and considering the lines of equilibrium and the barycentre of this configuration.

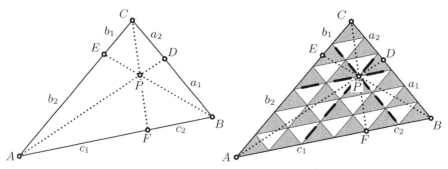

Fig. 4.11. Ceva's theorem and its *Stone Age proof*

Theorem 4.5 (Ceva). *In a triangle the lines AD, BE and CF (see Fig. 4.11, left) are concurrent if and only if*

$$\frac{a_1}{a_2} \cdot \frac{b_1}{b_2} \cdot \frac{c_1}{c_2} = 1 \, .$$

Proof. Joh. Bernoulli's proof uses Thales' theorem (see Exercise 6 on page 108) as in the proof above; Crelle's proof is in the style of Eucl. VI.2 (areas of the triangles, see Exercise 7).

---

[3] "Qui continentur ANEKΔOTA."

For readers who like to *see* the result by looking at a nice picture, we present a *Stone Age proof*, by supposing that P is a vertex of one of the small triangles of a subdivision, see Fig. 1.2 and Fig. 4.11, right.[4] Thales' theorem now shows that

$$\frac{a_1}{a_2} = \frac{3}{2}, \qquad \frac{b_1}{b_2} = \frac{1}{3}, \qquad \frac{c_1}{c_2} = \frac{2}{1}.$$

Here the numbers 1, 2 and 3 count the layers of small triangles which separate P from the three sides of the given triangle; they appear in pairs, one above and one below the division bar. So the product of all three quotients is 1. □

Remark. The apparent similarity of the conditions in Theorems 4.4 and 4.5 calls for an explanation. The reason will become clear in Chap. 11: if Figs. 4.10 and 4.11 are superimposed, one obtains a figure which is a complete quadrilateral. Hence by Theorem 11.10 on page 335 the points A, B, F, K are in *harmonic position*, with cross ratio -1. For this reason, it is somewhat nicer to write the condition of Menelaus' theorem with a minus sign, i.e. to take the distance BK in the negative sense.

The Gergonne point (Joseph Diaz Gergonne, 1771–1859).

> "... during the July Revolution [of 1830], when rebellious students began to whistle in his class, he regained their sympathy by beginning to lecture on the acoustics of the whistle."
> (Struik, quoted from MacTutor History of Mathematics archive[5])

Theorem 4.6. *In a triangle ABC, let D, E, F be the points where the inscribed circle touches the triangle, see Fig. 4.6, right. Then the lines AD, BE and CF are concurrent (in a point called the* Gergonne *point of the triangle).*

Proof. The lengths indicated in Fig. 4.6 (right) clearly satisfy the condition of Ceva's theorem. □

Remark. If D, E and F are the midpoints of the sides of ABC, so that $a_1 = a_2$, $b_1 = b_2$ and $c_1 = c_2$, Ceva's condition is again satisfied and P becomes the centroid G. In the case where D, E and F are the feet of the altitudes, the triangles AFC and AEB are similar triangles and we have by Thales $\frac{c_1}{b_2} = \frac{h_c}{h_b}$, and similarly $\frac{a_1}{c_2} = \frac{h_a}{h_c}$ and $\frac{b_1}{a_2} = \frac{h_b}{h_a}$, where h_a, h_b, h_c are the lengths of the altitudes. The hypothesis of Ceva's theorem is satisfied and we obtain another proof of Theorem 4.2.

4.5 The Theorems of Apollonius–Pappus–Stewart

A segment connecting a vertex of a triangle to a point on the opposite side is called a *cevian*, because of Ceva's theorem. We want to find its length.

---

[4] In other words, we suppose that P has rational *barycentric coordinates*.
[5] http://www-history.mcs.st-and.ac.uk/Biographies/Gergonne.html

Solution. We use Eucl. II.13 and Eucl. II.12 for the triangles FBC and AFC of Fig. 4.12 (a) (see formula (2.2) on page 38 and the footnote):

$$2un = w^2 + n^2 - a^2 \qquad \text{and} \qquad -2um = w^2 + m^2 - b^2. \tag{4.4}$$

In the case where $m = n$, the unknown u disappears if we add the two equations. We thus get for the *length of a median*:

Theorem 4.7 (Pappus, *Collection*,[6] Book VII, Prop. 122). *The length $CF = w$ of the median of a triangle with sides a, b and $c = 2n$ satisfies*

$$w^2 + n^2 = \frac{1}{2} a^2 + \frac{1}{2} b^2. \tag{4.5}$$

By extending the triangle ABC to a parallelogram $ADBC$ (see Fig. 4.12 (b)) with sides a and b and diagonals $d_1 = 2n$ and $d_2 = 2w$, we obtain from (4.5) the relation

$$d_1^2 + d_2^2 = 2a^2 + 2b^2. \tag{4.6}$$

This last formula is called the *parallelogram law*. Pappus' theorem is, according to Simson's restoration (1749, p. 152), part of the lost work *De locis planis* of Apollonius. It is also given in Giul. Fagnano (1750, vol. II, Appendice: *Nuova et generale proprietà de' Poligoni*, Lemma I), the same year when Euler published his generalisation to arbitrary quadrilaterals (see Exercise 3 on page 233).

Theorem 4.8 (M. Stewart, 1746, Proposition II). *For the length $CF = w$ of the cevian in Fig. 4.12 (a) we have the following generalisation[7] of (4.5)*

$$w^2 + nm = \frac{m}{n+m} a^2 + \frac{n}{n+m} b^2. \tag{4.7}$$

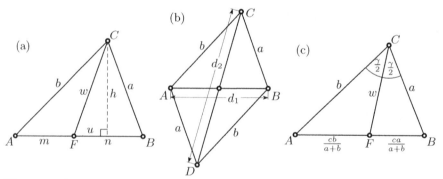

Fig. 4.12. Stewart's theorem for the length of a cevian (a); the parallelogram law (b), length of an angle bisector (c)

[6]The authors are grateful to Philippe Henry for this reference.

[7]Stewart wrote the result in the form $CA^2 \cdot FB + CB^2 \cdot AF - CF^2 \cdot AB = AB \cdot AF \cdot FB$, which may be nicer.

Proof. We multiply the first formula of (4.4) by m and the second one by n. By adding the resulting formulas, the unknown u disappears again and (4.7) is obtained. □

Corollary 4.9. *For the length $CF = w$ of the angle bisector in Fig. 4.12 (c) we have*

$$w^2 = ab\left(1 - \left(\frac{c}{a+b}\right)^2\right) = \frac{ab(a+b+c)(a+b-c)}{(a+b)^2}. \tag{4.8}$$

Proof. If CF is the angle bisector, we have $m = \frac{cb}{a+b}$ and $n = \frac{ca}{a+b}$ as a consequence of Eucl. VI.3 and $m + n = c$. This inserted into (4.7) leads after some simplifications to the stated result (the last identity by Eucl. II.5). □

4.6 The Euler Line and the Nine-Point Circle

> "Some of his [Euler's] simplest discoveries are of such a nature that one can well imagine the ghost of Euclid saying, 'Why on earth didn't I think of that?'" (H.S.M. Coxeter, 1961, p. 17)

The Euler line. Euler (1767a) discovered a *remarkable* property of four *remarkable points* of a triangle by analytical calculations (see Chap. 7). We note here with satisfaction that the combination of Gauss' proof of Theorem 4.2 with the medial reduction leads to an elegant proof in a few lines.

Theorem 4.10. *In any triangle the points H, G and O lie on a line (called the Euler line), and G divides the segment HO in the ratio $HG : GO = 2 : 1$.*

Proof. We follow Gauss' proof *backwards*, i.e. we exploit the similarity of a triangle ABC with its medial triangle $A'B'C'$ (see Fig. 4.13, left). As we know, the orthocentre H' of $A'B'C'$ coincides with the circumcentre O of ABC.

The triangle $A'B'C'$ is similar and parallel to the triangle ABC, shrunken by a factor $\frac{1}{2}$ with reversed directions. We thus have figures similar to those in Fig. 1.4, and see that all lines connecting corresponding points AA', BB', CC' (these are the medians) and HH' (this is the Euler line) must pass through the same point G. Further, G divides all these lines in the same ratio, which is $2 : 1$. □

The nine-point circle. The circumcircle of the triangle $A'B'C'$ is called the *nine-point circle* of the triangle ABC and has many interesting properties (see Fig. 4.13, right).

Theorem 4.11. *(a) The centre $O' = N$ of the nine-point circle is the midpoint between H and O and its radius is half of the radius of the circumcircle of ABC. Thus the circumcircle and the nine-point circle are in similarity position with ratio $2 : 1$ and similarity centre H (as in the Geneva duck theorem of Exercise 19 on page 26).*

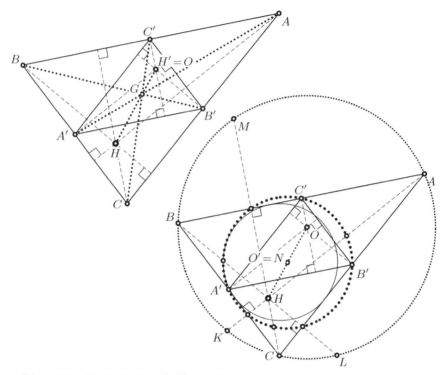

Fig. 4.13. The Euler line (left) and the nine-point (or Feuerbach) circle (right)

(b) The nine-point circle, which passes through $A'B'C'$ by definition, also passes through the feet of the altitudes AH, BH and CH.

(c) The nine-point circle further passes through the midpoints of the segments AH, BH and CH.

(d) The reflections K, L and M of the orthocentre H at the three sides of the triangle, which lie on the extended altitudes, lie on the circumcircle of ABC.

Proof. (a) This follows from the proof of Theorem 4.10, because by similarity the segment $ON = H'O'$ is half as long as HO.

(b) Because of (a), N is equidistant from H and O, hence also from the normal projections of these points onto the sides of the triangle.

(c) This follows from (a) because A, B and C lie on the circumcircle.

(d) This follows from (a) and (b) in a similar way. $\square$

There is still another remarkable property: the nine-point circle is *tangent to the incircle of ABC*, a result discovered by K.W. Feuerbach (see Theorem 7.23 below). This led to the name *Feuerbach circle*.

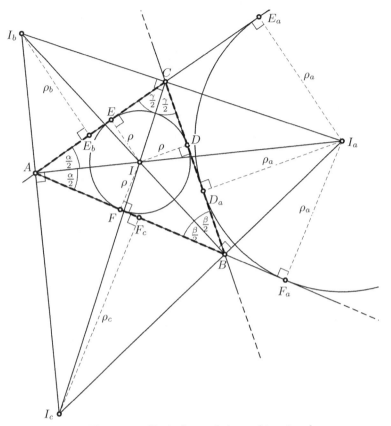

Fig. 4.14. Excircles and the orthic triangle

4.7 Excircles and the Nagel Point

The following results are among those discovered at the beginning of the 19th century by Simon Lhuilier (1810/11), A.L. Crelle[8] (in 1816), K.W. Feuerbach (in 1822) and C.H. von Nagel (in 1835). For a complete bibliography (with more than 300 references) we refer to P. Baptist (1992).

Excircles. Let a triangle ABC be given and produce its sides in both directions. Form a new triangle $I_a I_b I_c$ by drawing the lines through A, B, C orthogonal to the angle bisectors AI, BI, CI respectively (see Fig. 4.14). By Eucl. I.14 these lines are the angle bisectors of the *exterior* angles of the triangle. Precisely as in the proof of Eucl. IV.4, we conclude that, say, I_a, which is equidistant from AB and BC, and also from BC and CA, must be equidistant from AC and AB, hence must also lie on the interior angle bisector of BAC. We call I_a the centre of the *excircle* of radius ρ_a, touching the triangle from

---

[8]Today mostly remembered as the founder of *J. Reine Angew. Math.*

outside at the point D_a. Similarly there are two other excircles centred at I_b and I_c of radius ρ_b and ρ_c, respectively.

Remark. The original triangle ABC is the orthic triangle of $I_a I_b I_c$; this gives another proof of Theorem 4.3. Also, from Eucl. IV.4 applied to ABC, we obtain another proof of Theorem 4.2 for the triangle $I_a I_b I_c$.

Theorem 4.12 (Nagel). *The lines AD_a, BE_b and CF_c, joining the vertices of a triangle with the points of contact of the opposite excircles, intersect in a point, the* Nagel *point.*

Proof. By Eucl. III.36, $AF_a = AE_a$, $CD_a = CE_a$ and $BF_a = BD_a$. Therefore the broken lines ACD_a and ABD_a have the same length. This length must be s, the semi-perimeter, because together they form the entire perimeter. Subtracting $AB = c$ we obtain $BD_a = s - c$. Similarly,

$$CD_a = AF_c = s - b, \quad AE_b = BD_a = s - c, \quad BF_c = CE_b = s - a. \quad (4.9)$$

These values satisfy the condition of Ceva's theorem (Thm. 4.5). □

By Thales' theorem applied to the similar triangles AFI and $AF_a I_a$, and by remembering the values of Fig. 4.6, we obtain the following formulas, which will be of use later:

$$\rho_a = \frac{s \cdot \rho}{s - a}, \quad \rho_b = \frac{s \cdot \rho}{s - b}, \quad \rho_c = \frac{s \cdot \rho}{s - c}. \quad (4.10)$$

Kimberling's centre catalogue. In addition to I, O, G, H, the Gergonne point, the nine-point centre N and the Nagel point, more and more such "centres", i.e. points of a triangle defined in a certain natural way respecting symmetric exchanges of the vertices, have been discovered together with their interesting properties. The letters of the alphabet were quickly used up. Therefore, Kimberling (1994) started labelling these points as $X_1, X_2, \ldots$ by compiling a list of 100 such points. In particular, $I = X_1$, $O = X_3$, $G = X_2$, $H = X_4$, the nine-point centre $N = X_5$, the Gergonne point is X_7, the Nagel point is X_8, the Fermat–Torricelli point to be discussed in Chapter 7 is X_{13}, and so on. This list, together with supporting theory and hundreds of "central lines" (such as Euler's), was extended later in Kimberling (1998) to 400 centres. Now accessible on the internet, the list is even 10 times longer.

4.8 Miquel's Theorems

Auguste Miquel was a high school teacher in the French countryside (Nantua), which snobs in Paris call *la province*. And from this province the newly founded journal of Liouville[9] suddenly started receiving beautiful geometric

---

[9]Journal de mathématiques pures et appliquées

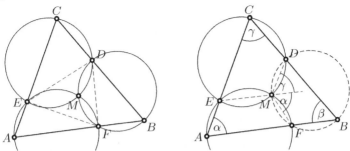

Fig. 4.15. Miquel's triangle theorem (left) and its proof (right)

discoveries (see Miquel, 1838a and 1838b). The first of these, concerning the
Miquel point, became the most famous (Theorem 4.13 below). But for Miquel
this was just an auxiliary result for the proof of his marvellous pentagon
theorem which will follow.

Theorem 4.13 (Miquel's triangle theorem). *Let D, E, F be arbitrary points
chosen on the sides BC, CA, AB, respectively, of a triangle (see Fig. 4.15,
left). Then the circumcircles of the triangles AEF, BDF and CDE are concurrent, in a point M called the* Miquel *point.*

Proof. Let M be the second intersection point (other than E) of the circle through A with the circle through C (see Fig. 4.15, right). Then, by
Eucl. III.22, the two angles denoted by α are equal, as are those denoted
by γ. By Eucl. I.32 we have $\alpha + \gamma + \beta = 2\llcorner$, which tells us, by the converse
of Eucl. III.22 applied to the quadrilateral $FMDB$, that M also lies on the
third circle. □

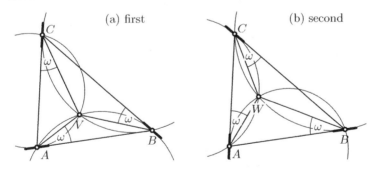

Fig. 4.16. Brocard points

An application: the Brocard points. If we move the points D, E and
F in Miquel's triangle theorem towards B, C and A, respectively, the circles
tend to circles which pass through *one* vertex of the triangle and are tangent
to the opposite side at a *second* vertex (see Fig. 4.16 (a)). The Miquel point
then tends to a point V which is called the *first Brocard point*. By applying
Eucl. III.32 (a variant of Eucl. III.21, see Exercise 17 on page 57), we find that

all three angles marked ω in this picture are equal. If we push the three points D, E and F to the *other* end of their respective sides, we obtain analogously the *second Brocard point* (Fig. 4.16 (b)), again with all three angles ω equal. It is more difficult to see that ω is the same in both cases. The Brocard points V and W are not centres in Kimberling's sense, because they lack symmetry with respect to the exchange of two vertices. The *midpoint* between them, however, is symmetric and has the number X_{39} in Kimberling's list.

Definition 4.14. *Let $ABDE$ be a convex quadrilateral (see Fig. 4.17). Produce the opposite sides until they meet in two points F and C. The figure obtained in this manner consists of four triangles and is called a* complete quadrilateral.

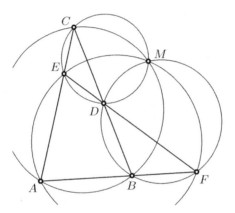

Fig. 4.17. Miquel's quadrilateral theorem

Theorem 4.15 (The quadrilateral theorem of Steiner and Miquel). *The circumcircles of all four triangles of a complete quadrilateral are concurrent in a point M.*

Proof. This result had already been announced earlier in two lines by Steiner (1827/1828, number 1°), but Miquel gave a detailed proof, which applied Euclid's theorems from Book III as in the preceding proof. However, we can easily see this theorem by using an idea which Poncelet called the *continuity principle* (see Chap. 11): in Theorem 4.13 we move the point F *outside* the interval AB and believe that this theorem still remains valid. We further move the point D of that theorem until E, D, F are collinear. Then the circumcircles of AFE, FBD and EDC meet in one point. By symmetry of the figure (exchange $F \leftrightarrow C$ and $E \leftrightarrow B$), the fourth circumcircle must also pass through the same point.   □

Theorem 4.16 (Miquel's pentagon theorem). *Let $ABCDE$ be a convex pentagon (see the upper picture of Fig. 4.18). Produce all sides until they meet*

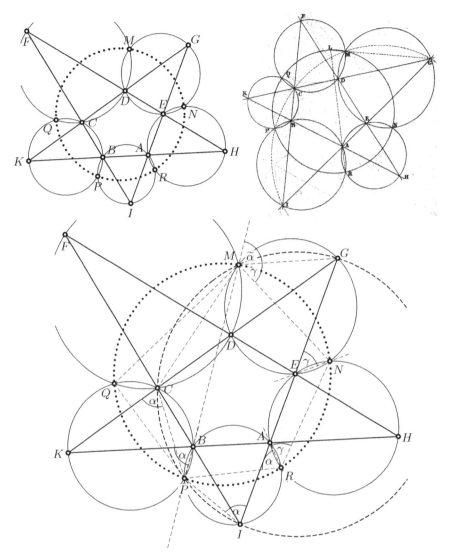

Fig. 4.18. Miquel's pentagon theorem (above; together with Miquel's original drawing); its proof (below)

in five points F, G, H, I, K and draw the circumcircles of the five triangles CFD, DGE, EHA, AIB and BKC. Then the second intersection points of these circles (other than A, B, C, D, E), namely the points M, N, R, P, Q, are concyclic.

Proof. Miquel proved this by using an auxiliary result, namely that the points I, P, C, M, G are concyclic, by repeated applications of Theorems 4.13 and

4.15. But we can also, as in A. Gutierrez' beautiful web-site,[10] reduce this proof exclusively to Euclid's Book III (see the lower picture of Fig. 4.18).

Using Eucl. III.22 once and Eucl. III.21 twice, we see that the four angles denoted by α are equal. Similarly, the three angles denoted by γ are equal. Applying the converse of Eucl. III.22 to the quadrilateral $IPCG$, we see that these four points are concyclic.

By symmetry of the configuration (exchange $A \leftrightarrow E$, $B \leftrightarrow D$, etc.) the fifth point M also lies on the same circle. Then by Eucl. III.22, applied to the quadrilateral $IPMG$, the angle denoted by $\widetilde{\alpha}$ is equal to α.

Finally we apply the converse of Eucl. III.22 to the quadrilateral $RPMN$ and see, because $\widetilde{\alpha} + \gamma = \alpha + \gamma$, that these four points are concyclic. Again, the fifth point Q lies on the same circle by symmetry of the configuration. A wonderful proof of a wonderful result.   □

4.9 Steiner's Circle Theorems

> "Gefunden Samstag den 10. Christmonat 1814, $3+3+4$ St. daran gesucht, des Nachts um 1 Uhr gefunden. [Found on Saturday Dec. 10th, 1814, after $3 + 3 + 4$ hours of efforts, at 1 o'clock in the night]."   (From Steiner's notes during his first month as a pupil in Yverdon's Pestalozzi school; quoted from J.-P. Sydler, *L'Enseignement Mathématique* 2e sér. vol. 11 (1965), p. 241.)

Jakob Steiner (1796–1863) has one of the most incredible biographies of a great mathematician: he was born in a small Swiss village (Utzenstorf close to Bern, see picture). His father forbade him to read, the village priest refused to let him write — he was not good enough in Catechism. At the age of 18 he entered, as the oldest pupil, the Pestalozzi school at Yverdon, where he began his education eagerly and with great energy (see the quotation). Later in Berlin, he was not allowed to teach higher mathematics at the Werden Gymnasium — he had not understood Hegel's philosophy well enough. So he survived as "Privatlehrer" [private teacher] and contributed five articles (No. 5, 18, 25, 31 and 32) to the first volume of the newly founded *Crelle Journal*. These articles include Steiner's first published long work (1826c), which we follow below. The book project Steiner (1826a), containing the same results, was published only in 1931.

Power of a point with respect to a circle. We recall from Eucl. III.36, with the notation of Fig. 2.17 (b) and (c) on page 40, that

$$EA \cdot EB = t^2 = (d + r)(d - r) = d^2 - r^2 \qquad (4.11)$$

is independent of the position of the points A and B on the circle, as long as they are aligned with E, and depends only on the radius r of the circle and

---

[10]Geometry step by step from the land of the Incas
http://agutie.homestead.com/

Fig. 4.19. Steiner's portrait from his *Gesammelte Werke* 1881 (left); Steiner's birthplace in Utzenstorf (right); photographs by Barbara Kummer (before the house was destroyed)

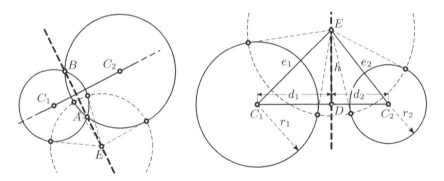

Fig. 4.20. Radical axis as line of equal powers (left); line of equal powers for nonintersecting circles (right)

the distance d of the point E from its centre. This quantity is called, since Steiner, *the power of the point E with respect to the circle.*

Line of equal powers for two circles. Let two circles be given with centres C_1, C_2 and radii r_1, r_2 (see Fig. 4.20). We look for the set of points E which have the same power with respect to both circles.

The answer is easiest if the two circles intersect (left picture): in this case we choose E on the line connecting A and B (the radical axis) and see that its power $EA \cdot EB$ is the same with respect to both circles. The tangents from

E to both circles have the same length $t = \sqrt{EA \cdot EB}$ and the circle centred at E with this radius *intersects both circles at right angles*.

Now let the distance of the centres be $a > r_1 + r_2$ (right picture). We first look for a point D on the line C_1C_2 having the same powers:

$$d_1^2 - r_1^2 = d_2^2 - r_2^2 \quad \text{and} \quad d_1 + d_2 = a. \tag{4.12}$$

We write this as $r_1^2 - r_2^2 = d_1^2 - d_2^2 = (d_1 + d_2)(d_1 - d_2) = a(2d_1 - a)$ and obtain

$$d_1 = \frac{a^2 + r_1^2 - r_2^2}{2a} = \frac{a}{2} + \frac{r_1^2 - r_2^2}{2a} \quad \text{and} \quad d_2 = \frac{a}{2} + \frac{r_2^2 - r_1^2}{2a}. \tag{4.13}$$

If we now place E on the line through D perpendicular to C_1C_2 at distance h from D and add h^2 to both sides of (4.12), we obtain by Pythagoras

$$h^2 + d_1^2 - r_1^2 = h^2 + d_2^2 - r_2^2 \quad \text{hence} \quad e_1^2 - r_1^2 = e_2^2 - r_2^2 \tag{4.14}$$

and E also has the same power with respect to each circle. That the converse is also true, i.e. that *every point which has the same power for both circles, and is the centre of a circle intersecting both circles at right angles, lies on DE, the "line of equal powers"*, was to trivial for Steiner to prove separately. It corresponds to the previous statement in the same way as Eucl. I.48 corresponds to Eucl. I.47.

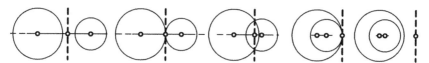

Fig. 4.21. Line of equal powers for one circle moving into the other

Remark. If the smaller circle moves into the larger, the line of equal powers first becomes the common tangent, then the radical axis, the common tangent again, and finally moves outside both circles (see Fig. 4.21).

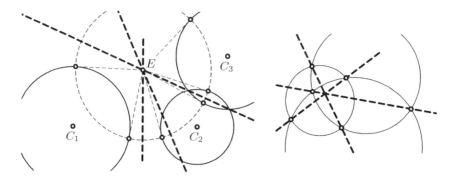

Fig. 4.22. Equal powers for three circles (left); the radical axes of three circles (right)

Point of equal powers for three circles. In the case of three circles, there are three lines of equal powers for the circles taken in pairs. *These three lines are concurrent in one point E, the "point of equal powers for the three circles"* (see Fig. 4.22, left). Furthermore, this point E is the centre of the circle intersecting all three circles at right angles. In particular, the three radical axes of three mutually intersecting circles are concurrent (Fig. 4.22, right).

The proofs of these statements rely on the fact that if among three powers any two are equal, then all three must be equal, a conclusion similar to that in the proofs of Eucl. IV.4 and IV.5.

Common power of two circles with respect to their similarity centre. Let two circles centred in C_1 and C_2 be given with E as similarity centre (see Fig. 4.23). Suppose $EC_2 = \theta \cdot EC_1$, i.e. the second circle is larger than the first by a factor θ. If a line through E intersects the circles in the points X', X, Y, Y' (in this order) then by Thales $EY = \theta \cdot EX'$ and $EY' = \theta \cdot EX$. If now $q = EX' \cdot EX$ is the power of E with respect to the first circle, then the power of E with respect to the second circle is $\theta^2 q$. In addition, we have that the "mixed products"

$$EX \cdot EY = EX' \cdot EY' = \theta q = p \qquad (4.15)$$

have the same value for all lines through E. In particular $EC \cdot ED = p$. Steiner calls this constant p the *common power of both circles with respect to their similarity centre.*

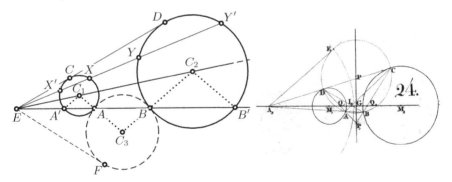

Fig. 4.23. Common power of two circles with respect to E (left); original illustration from Steiner 1826c (right)

Then let a third circle centred in C_3 be tangent to both circles at the points A and B, so that C_1, A, C_3 and C_3, B, C_2 are aligned. We define A' on the first circle such that $C_1 A'$ is parallel to $C_3 B$ and B' on the second one such that $C_2 B'$ is parallel to $A C_3$. We thus obtain three isosceles triangles $A'C_1 A$, $BC_3 A$ and $BC_2 B'$ which are similar and whose corresponding sides are parallel. Therefore, by Thales, A and B are aligned with E and we have from (4.15)

$$EA \cdot EB = EF^2 = p.\tag{4.16}$$

This means that *the common power of the two circles with respect to E is the (ordinary) power of E with respect to a circle tangent to both circles.*

Steiner's proof of Pappus' "ancient theorem". We can now present Steiner's elegant proof of a result, which already Pappus called an "ancient theorem":[11]

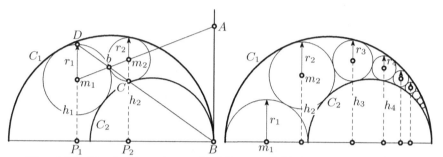

Fig. 4.24. The "ancient theorem" of Pappus (left); Steiner's proof (right)

Theorem 4.17 (Pappus *Collection*, Book IV, Props. 15, 16 and 18). *Let two semi-circles C_1 and C_2 be tangent at B (see Fig. 4.24, left) and two circles with centres m_i and radii r_i be tangent to C_1, C_2 and tangent to each other. Then*

$$\frac{h_2}{r_2} = \frac{h_1}{r_1} + 2\tag{4.17}$$

where h_i is the distance of m_i from the common diameter $P_1 P_2 B$. In particular (Pappus' Prop. 16, right picture), if we fill the space between C_1 and C_2 with an infinity of circles m_1, m_2, m_3, ... with m_1 on the common diameter, we have

$$\frac{h_1}{r_1} = 0, \quad \frac{h_2}{r_2} = 2, \quad \frac{h_3}{r_3} = 4, \quad \frac{h_4}{r_4} = 6, \quad \dots\tag{4.18}$$

If the initial circle is tangent to the common diameter (Pappus' Prop. 18), then the corresponding sequence of ratios is $1, 3, 5, 7, \dots$

Proof. For the proof of (4.17), of which (4.18) is a trivial consequence, we follow precisely the steps of Steiner's proof in (1826c):

(a) AB is the line of equal powers for the circles C_1 and C_2 (see Fig. 4.21);

(b) the similarity centre A for the circles m_1 and m_2 has the same power with respect to C_1 and C_2 (see (4.16)), hence lies on the common tangent at B;

(c) by Thales $P_1 B : P_2 B = r_1 : r_2$;

(d) AB^2 is the common power of m_1 and m_2 with respect to A (again by (4.16));

---

[11] "Circumfertur in quibusdam libris antiqua propositio huiusmodi."

(e) $Ab^2 = AB^2$ (by (4.15), b represents two points X and Y which collapse), hence $Ab = AB$;

(f) D, b, C are aligned; b, C, B are also aligned by Thales (because bm_2C and bAB are both isosceles and thus similar), hence D, b, C, B are aligned;

(g) by Thales and (c) we obtain

$$\frac{h_1 + r_1}{r_1} = \frac{h_2 - r_2}{r_2}$$

from which (4.17) follows.   □

Once a simple proof has been found for a famous result, the way is open for numerous extensions and generalisations, which Steiner then pursued with considerable energy.

Steiner's Porism.

> "A porism is a mathematical 'proposition affirming the possibility of finding such conditions as will render a certain problem indeterminate or capable of innumerable solutions'."
>
> (J. Playfair, 1792)

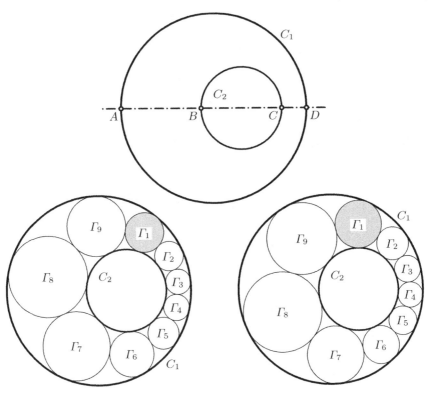

Fig. 4.25. Steiner's porism

Theorem 4.18 (Steiner, 1826c, second part, §22). *Let C_1 and C_2 be two circles, C_2 inside C_1. Suppose that one fills the space between the two circles by a chain of circles Γ_1, Γ_2, Γ_3 etc., each tangent to its neighbours and to C_1 and C_2. Then, if $\Gamma_n = \Gamma_1$ for some integer n (see the second picture of Fig. 4.25), the same property holds for the same n for any choice of Γ_1 (see the third picture).*

Steiner obtained this result after a long struggle. Fig. 4.26 gives some idea of his original thoughts. We will see in the next chapter (Section 5.5) that this result can be obtained in a very elegant manner by using the stereographic projection.

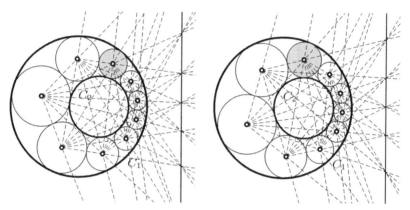

Fig. 4.26. Connecting the circle centres in Steiner's porism

4.10 Morley's Theorem

"Morley's theorem is startling, difficult to prove, and utterly beautiful."

(W. Dunham, *Euler, the master of us all.* Math. Ass. Amer. 1999)

"... on s'empressa ... de rechercher une démonstration aussi courte et aussi élégante que l'énoncé ... A mon avis, de tels désirs ne sauraient être satisfaits. [... one eagerly sought a proof as simple and elegant as the statement ... In my opinion such desires cannot be satisfied.]"

(H. Lebesgue, *L'Enseignement Mathématique* 38 (1939), p. 39)

"Much trouble is experienced if we try a direct approach, but the difficulties disappear if we work backwards, ..."

(H.S.M. Coxeter, 1961, p. 24)

Theorem 4.19. *For a given triangle ABC, let PQR be the triangle formed by the intersections of the angle trisectors of ABC, see Fig. 4.27. Then, PQR is equilateral.*

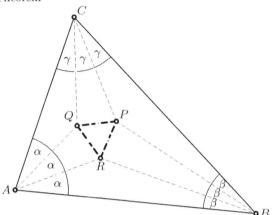

Fig. 4.27. Morley's triangle

Remark. Despite the simplicity and elegance of its statement, this theorem, discovered in 1904, escaped the attention of even the most ingenious geometers for more than two thousand years. At first sight it seems very difficult to prove (see the quotations; for a detailed historical account, see Loria (1939, p. 367), Coxeter (1961, p. 24), Coxeter and Greitzer (1967, p. 47) and the particularly rich bibliography of Oakley and Baker (1978). Surprisingly, the proof becomes simple if one proceeds *backwards* (see the last quotation). The proof given below, as elaborated in Wanner (2004), uses only Euclid's Books I, III and VI.

Proof. We forget about the triangle ABC in Fig. 4.27 and keep in mind only the values of the angles α, β and γ with

$$\alpha + \beta + \gamma = \frac{2}{3}\,\llcorner \tag{4.19}$$

(Eucl. I.32). We start from an equilateral triangle PQR, say of side length 1 and fixed in Fig. 4.28. We then try to reconstruct a triangle similar to ABC (again denoted by the same letters) satisfying Morley's theorem. By the uniqueness of the construction of P, Q, R, the original triangle must then also satisfy Morley's theorem.

With Eucl. III.20 in mind, we draw the three circles with centres K, L, M and central angles 2α, 2β, 2γ, respectively, that have the sides of the triangle PQR as chords. We then add four additional chords QD, QE, RF and RG, all of length 1. This creates several isosceles triangles with vertices at K, L, M, Q and R respectively. We next compute all angles of these triangles using Eucl. I.5 and Eucl. I.32. The angles around Q must sum to $4\,\llcorner$,[12] whence

$$\delta = 2\alpha + 2\gamma - \frac{2}{3}\,\llcorner\,, \qquad \text{and at } D, \qquad \varepsilon = \llcorner - \frac{\delta}{2} = \frac{4}{3}\,\llcorner - \alpha - \gamma. \tag{4.20}$$

---

[12]The authors are grateful to Christian Aebi for a suggestion concerning this part of the proof.

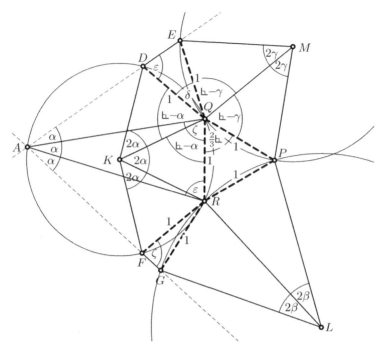

Fig. 4.28. Proof of Morley's theorem

Similarly, for the angle ζ at F (exchange β and γ) we obtain

$$\zeta = \frac{4}{3}\llcorner - \alpha - \beta \qquad \text{whence, with (4.19) and (4.20),} \qquad \alpha + \varepsilon + \zeta = 2\llcorner .$$

This means that by constructing the triangle QAR on RQ using the angles ε and ζ (version ASA of Eucl. I.26), we find a point A which lies on the circle centred at K (Eucl. I.32 and converse of Eucl. III.20). Then the quadrilaterals $ARQD$ and $AQRF$ are inscribed in the same circle and A, D, E and A, F, G are collinear (Eucl. III.22).

The proof is concluded by constructing B and C in the same way. □

Another backwards proof (see the last quotation) due to R. Penrose (1953) is given in Exercise 12 below.

4.11 Exercises

1. (*The cissoid of Diocles*) Prove the following result of Pappus (*Collection*, Book III, Chap. 10): *Let $AB\Gamma$ be a semicircle with centre Δ and let ΔE be a given distance (see Fig. 4.29, left). Produce ΓE to Z on the circle. Determine a line $AH\Theta K$ where K is on the arc $B\Gamma$, H on ΓZ and Θ on ΔB in such a way that $H\Theta = \Theta K$. Then $E\Delta$ is to $B\Delta$ as the cube*

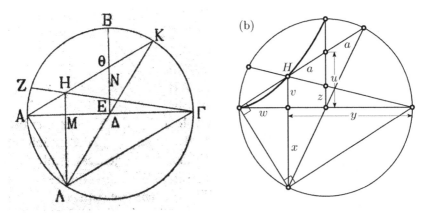

Fig. 4.29. The doubling of the cube by Diocles and Pappus; reproduced from Ver Eecke's translation (1933, left); in modern notation (right)

of $\Theta\Delta$ is to the cube of $B\Delta$. In other words, the quantity $\Theta\Delta$ solves the problem of multiplying the cube by any given ratio $E\Delta/B\Delta$.

Hint. Pappus suggests to complete the semicircle to a full circle and to draw the diameter $K\Delta\Lambda$ so that $AK\Gamma\Lambda$ is a rectangle. By hypothesis and Thales, $HM\Lambda$ is parallel to $\Theta\Delta$. Pappus' proof then requires an entire page of prose. In modern notation (see Fig. 4.29 (b)), supposing that the circle has radius 1, the required result becomes $z = u^3$, or $u = \sqrt[3]{z}$, which can be proved in one line.

Remark. In Pappus' version, the line through A must be placed by trial and error, until the segments $H\Theta$ and ΘK are equal. Another possibility is to determine for *all* lines through A the point H as the point of distance ΘK from Θ. This curve is called the *cissoid of Diocles* (drawn in Fig. 4.29 (b)). Once found, it solves the problem of multiplying the cube by drawing two lines, first ΓEH, then $AH\Theta$.

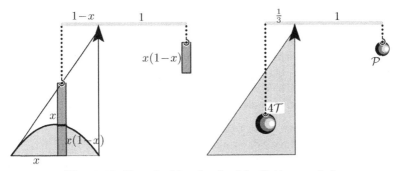

Fig. 4.30. How Archimedes "weighed" the parabola

2. Explain the beautiful idea of Archimedes, displayed in Fig. 4.30, of placing the parabola

$$y = \frac{1}{4} - \left(x - \frac{1}{2}\right)^2 = x(1 - x)$$

"on a balance" in order to determine its area.[13] This was Archimedes' first approach to the formula $\mathcal{P} = \frac{4}{3}\mathcal{T}$.

3. Let D, E, F of Fig. 4.15 be the feet of the altitudes of the triangle ABC. Show that the Miquel point of the corresponding circumcircles then coincides with the orthocentre. Therefore Miquel's result provides another proof of Theorem 4.2.

4. Reconstruct Newton's proof of Theorem 4.2 (see Fig. 4.31 (a)) by using Thales' theorem to compute y twice: by considering first the intersection of FC and AD, and then the intersection of FC and BE (see the, at that time, unpublished manuscript Newton 1680, p. 454).

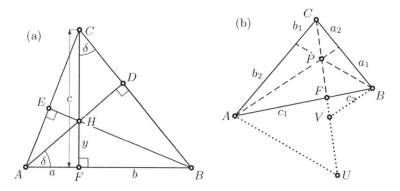

Fig. 4.31. Newton's proof for the orthocentre (a); Joh. Bernoulli's proof of Ceva's theorem (b)

5. Show: *If H is the orthocentre of the triangle ABC, then C is the orthocentre of the triangle ABH.* A map which is its own inverse (such as $H \mapsto C$) is called an *involution*.

6. Reconstruct Joh. Bernoulli's proof of Ceva's theorem by producing the line CPF and drawing AU parallel to PB and BV parallel to AP (see Fig. 4.31 (b)). You obtain several pairs of similar triangles for which you can play around with Thales' theorem. Readers familiar with Bernoulli's work will not be surprised by the elegance of his proof.

7. Give a proof of Ceva's theorem based on the areas of the six triangles in Fig. 4.11 (left) on page 88, similar to Euclid's proof of Eucl. VI.2.

---

[13]The authors owe this exercise to a suggestion of Martin Cuénod, Geneva.

8. (a) Consider a triangle ABC (see Fig. 4.32 (a)), whose area we denote by Δ. The two medians AD and CF determine four polygons of areas $\mathcal{A}, \mathcal{B}, \mathcal{C}$ and $\mathcal{D}$. How do these areas depend on Δ?

(b) The segments AD, BE and CF divide the opposite sides in the ratio $1:2$ and create a triangle of area $\mathcal{T}$ (see Fig. 4.32 (b) and Fig. 4.33 from Steiner (1828b)). How does $\mathcal{T}$ depend on Δ?

Remarks. Exercise (a) is among the first problems of the very first geometry lessons given by Mr. Maurer to the young Jakob Steiner in 1814 at Yverdon's Pestalozzi School. Throughout his life Steiner kept these lessons in high esteem. Exercise (b) was solved by Clausen (1828) using complicated trigonometric formulas. Steiner answered in Steiner (1828a) that such problems were treated extensively in Pestalozzi's School "wegen mancherlei pädagogischer Vorzüge [due to several pedagogical advantages]" and led to very elegant solutions. These results were later rediscovered and are called the *Dudeney–Steinhaus theorem* or also *Routh's theorem* (see Elem. Math. 22 (1967), p. 49).

9. (*The Torricelli–Fermat point and Napoleon's theorem*) Consider a triangle ABC with all angles less or equal $120°$. Construct an equilateral triangle outwards on each side of ABC, and call the outside vertices D, E and F. Then prove (see Fig. 4.34 (a) and (b)) that

(a) the three lines AD, BE, CF meet in one point P;

(b) the circumcircles of these three triangles meet in the same point P;

(c) all six angles at P are $60°$;

(d) the centres of these three circumcircles form an equilateral triangle;

(e) the segments AD, BE and CF are all equal to $PA + PB + PC$.

The origin of these results was a challenge of Fermat (see Sect. 7.4 below), which Father Mersenne, on his way to Rome in 1644, presented to E. Torricelli[14] during a stop-over in Florence ("Questi tre Problemi ... sono

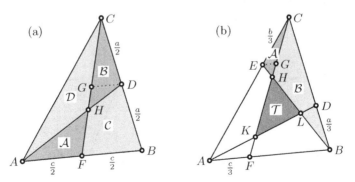

Fig. 4.32. Pestalozzi problems

---

[14]Evangelista Torricelli (1608–1647), famous for inventing the barometer.

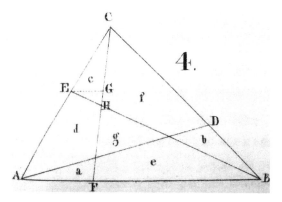

Fig. 4.33. Jakob Steiner's original picture in Steiner (1828b)

di Monsù de Fermat, Senatore di Tolosa [these three problems ... are by Monsieur de Fermat, senator from Toulouse]"). Torricelli included his solution in his manuscript *De Maximis et Minimis* (1646), Opere vol. I, parte sec., pp. 90–97. Today the point $P = X_{13}$ is called the *Torricelli–Fermat point*. Result (a), usually called *Napoleon's theorem*, was first published by Thomas Simpson, *The Doctrine and Application of Fluxions*, London 1750. Result (b) was Mersenne's tool for solving the problem, result (c) is the main reason for the minimality property, and result (e) is due to F. Heinen, 1834. For bibliographic details and generalisations, see *Encyklopädie der Math. Wiss.*, vol. III.1.2, p. 1129.

10. *Reflection of the circumcentre O in the sides of ABC.* Given a triangle ABC, define a new triangle $A'B'C'$ by reflecting the point O in the sides BC, CA and AB respectively (see Fig. 4.35). Prove the following properties, which were discovered by Odehnal (2006) using analytical calculations:

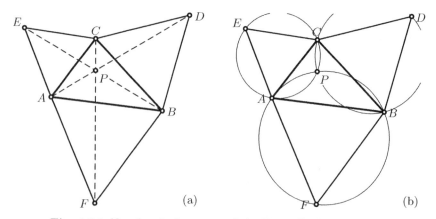

Fig. 4.34. Napoleon's theorem and the Torricelli–Fermat point

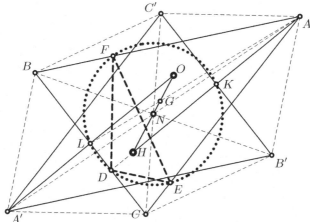

Fig. 4.35. Reflection of the circumcentre O in the sides of ABC

(a) One obtains the triangle $A'B'C'$ by rotating the triangle ABC through 180° around the nine-point centre.

(b) The points A', B' and C' are the circumcentres of the triangles HBC, HCA and HAB, respectively.

(c) The lines $A'B$, $A'C$, and $A'H$ are perpendicular to the sides ED, DF, and FE, respectively, of the orthic triangle DEF; and similarly for B' and C'.

11. Inspired by a drawing of Dürer, see Fig. 4.36, we ask to fill the lens-shaped space between two circles having the same radius R, and whose centres have distance $2a$, by an infinity of circles which touch each other and which touch the two circles. Compute the positions of the intersection points of these circles with the radical axis.

Dürer did not explain how these points, i.e. given d find e, can be constructed with ruler and compass, although he had "mit dem zirckel vnd richtschent" [with ruler and compass] in the title of his book. Repair this omission.

12. Elaborate Roger Penrose's backwards proof of Morley's theorem, as given in Coxeter (1961),[15] using the following idea: *Assume* that the triangle PQR is equilateral; then, since the point R in Fig. 4.37 (a) is the incentre of the triangle ABW, the triangle QPW must be isosceles. Therefore, start from an equilateral triangle and attach three isosceles triangles; produce their sides to obtain the points A, B and C. Determine all the angles correctly.

---

[15]Coxeter presents Penrose, who was at that time 22 years old, as follows: "Roger Penrose is a son of Professor Lionel Penrose the geneticist, and a brother of Jonathan Penrose the chess champion". In the meantime, *Sir* Roger Penrose has become famous in his own right.

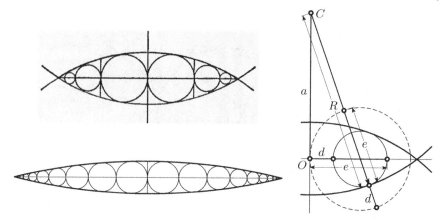

Fig. 4.36. Filling the intersection of two discs with circles (A. Dürer, Underweysung der messung, 1525, book 2)

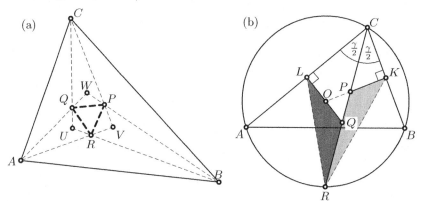

Fig. 4.37. Roger Penrose's backwards proof of Morley's theorem (a); the Olympiad problem (b)

13. (A problem from the Int. Math. Olympiad, Hanoi 2007) In triangle ABC the bisector of angle BCA intersects the circumcircle again at R, the perpendicular bisector of BC at P, and the perpendicular bisector of AC at Q. The midpoint of BC is K and the midpoint of AC is L (see Fig. 4.37 (b)). Prove that the triangles RPK and RQL have the same area.

14. (From a note by Indika Shameera Amarasinghe, Math. Spectrum, 2011) Extend the cevian CF in Fig. 4.12 (a) to a point D on the circumcircle of the triangle ABC and deduce Stewart's theorem from Ptolemy's lemma (see page 114 below), Thales and Eucl. III.35.

Trigonometry

τρεῖς, τρία, three, ...
(Liddel and Scott, *Greek-English Lexicon*, Oxford)

γωνία, corner, angle, ...
(Liddel and Scott, *Greek-English Lexicon*, Oxford)

"I have concentrated on trigonometry."
(Mister Bean in *The Exam*, 1989)

5.1 Ptolemy and the Chord Function

The next giant of Greek science was Ptolemy, who lived around 150 A.D. and was famous both as a great astronomer and as the author of his *Geographia*. Ptolemy's astronomical achievements are collected in his masterpiece μαθη-ματική σύνταξις, later called μεγάλη σύνταξις,[1] the great treatise. Arabic astronomers gave it the Arabic-Greek title al-μεγίστη, and so the book became the *Almagest*, translated into Latin by Regiomontanus. It was the *second* scientific book to be printed (in 1496), after Euclid's *Elements* (in 1482); see Fig. 5.1, left.

All of Ptolemy's measurements were based on the chord function, cord α (see Fig. 5.1, right),[2] which measures the length of a chord in a circle of radius 1 (or 60) as a function of the corresponding central angle α. A table of chords is given in Ptolemy's Almagest for angles between $\frac{1}{2}^\circ$ and 180° in steps of $\frac{1}{2}^\circ$ (see Fig. 5.2). The chords are given for radius 60, using the Babylonian sexagesimal system, refined by adding two sexagesimal digits after the sexagesimal point (usually all correct, with an error of at most 1 sec). He called these digits *partes minutae primae* (first small parts) and *partes minutae secondae* (second small parts), which is the origin of our *minutes* and *seconds*.

---

[1]Megabyte, megaflops, megawatt, megalith, megalomania, ...; a work starting with *mega* is certainly not a bagatelle ...

[2]Ptolemy used εὐτεῖα (straight, direct); "chord" comes from χορδή (bowel, string of gut), Latin "chorda", which was used to produce strings of musical instruments.

A. Ostermann and G. Wanner, *Geometry by Its History,*
Undergraduate Texts in Mathematics, DOI: 10.1007/978-3-642-29163-0_5,
© Springer-Verlag Berlin Heidelberg 2012

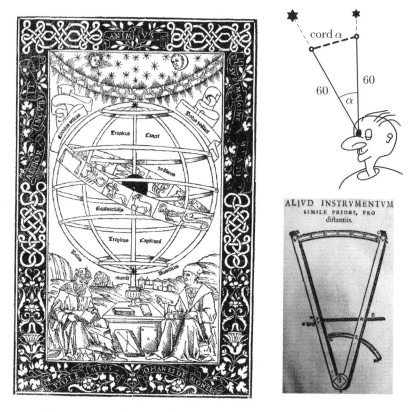

Fig. 5.1. Frontispiece of *Almagest* printed in 1496 (Ptolemy and Regiomontanus seated facing each other, left); Ptolemy's chord function (top right); Tycho Brahe's instrument from 1586 for measuring angles (bottom right)

How did Ptolemy compute his table? For particular angles like $36°$ or $60°$, the lengths of the corresponding chords can easily be determined from the regular hexagon or the regular decagon (we see from Table 1.1 on page 18 that $\operatorname{cord} 60° = 1$ and $\operatorname{cord} 36° = \frac{1}{\Phi}$). Once $\operatorname{cord} \alpha$ was known, he obtained $\operatorname{cord} \frac{\alpha}{2}$ in the same way as Archimedes (see Exercise 22 on page 58). For sums and differences of angles, he used the following identity for the chord function:

$$2 \operatorname{cord}(\alpha + \beta) = \operatorname{cord} \alpha \operatorname{cord}(180° - \beta) + \operatorname{cord}(180° - \alpha) \operatorname{cord} \beta. \tag{5.1}$$

Ptolemy thus arrived successively at $\operatorname{cord} 3°$, $\operatorname{cord} 1.5°$ and $\operatorname{cord} 0.75°$. However, $\operatorname{cord} 1°$ was inaccessible (trisection of an angle). Therefore, Ptolemy calculated $\operatorname{cord} 1°$ by brute interpolation, correctly to the given precision.

The original proof of (5.1) by Ptolemy (see Fig. 5.3) is based on the following lemma.

Lemma 5.1 (Ptolemy). *Let a quadrilateral with sides a, b, c, d be inscribed in a circle. Then the diagonals δ_1, δ_2 satisfy $ac + bd = \delta_1 \delta_2$.*

TABLE DES DROITES INSCRITES DANS LE CERCLE.							
ARCS.		CORDES.		TRENTIÈMES DES DIFFÉRENCES.			
Degrés	Min.	Part. .lu Diam.	Prim	Secon.	Part.	Prim.	Secon.	Tierc.
0	3o	0	31	25	0	1	2	5o
1	0	1	2	5o	0	1	2	5o
1	3o	1	34	15	0	1	2	5o
2	0	2	5	40	0	1	2	5o
2	3o	2	37	4	0	1	2	48
3	0	3	8	28	0	1	2	48
3	3o	3	39	52	0	1	2	48
4	0	4	11	16	0	1	2	47
4	3o	4	42	40	0	1	2	47
5	0	5	14	4	0	1	2	46
5	3o	5	45	27	0	1	2	45
6	0	6	16	49	0	1	2	44

ΚΑΝΟΝΙΟΝ ΤΩΝ ΕΝ ΚΥΚΛΩ ΕΥΘΕΙΩΝ.							
ΠΕΡΙΦΕ-ΡΕΙΩΝ.		ΕΥΘΕΙΩΝ.			ΕΞΗΚΟΤΩΝ.		
Μοιρῶν.		Μ.	Π.	Δ.	Μ.	Π.	Δ.	Τ.
ō	ς	ō	λα	κε	ō	α	β	ν
α	ō	α	β	ν	ō	α	β	ν
α	ς	α	λδ	ιε	ō	α	β	ν
β	ō	β	ε	μ	ō	α	β	ν
β	ς	β	λζ	δ	ō	α	β	μη
γ	ō	γ	η	κη	ō	α	β	μη
γ	ς	γ	λθ	νβ	ō	α	β	μη
δ	ō	δ	ια	ις	ō	α	β	μζ
δ	ς	δ	μβ	μ	ō	α	β	μζ
ε	ō	ε	ιδ	δ	ō	α	β	μς
ε	ς	ε	με	κζ	ō	α	β	με
ς	ō	ς	ιπ	μθ	ō	α	β	μγ

0	30	0	31	25
1	0	1	2	50
1	30	1	34	15
2	0	2	5	**39**
2	30	2	37	4
3	0	3	8	28
3	30	3	39	**53**
4	0	4	11	**17**
4	30	4	42	40
5	0	5	14	4
5	30	5	45	27
6	0	6	16	49

Fig. 5.2. Beginning of Ptolemy's chord table (left as published in Paris 1813); correct values (right)

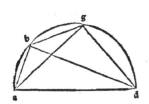

Propofitio   iiij.

Ὅτις chordis inequalium arcuum in femicirculo: arcus quo maior minore fuperat chorda nota fiet. ⟨Ut in femicirculo.a.b.d.fupra diametru.a.d.note fint chor de.a.b.a.g.Dico notam fieri chorda:dam.b.g.nam per correla/ rium prime huius note etiam fient chorde.b.d.z.g.d. ⟨Sint in quadrilatero.a.b.g.d.diametri.a.g.z.b.d.note.funt z late a.a.b.z.g.d.oppofita nota.igif per premiffam quod fit er.a.d.in.b.g.notū fiet.Sed.a.d.eft nota:quia diameter circuli.ideo.b.g.nota fiet: ô querebat. Per hâc plurimor arcuū chordas cognofces. Repies eni chorda arcus quo qĩnta pars circūferentie fertā fupat.f.chorda arcus.12.graduū:z fic de alijs.

Fig. 5.3. Proof of (5.1) using Ptolemy's lemma in the *Almagest*, 1496

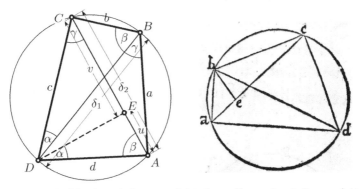

Fig. 5.4. Proof of Ptolemy's lemma; right: from Copernicus' *De revolutionibus*

Proof. Let E be the (unique) point on AC such that the angle EDA equals the angle CDB, see Fig. 5.4. By Eucl. III.21, the two angles denoted by β are equal (as well as those denoted by γ). Therefore, the triangles EDA and CDB (as well as DCE and DBA) are similar, whence

$$\frac{b}{\delta_1} = \frac{u}{d} \quad \text{and} \quad \frac{a}{\delta_1} = \frac{v}{c} \quad \Rightarrow \quad bd + ac = \delta_1(u + v) = \delta_1\delta_2. \qquad \square$$

5.2 Regiomontanus and Euler's Trigonometric Functions

During the Greek period, only the chord function was used. Later (Brahmagupta circa 630 A.D., Regiomontanus 1464, printed in 1533) it became clear that sines and cosines are better adapted to calculations with triangles (see Fig. 5.5).

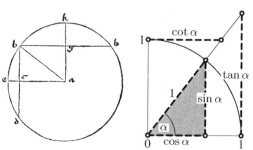

quoad fatis eft prolongatū in e punƈto. Diſco quòd latus b c angulo b a c oppofitum eft finus arcus b e diƈtum angulum fubtendentis. Latus autem tertium, fcilicet a c, æʒquale eft finui reƈto complementi arcus b e. Extendatur enim latus b c occurrendo circumferentiæ circuli in punƈto d . à punƈtis autem a quidem centro circuli exeat femidiameter a k æquediftans lateri b c.& à punƈto b corda b h æquediftans lateri a c. feʒcabunt autem fe neceffario duæ lineæ b h & a k, angulis a b h & b a k acutis exiftentibus, quod fiat in punƈto g. Quia itaʒ femidi

Fig. 5.5. The "trigonometric circle" and definitions of the trigonometric functions. Left: First publication by Regiomontanus, 1464, printed in 1533

$\dfrac{(: BC) \div (AB \doteq AC) \text{ in } \square\, R}{\underset{\text{--- Sin ---}}{\mid AB, \text{ in } AC \mid}} = (: A),$	J. J. Stampioen (Leyden 1632)
cosinum anguli ad A fore $= \dfrac{rq - Cc}{Ss}\, r$, sinu cruris $AB = S$, cosinus eiusdem $= C$, sinu cruris $AC = s$ et cosinu $= c$, cosinu baseos $BC = q$, et radio $= r$;	F. C. Maier (St. Petersburg, 1727)
$\cos : \text{anguli } A = \dfrac{\cos : BC - \cos : AB \cdot \cos : AC}{s\, AB \cdot s\, AC}$, posito radio vel sinu toto 1.	L. Euler (E14, 1735)
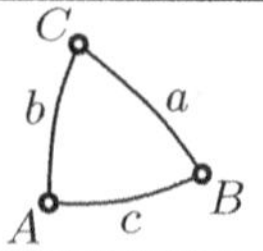 $\cos A = \dfrac{\cos a - \cos b \cdot \cos c}{\sin b \cdot \sin c}$	L. Euler (E214, 1755)

Fig. 5.6. Development of the notation for the trigonometric functions; Stampioen's notation is taken from A. v. Braunmühl, *Bibliotheca Mathematica* 1 (1900), p. 73

However, another two centuries of effort were needed to turn all this Latin text into satisfactory mathematical notation. We show in Fig. 5.6 the development of the notation, each time for the same theorem, the cosine rule of spherical trigonometry, see Formula (5.35) below. The major steps of this development were taken by Euler, who established the sines and cosines, not only as some numbers in a table, but as true functions, allowing all algebraic operations, differentiations, integrations without obstacles. Euler contributed throughout his life in establishing more and more properties and formulas of these functions. A classical treatise on trigonometric functions is Hobson (1891).

To *define* the trigonometric functions, we consider a right-angled triangle placed in a circle of radius 1 as shown in Fig. 5.5. Then the length of the leg *opposite* the angle α is called $\sin \alpha$, whereas the length of the *adjacent* (horizontal) leg is $\cos \alpha$. One further defines

$$\tan \alpha = \frac{\sin \alpha}{\cos \alpha} \quad \text{and} \quad \cot \alpha = \frac{\cos \alpha}{\sin \alpha}. \tag{5.2}$$

By Pythagoras' theorem,

$$\sin^2 \alpha + \cos^2 \alpha = 1. \tag{5.3}$$

The sine function is related to the chord function by

$$\sin \alpha = \frac{1}{2} \operatorname{cord} 2\alpha. \tag{5.4}$$

With the help of Thales, these definitions apply immediately to right-angled triangles of arbitrary size:

$$a = c \cdot \sin \alpha, \quad b = c \cdot \cos \alpha, \quad a = b \cdot \tan \alpha,$$
$$\sin \alpha = \frac{a}{c}, \quad \cos \alpha = \frac{b}{c}, \quad \tan \alpha = \frac{a}{b}. \tag{5.5}$$

Theorem 5.2 (addition formulas). *The following identities hold:*

$$\sin(\alpha + \beta) = \sin \alpha \cos \beta + \cos \alpha \sin \beta,$$
$$\cos(\alpha + \beta) = \cos \alpha \cos \beta - \sin \alpha \sin \beta, \tag{5.6}$$
$$\tan(\alpha + \beta) = \frac{\tan \alpha + \tan \beta}{1 - \tan \alpha \tan \beta}.$$

Proof. The first formula is the same as (5.1) together with (5.4). A geometric proof (by Viète, based on Eucl. III.21) for the first two identities is given in Exercise 3 below. Fig. 5.7 (left) shows today's standard geometric proof of these equations based on the relations (5.5). The third formula is obtained on dividing the first by the second.

However, for this last formula connecting three values of tan, one might also like to see a direct geometric proof. This is given in Fig. 5.7 (right). We see from (5.5), applied to the triangle ABE, that $AB = \tan \alpha \tan \beta$, hence $OA = 1 - \tan \alpha \tan \beta$. Thales' theorem shows that

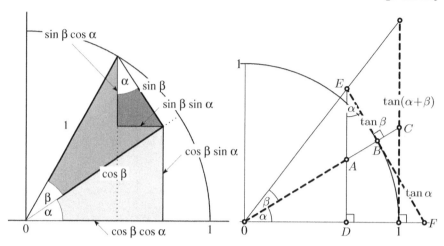

Fig. 5.7. Proof of addition formulas for sin and cos (left), for tan (right)

$$\tan(\alpha + \beta) = \frac{ED}{OA} \cdot OC \quad \text{and} \quad \frac{OC}{1} = \frac{EF}{ED} \quad \Rightarrow \quad \tan(\alpha + \beta) = \frac{EF}{OA},$$

which is the stated result.[3]   □

Later, one also accepted negative angles and defined

$$\cos(-\alpha) = \cos\alpha, \qquad \sin(-\alpha) = -\sin\alpha.$$

With this, we at once derive from (5.6) the identities

$$\begin{aligned}
\sin(\alpha - \beta) &= \sin\alpha\cos\beta - \cos\alpha\sin\beta, \\
\cos(\alpha - \beta) &= \cos\alpha\cos\beta + \sin\alpha\sin\beta.
\end{aligned} \tag{5.7}$$

Relations for double and half angles. Putting $\alpha = \beta$ in (5.6) gives

$$\begin{aligned}
\sin 2\alpha &= 2\sin\alpha\,\cos\alpha, \\
\cos 2\alpha &= \cos^2\alpha - \sin^2\alpha = 1 - 2\sin^2\alpha = 2\cos^2\alpha - 1.
\end{aligned} \tag{5.8}$$

Finally, replacing in the last formula α by $\alpha/2$ we obtain

$$\sin\frac{\alpha}{2} = \pm\sqrt{\frac{1 - \cos\alpha}{2}}, \qquad \cos\frac{\alpha}{2} = \pm\sqrt{\frac{1 + \cos\alpha}{2}}, \tag{5.9}$$

(for a geometric proof see Fig. 5.8, left).

Some values of sine and cosine. The proportions of the equilateral triangle, the square, the pentagon and the decagon provide *sine* and *cosine* for the

---

[3]Another geometrical proof of this formula, not based on Thales, but on Eucl. III.21 and III.35, is given in Hobson (1891, Art. 54) and attributed to "Mr Hart in the *Messenger of Mathematics*, vol. IV".

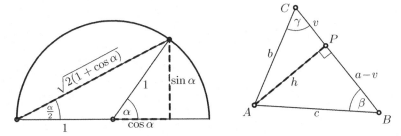

Fig. 5.8. Half angles (left); the law of cosines and the law of sines (right)

angles 60°, 45°, 36°, 30° and 18°, respectively. Once the sine is known, one finds the cosine by Pythagoras (5.3). Then, taking half angles gives the result for 15°, and with the help of differences, we arrive at 3°. Using the identities (5.6) we can finally calculate sine and cosine for $\alpha = 3°, 6°, 9°, 12°, \ldots$, see Table 5.2 in Exercise 4 below. The complete list of algebraic expressions was given by Lambert (1770). See also Hobson (1891, Art. 66).

5.3 Arbitrary Triangles

The trigonometric functions have been defined for right-angled triangles. We will now derive trigonometric relations for arbitrary triangles.

The law of cosines. When ACB is not a right angle, we consider the point P on BC such that APB is a right angle (see Fig. 5.8, right).[4] The *law of cosines* (cosine rule) is the same as Eucl. II.13, because if we insert into formula (2.17) on page 57 the value $v = b \cos \gamma$, which is (5.5) applied to the triangle CPA, we obtain

$$c^2 = a^2 + b^2 - 2ab \cos \gamma \quad \text{or} \quad \cos \gamma = \frac{a^2 + b^2 - c^2}{2ab}. \qquad (5.10)$$

Cyclic permutations of sides and angles $a \to b$, $b \to c$, $c \to a$ give four additional formulas. They all reduce to Pythagoras' theorem if the cosine is zero, i.e. if the angle is right.

The first formula allows one to compute the third side of a triangle in the SAS-case of Fig. 2.9; the second formula allows one to find the three angles in the SSS-case.

The law of sines. Computing h in Fig. 5.8 (right) with the help of (5.5), we find $h = b \sin \gamma$ and also $h = c \sin \beta$. Dividing one formula by the other (and applying cyclic permutations), we find the *law of sines* (sine rule)

---

[4]The reader might see a problem if γ is an obtuse angle. In this case the point P moves outside the triangle and u becomes negative; in contrast to the Ancients, we are no longer afraid of such quantities.

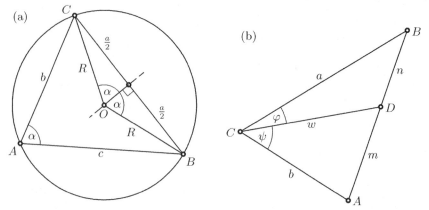

Fig. 5.9. Proof of the law of sines (left) and length of a cevian (right)

$$\frac{\sin \alpha}{a} = \frac{\sin \beta}{b} = \frac{\sin \gamma}{c} . \tag{5.11}$$

This rule is useful if one knows one side and two angles (and hence all three angles).

Relations involving the circumcircle. We see from Fig. 5.9 (a), which is the same as Fig. 4.7 (b) on page 84 and is justified by Eucl. III.20, that $\frac{a}{2} = R \sin \alpha$ and similarly for the other angles. This proves the sine rule once again, but also gives more information:

$$\frac{\sin \alpha}{a} = \frac{\sin \beta}{b} = \frac{\sin \gamma}{c} = \frac{1}{2R} \quad \text{or} \quad \begin{array}{l} a = 2R \sin \alpha , \\ b = 2R \sin \beta , \\ c = 2R \sin \gamma , \end{array} \tag{5.12}$$

where R is the radius of the circumcircle. These last expressions transform the law of cosines (5.10) into the following identity involving the three angles of a triangle (Sturm, 1823/24):

$$\sin^2 \alpha + \sin^2 \beta - \sin^2 \gamma = 2 \sin \alpha \sin \beta \cos \gamma . \tag{5.13}$$

Relations involving the area of the triangle. Inserting $h = c \sin \beta$ into Eucl. I.41 we obtain a formula for the area of the triangle in Fig. 5.8,

$$\mathcal{A} = \frac{ac}{2} \sin \beta . \tag{5.14}$$

This formula divided by abc yields a symmetric expression which allows one to conclude that

$$\frac{2\mathcal{A}}{abc} = \frac{\sin \alpha}{a} = \frac{\sin \beta}{b} = \frac{\sin \gamma}{c} , \tag{5.15}$$

a *third* proof of the law of sines. Comparing this with (5.12) leads to two interesting formulas,

$$\mathcal{A} = \frac{abc}{4R} \quad \text{and} \quad \mathcal{A} = 2R^2 \sin \alpha \sin \beta \sin \gamma. \tag{5.16}$$

This area is also the sum of the areas of three triangles. One of these, BOC, is drawn in Fig. 5.9 (left); its area is $R^2 \cos \alpha \sin \alpha$. Comparing this with (5.16), we obtain another identity for the three angles of a triangle:

$$\sin \alpha \cos \alpha + \sin \beta \cos \beta + \sin \gamma \cos \gamma = 2 \sin \alpha \sin \beta \sin \gamma. \tag{5.17}$$

Since the area of our triangle is the sum of the areas of the triangles BCI, CAI and ABI of Fig. 4.6 on page 83, we have

$$\mathcal{A} = \frac{a+b+c}{2} \cdot \rho, \quad (\rho \text{ radius of the incircle}). \tag{5.18}$$

We finally insert (5.12) for a, b, c and compare with (5.16) to obtain

$$\sin \alpha + \sin \beta + \sin \gamma = \frac{2R}{\rho} \cdot \sin \alpha \sin \beta \sin \gamma. \tag{5.19}$$

Length of a cevian and of an angle bisector. In addition to the results of Section 4.5, we here derive trigonometric formulas for the length of a cevian. Choose the vertex C, call the cevian CD, and denote the angles ACD and DCB by ψ and φ respectively (see Fig. 5.9 (b)). Then for the length w of CD we have the formula

$$\frac{\sin(\varphi + \psi)}{w} = \frac{\sin \psi}{a} + \frac{\sin \varphi}{b}, \tag{5.20}$$

which in a certain sense extends the sine rule. To prove this formula, we multiply it by abw, which gives $ab \sin(\varphi + \psi) = bw \sin \psi + aw \sin \varphi$ and reduces with (5.14) to the relation $\mathcal{A} = \mathcal{A}_1 + \mathcal{A}_2$ for the areas of the three triangles in Fig. 5.9.

The length w of the angle bisector through C satisfies

$$\frac{2 \cos \frac{\gamma}{2}}{w} = \frac{1}{a} + \frac{1}{b}. \tag{5.21}$$

This follows from formula (5.20) by choosing $\varphi = \psi = \frac{\gamma}{2}$ and using (5.8). Apart from the factor $\cos \frac{\gamma}{2}$, w is the *harmonic mean* between a and b (see equation (7.59) on page 223), a result discovered by Pappus (*Collection*, Book III, Prop. 9). We will return to this mysterious resemblance in Chap. 11 (Exercise 5 on page 342).

5.4 Trigonometric Solution of Malfatti's Problem

Problem. Inscribe in a given triangle ABC three circles, each tangent to the other two and to two sides of the triangle (see Fig. 5.10 (b)).

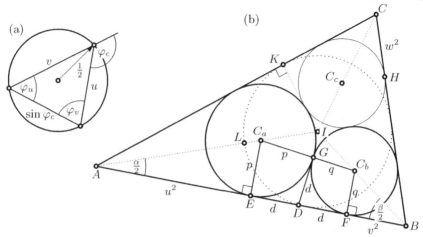

Fig. 5.10. Malfatti's problem and Schellbach's solution (b); auxiliary triangle (a)

This problem appeared in several places towards the beginning of the 19th century and had for a long time the reputation of being very difficult. Since the centres C_a, C_b and C_c of the Malfatti circles must lie on the angle bisectors, we have solved the problem once we have found the points of contact E, F and H with sides of the triangle.

Malfatti's solution. After the problem had been published in the *Gergonne Journal* vol. 10 (1810/11), the "rédacteurs des *Annales*" were informed by a letter from Italy that "M. Malfatti, géomètre italien très-distingué" had obtained a solution already in 1803, after analytic calculations which were to long for a letter to be reproduced, as (in the notations of Fig. 5.10 (b))

$$2AE = AB + KC + AL - CI - BI.$$

Steiner's construction. Steiner, in (1826c), then gave the elegant geometric construction which is given in Fig. 5.11. It is based on the incircles of the triangles IAB, IBC and ICA. Steiner claims, without proof, that the tangents from D, the point of contact of one of these circles with the side AB, to the other two circles is one and the same line, which determines the points E and F. He then claims, again without proof, that the incircles of the triangles ADE and DBF touch the line DEF at the same point. Later reconstructions of Steiner's proofs (for example in Carrega 1981, pp. 101–106) are long, complicated and without any insight into Steiner's way of discovery.

Schellbach's solution. For the elegant solution by K. Schellbach (1853) it will turn out to be preferable to denote the unknown distances AE, BF and CH by u^2, v^2 and w^2 respectively (instead of u, v and w). It will also be useful to normalise the semi-perimeter $s = \frac{a+b+c}{2}$ to $s = 1$.

We denote by D (see Fig. 5.10 (b)) the intersection of the common tangent of two of the Malfatti circles with the side AB. The tangents DE, DG and

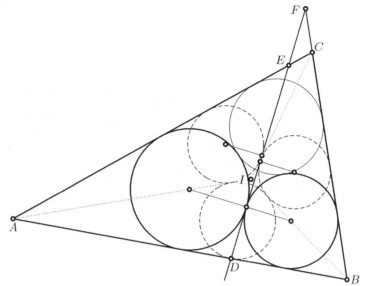

Fig. 5.11. Steiner's construction for Malfatti's problem

DF are equal and we write $DE = DG = DF = d$. The quadrilaterals $C_a GDE$ and $DGC_b F$ are similar, hence $\frac{p}{d} = \frac{d}{q}$ or

$$d = \sqrt{pq} = \sqrt{\tan \tfrac{\alpha}{2} \cdot \tan \tfrac{\beta}{2}} \cdot uv \quad \text{because} \quad p = u^2 \tan \tfrac{\alpha}{2}, \ q = v^2 \tan \tfrac{\beta}{2}.$$

We simplify this expression by using a formula from (5.60) in Exercise 5 below and obtain

$$\tan \tfrac{\alpha}{2} \cdot \tan \tfrac{\beta}{2} = \sqrt{\tfrac{(1-c)(1-b)}{1-a} \tfrac{(1-a)(1-c)}{1-b}} = 1 - c \quad \text{hence} \quad d = \sqrt{1-c} \cdot uv.$$

Adding up the segments on the side AB in our triangle, we get finally

$$c = u^2 + 2d + v^2 = u^2 + 2\sqrt{1-c} \cdot uv + v^2 \tag{5.22}$$

and similar equations for the sides b and a. The solution of this system of three quadratic equations is the main difficulty of the problem. The elegant idea of Schellbach is the following: determine an angle φ_c such that $c = \sin^2 \varphi_c$ (we know that $c < 1$ because $s = 1$). Then $\sqrt{1-c} = \cos \varphi_c$ and (5.22) becomes

$$\sin^2 \varphi_c = u^2 + 2uv \cos \varphi_c + v^2$$

which has a striking similarity with the law of cosines in (5.10). Thus the triangle with sides $u, v, \sin \varphi_c$, sketched in Fig. 5.10 (a), has exterior angle of φ_c (because of the +-sign at the cos-term). Then from the three formulas of the law of sines in (5.12) we obtain first $2R = 1$, and then $u = \sin \varphi_u$ and $v = \sin \varphi_v$. Therefore we have from (1.2)

Fig. 5.12. Ptolemy's map; the line reproduced from Ptolemy's *Geographia* gives the coordinates of "polis Parision Loukotekia" as $23\frac{1}{2}^\circ$ and $48\frac{1}{2}^\circ$

$$\varphi_c = \varphi_u + \varphi_v, \quad \text{and similarly} \quad \varphi_b = \varphi_w + \varphi_u, \quad \varphi_a = \varphi_v + \varphi_w.$$

As in the remark to Eucl. IV.4 on page 83 we obtain

$$\varphi_u = \sigma - \varphi_a, \quad \varphi_v = \sigma - \varphi_b, \quad \varphi_w = \sigma - \varphi_c, \quad \text{where } \sigma = \frac{\varphi_a + \varphi_b + \varphi_c}{2},$$

which determine the positions (and radii) of the three circles. All operations can be turned into constructions by ruler and compass, so that finally no sine tables or calculators are required.

5.5 The Stereographic Projection

Ptolemy was also the architect of *spherical trigonometry*, which he developed mainly for applications in astronomy and geography. This will be our subject in the next sections. Ptolemy's monumental work *Geographia* (eight books) contained longitudes and latitudes of eight thousand locations of the known world (see Fig. 5.12). *One* location, that of "Parision", is reproduced in the second picture of Fig. 5.12 with latitude $48\frac{1}{2}$ degrees, a value which is correct

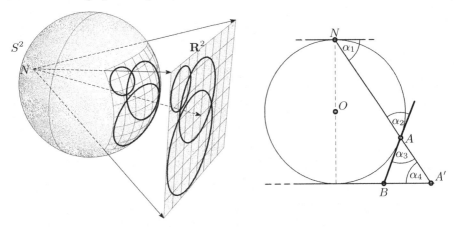

Fig. 5.13. Stereographic projection (left); preservation of angles (right)

to within one third of a degree. Longitudes, however, were difficult to measure with the instruments of that time. The errors near India are more than $30°$. The resulting map spans half of the globe and remained the standard reference until the 16th century.

In order to map the sphere onto a flat sheet, Hipparchus and Ptolemy invented the projection from the antipodal point N onto a plane tangent to the sphere, see Fig. 5.13, left. Since 1613 this projection is called a *stereographic projection* (see Cantor, 1894, vol. I, p. 395). For his terrestrial maps, however, Ptolemy used instead conic-like projections.

Theorem 5.3 (Halley 1696). *The stereographic projection is conformal, i.e. preserves all angles.*

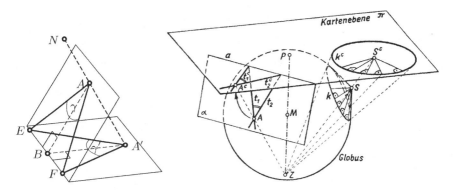

Fig. 5.14. The stereographic projection preserves angles and circles (illustration on the right taken from "Darstellende Geometrie" by E. Ludwig and J. Laub, Wien 1956)

Proof. In the picture on the right of Fig. 5.13 one verifies that $\alpha_1 = \alpha_2$ (isosceles triangle), $\alpha_2 = \alpha_3$ (opposite angles), and $\alpha_1 = \alpha_4$ (parallel angles). Thus $\alpha_3 = \alpha_4$, so $AB = A'B$ and the triangle ABA' is *isosceles*. An angle at A, formed by two lines in the tangent plane, is mapped onto the same angle at A', since the intersection points E and F of these two lines with the horizontal plane form identical triangles with A and A', see the picture on the left of Fig. 5.14. □

Theorem 5.4 (Ptolemy, rediscovered by Miquel (1838b, Thm. IV)).   *The stereographic projection maps circles onto circles.*

Proof. Let a circle be given on the sphere. An elegant idea consists in adding to this circle the cone formed by all tangents to the sphere that are orthogonal to the circle, see the picture on the right of Fig. 5.14. Mapping everything by the stereographic projection gives the image k^c of the circle together with a bundle of lines. The lines all pass through the same point (which is the image of the vertex of the cone) and, by Theorem 5.3, are *orthogonal* to the curve k^c. Therefore, k^c must be a circle. □

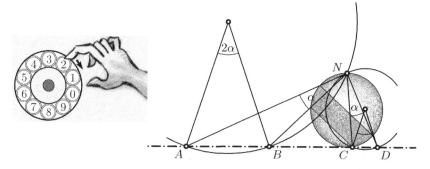

Fig. 5.15. Proof of Steiner's porism

Proof of Steiner's porism. As we have seen in Section 4.9, Steiner obtained Theorem 4.18 after a long struggle. He never mentioned with a word — either he did not know, or he wanted to keep it as his secret — that this result becomes nearly evident with the use of the stereographic projection. Firstly, the statement is trivial if C_1 and C_2 are concentric (see Fig. 5.15, left). Otherwise we look for a stereographic projection which transforms C_1 and C_2 into concentric circles on the sphere. For this we remember Eucl. III.20, draw the line of symmetry in the first picture of Fig. 4.25, cutting the circles in four points A, B, C and D, and construct on each of the segments AB and CD (similar) isosceles triangles with the same angle 2α at the vertex (see Fig. 5.15, right).[5] We then choose for N one of the intersection points of the two circles drawn around these vertices as centres. The inscribed angles ANB and CND are

---

[5] This angle should be chosen not too large.

equal to α and the (inverse) stereographic projection from this point transforms the original circles C_1 and C_2 into concentric ones. All the small circles are transformed into circles with the same radius, and the porism is also clear in the second case. □

5.6 The Spherical Trigonometry of Right-Angled Triangles

On the sphere, the analogue of a straight line becomes a circle whose centre coincides with that of the sphere (geodesic line). A circle on the sphere with this property is called a *great circle*. A *segment* on the sphere is a connected part of a great circle. The length of a segment is measured *in radians* by its angle at the centre.

Thus, a *spherical triangle* is composed of three great circles, whose side lengths a, b, c are the corresponding angles at the centre, and three angles α, β, γ, which express the inclination of the three great circles taken two at a time. If one of the angles, say γ, is a right angle, then the spherical triangle is called *right-angled*.

Spherical trigonometry consists in finding all possible relations between these quantities. The first ones were found by Ptolemy in his *Almagest* (Book I, Chap. 11). He considered right-angled triangles and used a complicated theory that goes back to Menelaus. Spherical trigonometry was further developed by mathematicians of the Islamic world,[6] and taken up again by Napier and later by Euler. The latter developed the general case in a straightforward way, "ex primis principiis breviter et dilucide derivata" (see Euler, 1782).[7] We start by applying Euler's ideas to *right-angled* triangles in an even more *brevis et dilucidus* way.

Idea. Take a spherical triangle ABC with a right angle at C and the angle β at B *and project it from the centre O onto the tangent plane at B*. In this way we obtain a right-angled triangle $A'BC'$ which *again* has the angle β at B, see Fig. 5.16.

We assume that $OB = 1$ and find from the right-angled triangle OBC' the quantities $BC' = \tan a$ and $OC' = \frac{1}{\cos a}$, similarly from the right-angled triangle OBA' the quantities $BA' = \tan c$ and $OA' = \frac{1}{\cos c}$, and from the triangle

---

[6]In the 10th century Abū'l-Wafā' Būzjānī discovered the law of sines for spherical triangles, see Berggren (1986), p. 175. In the 11th century al-Jayyānī wrote an influential treatise on spherical trigonometry with the title *The book of unknown arcs of a sphere*.

[7]The same idea had been used in a short note by Francis Blake in 1752. It was later discovered that Newton had also used the same figures in an unpublished manuscript from 1684 (see Newton, Mathematical Papers, vol. IV, p. 174, note (9) for more details).

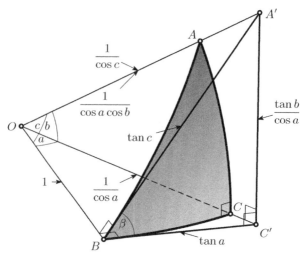

Fig. 5.16. A right-angled spherical triangle

$OC'A'$ the quantities $C'A' = \frac{\tan b}{\cos a}$ and $OA' = \frac{1}{\cos a \cos b}$. The two values obtained for OA', which must be equal, provide a first interesting formula, the *right-angled law of cosines* (cosine rule for sides):

$$\cos c = \cos a \cdot \cos b. \qquad\qquad (5.23)$$

A right-angled spherical triangle has five unknowns a, b, c, α and β. Formula (5.23) relates the three side lengths a, b, c in such a way that *one* can be found if the *other two* are known. This is symbolised by the small forget-me-not.

Remark. If the angles a, b, c become small, i.e. if the triangle tends to a planar triangle, then we may neglect the terms a^4, b^4, c^4 and a^2b^2, and their products, in the series for $\cos c = 1 - \frac{c^2}{2} + \frac{c^4}{24} - \ldots$ The law of cosines then becomes

$$1 - \frac{c^2}{2} \approx \cos c = \cos a \cdot \cos b \approx (1 - \frac{a^2}{2})(1 - \frac{b^2}{2}) \approx 1 - \frac{a^2}{2} - \frac{b^2}{2} \qquad (5.24)$$

which, on subtracting 1 from both sides and multiplying by -2, reduces to Pythagoras' theorem.

Formulas for the angles. Consider now the plane triangle $A'BC'$. We find here the following three relations (and analogous ones by exchanging $a \leftrightarrow b$, $\alpha \leftrightarrow \beta$):

$$\sin \beta = \frac{\tan b}{\cos a \cdot \tan c} \overset{(5.23)}{=} \frac{\sin b}{\sin c} \qquad\qquad \sin \alpha = \frac{\sin a}{\sin c}$$

$$(5.25)$$

$$\cos\beta = \frac{\tan a}{\tan c} \qquad\qquad \cos\alpha = \frac{\tan b}{\tan c} \tag{5.26}$$

and

$$\tan\beta = \frac{\tan b}{\cos a \cdot \tan a} = \frac{\tan b}{\sin a} \qquad\qquad \tan\alpha = \frac{\tan a}{\sin b}. \tag{5.27}$$

Dividing a formula on one side of (5.26) by that on the other side of (5.25) gives

$$\frac{\cos\beta}{\sin\alpha} = \frac{\tan a \cdot \sin c}{\tan c \cdot \sin a} \overset{(5.23)}{=} \cos b \qquad\qquad \frac{\cos\alpha}{\sin\beta} = \cos a. \tag{5.28}$$

Inserting the last expressions into (5.23) then yields

$$\cos c = \frac{1}{\tan\alpha \cdot \tan\beta}. \tag{5.29}$$

Finally, we have a nice harvest — without too much calculational effort. In order to memorise this long list, clever rules have been devised (e.g. Napier's rules). Instead, we suggest that you memorise the proof rather than the particular formulas.

Example 1. The shortest day in Wrocław. Hugo Steinhaus (1887–1972), one of Poland's most distinguished mathematicians, felt that "a few years after the [second world] war the inadequacy of mathematical education in our high schools became evident (...) and a closer collaboration between mathematicians and school teachers could no longer be postponed". In order to "stimulate interest in mathematics", Steinhaus published the booklet Steinhaus (1958). Problem number 76 asks for the length of the shortest day in Wrocław, Steinhaus' hometown, situated at the latitude $\varphi = 51°07'$. On the shortest day (tropic of Capricorn), the sun is at the latitude $\delta = -23°27'$. We are asked to find the length of this day.

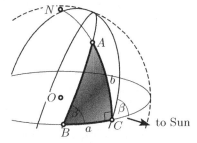

Fig. 5.17. Shortest day in Wrocław

Solution. On the 21st of December, the limit between shadow and light is formed by a great circle making a constant angle of $\beta = 90° - \delta = 66°33'$ with the Equator and rotating from east to west. This circle crosses the intersection of the "meridian of Wrocław" with the Equator (the point C in Fig. 5.17 (a)) at precisely 6 a.m. "local Wrocław time". The circle then moves on along the Equator through a certain angle a, until its intersection with the meridian of Wrocław reaches the point A at latitude $b = \varphi$. We have to solve a right-angled triangle with known β, b and unknown a. The solution is given by (5.27) and yields

$$\sin a = \frac{\tan b}{\tan \beta} = \frac{\tan 51°07'}{\tan 66°33'} = 0.5379034896 \quad \Rightarrow \quad a = 32.541034°. \quad (5.30)$$

For the corresponding quantity measured in hours we divide this value by $\frac{360}{24} = 15$ and obtain $2.169402268 = 2\text{h}\,10\text{min}\,10\text{sec}$, so that sunrise occurs at $8\text{h}\,10\text{min}\,10\text{sec}$. The same amount of time is lost in the evening, so that $7\text{h}\,39\text{min}\,40\text{sec}$ remain for the the total length of the day. Needless to say that this wonderfully precise value does not take into account that the data for φ and δ are not so precise, that the sun is not a perfect point, that the earth is not a perfect sphere and that the light is refracted.

Example 2. Platonic solids. Following in Euler's footprints (1781), we now want to use the above formulas to discover some secrets of the Platonic solids.

Problem. Under which angle do adjacent faces of a Platonic solid meet?

Solution. Let the solid be composed of regular k-gons, of which ℓ meet at each vertex. One places the centre of a small sphere at one vertex of the solid (Fig. 5.18 (a) illustrates this idea for the octahedron). The edges then cross this sphere perpendicularly and the faces create a regular ℓ-gon, which we cut into 2ℓ right spherical triangles such as EAN in this figure. Since $2\ell\alpha = 360°$, we have $\alpha = \frac{180°}{\ell}$. The sides $2a$ of this ℓ-gon are equal to the angles formed by regular plane k-gons, i.e. $180° \cdot \frac{k-2}{k}$. Thus we have for each Platonic solid

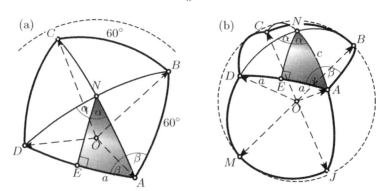

Fig. 5.18. Angles between adjacent faces of an octahedron (a); angles of the edges of a cube seen from the centre (b)

the values of α and a, while the angle 2β is the required answer. A discreet glance at our forget-me-nots above shows that we must use the formula on the right of (5.28), i.e.

$$\sin \beta = \frac{\cos \alpha}{\cos a}, \tag{5.31}$$

which allows the calculations of β and 2β. The results are:

solid	k	ℓ	α	a	2β
tetrahedron	3	3	60°	30°	70°31′43″36‴20⁗35‴‴
cube	4	3	60°	45°	90°00′00″00‴00⁗00‴‴
octahedron	3	4	45°	30°	109°28′16″23‴39⁗25‴‴
dodecahedron	5	3	60°	54°	116°33′54″11‴03⁗15‴‴
icosahedron	3	5	36°	30°	138°11′22″51‴58⁗57‴‴

$$(5.32)$$

Euler computed the division in (5.31) with logarithms and obtained the values of β with correct degrees, minutes and seconds, to which we proudly add the terces, quartes and quintes. The result of 90° for the cube ("scilicet hic angulus ipse est rectus") is no surprise.

Problem. Under which angle do the edges of a Platonic solid appear when seen from the centre?

Solution. Let again the solid be composed of regular k-gons, with ℓ at each vertex. Projecting these k-gons from the centre onto the circumscribed sphere produces spherical k-gons (see Fig. 5.18 (b) for the example of the cube). We decompose these k-gons into $2k$ right spherical triangles such as NEA in this figure. This time we know the angles $\alpha = \frac{\pi}{k}$ and $\beta = \frac{\pi}{\ell}$, while the arc a is requested (because $2a$ is the answer of our problem). We obtain a directly from the formula on the right of (5.28). This angle also allows one to compute the radius R of the circumscribed sphere of the Platonic solid. If we normalise the edges to 1, the required value is just the chord of $2a$, since we have $1 = 2R \sin a$. The results are:

solid	k	ℓ	$2a$	R	ρ/R
tetrahedron	3	3	109°28′16″24‴	0.612372436	0.333333333
cube	4	3	70°31′43″36‴	0.866025404	0.577350269
octahedron	3	4	90°00′00″00‴	0.707106781	0.577350269
dodecahedron	5	3	41°48′37″08‴	1.401258538	0.794654472
icosahedron	3	5	63°26′05″49‴	0.951056516	0.794654472

$$(5.33)$$

For most of these results, we don't really need spherical trigonometry. We have from Pythagoras that $R = \frac{\sqrt{6}}{4}$ for the tetrahedron, $R = \frac{\sqrt{2}}{2}$ for the octahedron and $R = \frac{\sqrt{3}}{2}$ for the cube. We also know from Eucl. XIII.17 (see Fig. 2.37) that $R = \frac{\Phi\sqrt{3}}{2}$ for the dodecahedron. This serves as a control for our formulas.

Problem. Find the radius of the inscribed spheres of the Platonic solids.

Solution. The inscribed spheres touch a face of the solid in its midpoint. For example the plane through $ABCD$ in Fig. 5.18 (b) is touched in the orthogonal projection of A onto ON. Hence $\rho = R \cdot \cos c$, where c is the arc AN. Inserting formula (5.29) we obtain

$$\rho = \frac{R}{\tan \alpha \cdot \tan \beta} \tag{5.34}$$

which leads to the last column in the table of (5.33). The symmetry of this formula shows that the dual solids cube and octahedron as well as icosahedron and dodecahedron have the same radius for the inscribed sphere.

Kepler's first cosmological model. From the similitude of the ratios in the above table with the ratios of the semi-major axes of the planet's orbits known at that time

$$a_{\text{Jupiter}}/a_{\text{Saturn}} = 0.545 \qquad a_{\text{Mars}}/a_{\text{Jupiter}} = 0.293$$
$$a_{\text{Earth}}/a_{\text{Mars}} = 0.657 \qquad a_{\text{Venus}}/a_{\text{Earth}} = 0.723$$
$$a_{\text{Mercury}}/a_{\text{Venus}} = 0.536$$

Kepler was convinced (*Mysterium cosmographicum* 1596) that the Creator of the best of all Universes wanted his planets to move on inscribed and superscribed spheres of Platonic solids (see Fig. 5.19). Throughout his live, Kepler considered this to be the greatest of all his discoveries.

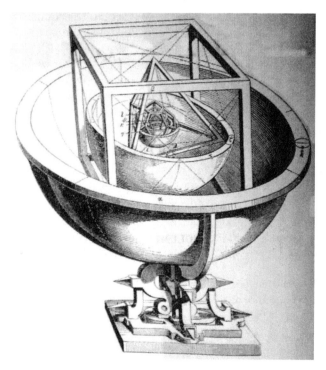

Fig. 5.19. Kepler's cosmological model

5.7 The Spherical Trigonometry of General Triangles

In the case of arbitrary triangles we can proceed in two ways: either we divide the spherical triangle ABC into two right-angled spherical triangles and apply, as in Sect. 5.3, the above formulas for right-angled triangles. This will be done in Exercise 16 below. Euler's original proof, however, leads directly to the general cosine and sine rules and to many additional results. It uses a decomposition of the *projected planar triangle*. We now follow precisely this proof of Euler (1782), full of admiration for this 74-year-old man and what he "saw" in the dark night of total blindness.

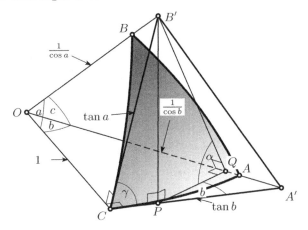

Fig. 5.20. Euler's proof for the general spherical triangle

Idea. We project the spherical triangle ABC from the centre O onto the tangent plane at one of the vertices, which we choose to be C. In this way we obtain a planar triangle $A'B'C$ which again has the angle γ at C, see Fig. 5.20. Then, as in Sect. 5.3, we let P be the orthogonal projection of B' onto the plane OAC. This yields the right-angled planar triangle CPB'. Finally, we let Q be the orthogonal projection of P onto the line OA. This produces the right-angled planar triangles OQB' and $B'PQ$, *where the angle α is preserved*, because OA is orthogonal to the plane $B'PQ$. Furthermore, the angle $A'PQ$ is equal to b (angle orthogonal to the angle b at the centre). Now we are in the position to compute the required distances in Fig. 5.20. The results are collected in Table 5.1.

The law of cosines. We see from Fig. 5.20 that $OA' - QA' = OQ$. By inserting the values of Table 5.1, we obtain after simplification our first formula,

$$\cos c = \cos b \cos a + \sin b \sin a \cos \gamma \quad \text{or} \quad \cos \gamma = \frac{\cos c - \cos b \cos a}{\sin b \sin a}, \quad (5.35)$$

the *law of cosines for arbitrary spherical triangles*. For another proof, see Exercise 16 below.

Table 5.1. Lengths of the edges of the triangles in Fig. 5.20

length	justification
$OC = 1$	by definition
$OA' = \dfrac{1}{\cos b}$	$OC = OA' \cdot \cos b$
$CA' = \tan b$	$CA' = OC \cdot \tan b$
$OB' = \dfrac{1}{\cos a}$	$OC = OB' \cdot \cos a$
$CB' = \tan a$	$CB' = OC \cdot \tan a$
$CP = \tan a \cos \gamma$	$CP = CB' \cdot \cos \gamma$
$B'P = \tan a \sin \gamma$	$B'P = CB' \cdot \sin \gamma$
$B'Q = \dfrac{\sin c}{\cos a}$	$B'Q = OB' \cdot \sin c$
$OQ = \dfrac{\cos c}{\cos a}$	$OQ = OB' \cdot \cos c$
$PA' = \tan b - \tan a \cos \gamma$	$PA' = CA' - CP$
$PQ = \sin b - \tan a \cos b \cos \gamma$	$PQ = PA' \cdot \cos b$
$QA' = \dfrac{\sin^2 b}{\cos b} - \tan a \sin b \cos \gamma$	$QA' = PA' \cdot \sin b$

Remark. As in (5.24), if a, b, c become small, the above cosine theorem tends to the planar cosine rule (5.10).

The law of sines. We next compute $\sin \alpha$ from the right-angled triangle $B'PQ$ and obtain with the values of Table 5.1

$$\sin \alpha = \frac{B'P}{B'Q} = \frac{\sin a \sin \gamma}{\sin c} \quad \Rightarrow \quad \frac{\sin \alpha}{\sin a} = \frac{\sin \beta}{\sin b} = \frac{\sin \gamma}{\sin c}, \qquad (5.36)$$

the *law of sines for arbitrary spherical triangles*. These identities were already discovered by Abū'l-Wafā' Būzjānī in the 10th century.

Cotangent theorems. We finally compute $\cos \alpha = \frac{PQ}{B'Q}$ and obtain with the values of Table 5.1

$$\cos \alpha = \frac{\cos a \sin b - \sin a \cos b \cos \gamma}{\sin c}. \qquad (5.37)$$

If this is divided by $\sin \alpha = \frac{\sin \gamma \sin a}{\sin c}$ (see (5.36)), we obtain a first interesting formula,

$$\cot \alpha = \frac{\cos a \sin b - \sin a \cos b \cos \gamma}{\sin a \sin \gamma}. \qquad (5.38)$$

This is used to compute the remaining angles in the SAS-case.

Incessantly, Euler continued to transform the formulas in all possible manners: we write (5.37) in the form

$$\cos \alpha \sin c = \cos a \sin b - \sin a \cos b \cos \gamma$$

and multiply the three terms by $\frac{\sin \gamma}{\sin c}$, $\frac{\sin \beta}{\sin b}$, $\frac{\sin \alpha}{\sin a}$ respectively; this leads to an "aequatio memorabilis", which, after division by $\sin \beta$ and the permutations $b \leftrightarrow c$, $\beta \leftrightarrow \gamma$ becomes

$$\cos a = \frac{\cos \alpha \sin \beta + \sin \alpha \cos \beta \cos c}{\sin \gamma}, \qquad (5.39)$$

a formula curiously similar to (5.37). We finally divide, as above, by $\sin a = \frac{\sin \alpha \sin c}{\sin \gamma}$ and obtain the formula

$$\cot a = \frac{\cos \alpha \sin \beta + \sin \alpha \cos \beta \cos c}{\sin \alpha \sin c}, \qquad (5.40)$$

analogous to (5.38). This formula is useful in the ASA-case to compute the remaining sides "ex datis duobus angulis α, β cum latere intercepto c".

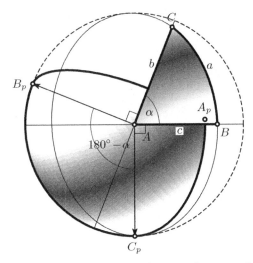

Fig. 5.21. Formation of the polar triangle; in order to make the values of the opposite angles more visible, the projection point is chosen vertically above the point A

Duality. After another half page of calculations, Euler finally managed to master the "casus alias difficillimus", to calculate a side from "datis tribus angulis". The result is

$$\cos \gamma = - \cos \beta \cos \alpha + \sin \beta \sin \alpha \cos c \quad \text{or} \quad \cos c = \frac{\cos \gamma + \cos \beta \cos \alpha}{\sin \beta \sin \alpha}. \tag{5.41}$$

All this beautiful duality between (5.35) and (5.41), between (5.37) and (5.39), between (5.38) and (5.40), calls for an elegant explanation. The explanation is the following THEOREMA: *if a spherical triangle ABC with sides a, b, c and angles α, β, γ is replaced by the triangle with vertices A_p, B_p, C_p, which are the* poles *of the great circles on which the sides a, b, c lie,[8] then the roles of the angles and of the sides are interchanged* and become the opposite angles

$$\alpha_p = 180° - a, \quad \beta_p = 180° - b, \quad \gamma_p = 180° - c,$$
$$a_p = 180° - \alpha, \quad b_p = 180° - \beta, \quad c_p = 180° - \gamma,$$

see Fig. 5.21. Replacing the angles by the opposite angles changes the signs of the cosines, but not those of the sines. This explains the sign changes in the above dual formulas.

Example 1. Cardan joint. We are interested in solving the following problem: Find the relation between the rotation angles of the shafts (which we call a and b) of a Cardan joint, when the bend angle γ is given (see Fig. 5.22, left).

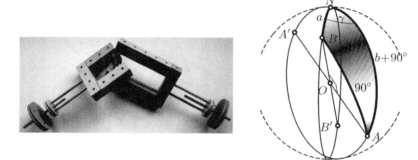

Fig. 5.22. A high-tech Cardan joint (left), and its spherical triangle (right)

Solution. We place a sphere at the centre of the joint. Then the endpoints of the cross shaft, say A and its opposite A', as well as B and its opposite B', rotate on two great circles which intersect at an angle γ. The cross shaft always keeps AA' and BB' at right angles to each other. We express the required rotation angles by prescribing the angles at A and B with the North Pole N to be respectively a and $b + 90°$ (see Fig. 5.22, right). This produces a spherical triangle BNA to which we apply the law of cosines (5.35), and require that $\cos c = \cos 90° = 0$. This gives

$$0 = \cos(90°+b)\cos a + \sin(90°+b)\sin a \cos\gamma \quad \text{or} \quad \tan b = \cos\gamma \cdot \tan a, \quad (5.42)$$

---

[8] "... formari ex Polis trium laterum ..."; there are always two poles to choose from, we respect the orientation and choose the one which points *outside* the triangle. The triangle $A_p B_p C_p$ is called the polar triangle of ABC.

which is the desired relation. We see that the more the bend angle γ differs from zero, the smaller $\cos \gamma$ becomes, the more the angle b deviates from a, and the more the joint suffers from jerky rotation.

Example 2. The spherical distance of two points. The admirers of Goscinny are familiar with "the most prestigious cities in the universe", Rome (*Asterix and the Laurel Wreath*, p. 1) and Lutetia (p. 2); the admirers of Ptolemy can find in his *Geographia* the following longitudes and latitudes: $36\frac{2}{3}°, 41\frac{2}{3}°$ for the one, and $23\frac{1}{2}°, 48\frac{1}{2}°$ (see Fig. 5.12) for the other. He established his zero meridian at the *Fortunate Islands* (now roughly the Canaries), the western-most point of the world known to him, see Brown (1949, p. 75); the admirers of Eratosthenes know that the radius of the Earth is about 6360 km. We want to calculate the shortest distance (measured along the corresponding great circle) between these two famous cities.

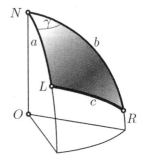

Solution. The idea is to make the great circle LR a side of the spherical triangle LNR where N is the North Pole (see figure). Thus, one knows $a = 90 - 48\frac{1}{2} = 41.5°$ and $b = 90 - 41\frac{2}{3} = 48.33°$. The angle γ is the difference of the longitudes $\gamma = 36\frac{2}{3} - 23\frac{1}{2} = 13.17°$. Hence, the third side can be calculated by the cosine rule (5.35). One obtains $\cos c = 0.979885$, $c = 11.5°$, and finally $d = 1275.8$ km. Note, however, that the error of one third of a degree in latitude corresponds to 37 km on the meridian. One should therefore not put too much faith in the precision obtained.

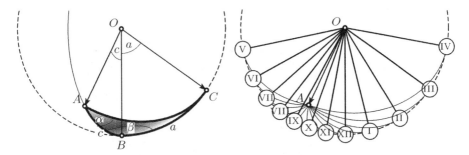

Fig. 5.23. A sundial for Geneva

Example 3. Sundials. Good bye, Wrocław, welcome Geneva at the latitude $\varphi = 46°12'$. We want to create a sundial on a wall, whose normal deviates from the south in direction of the east by an angle of $\sigma = 30°45'$. The sundial should measure the *apparent solar time*, without any of the corrections required by modern life (compensation for the nonuniform Kepler motion, time zones). We plant at a point O on the wall a *gnomon* OA parallel to the Earth's axis (see Fig. 5.23, left). The sun, together with its shadow, rotates around this gnomon OA at the constant angular speed of $15°$ per hour. At noon, the sun is precisely in the south and the shadow is in the vertical position OB. We want to determine the position of the shadow, i.e. the angle a, at any other time h.

Solution. We imagine half of a sphere of radius 1 fastened to the wall, with its centre at O. The point A of the gnomon makes an arc of $c = 90° - \varphi = 43°48'$ with B on the wall. The great circle AB makes an angle of $\beta = 90° + \sigma = 120°45'$ with the wall (see again Fig. 5.23, left). The great circle AC represents the shadow of the gnomon; the angle $\alpha = 0$ at noon and increases by $15°$ every hour, hence we have $\alpha = (h - 12) \cdot 15°$. ABC is a spherical triangle with one side c and two adjacent angles α and β known. We are clearly in the ASA-case, and formula (5.40) can readily be applied to obtain the required side a. Rewritten for the above data this formula becomes

$$\cot a = \frac{\cos\alpha\cos\sigma - \sin\alpha\sin\sigma\sin\varphi}{\sin\alpha\cos\varphi}. \tag{5.43}$$

This formula gives for the above sundial the values

VII	VIII	IX	X	XI	XII	I	II	III	IV
$-49°7'$	$-38°40'$	$-29°24'$	$-20°26'$	$-10°57'$	$0°0'$	$13°42'$	$31°44'$	$54°41'$	$79°35'$

(see Fig. 5.23, right). Alsatian painters, of course, make sundials much more poetically (see Fig. 5.24).

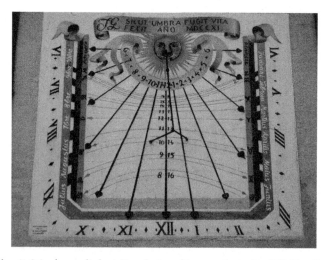

Fig. 5.24. A sundial at Bergheim, Alsace; photo by J.P. Kauthen

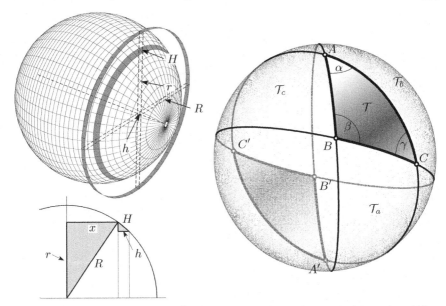

Fig. 5.25. Area of the sphere (left) and of a spherical triangle (Girard, right)

5.8 The Area of a Spherical Triangle

The following theorem is due to Archimedes (*On the sphere and cylinder I*, Props. 25 and 30).

Theorem 5.5. *The area of a sphere is four times that of its great circle, i.e.*

$$\mathcal{A} = 4R^2\pi, \qquad \text{where } R \text{ is the radius of the sphere.} \qquad (5.44)$$

We will prove, in fact, the following extension of this theorem.

Theorem 5.6 (Lambert 1772). *The cylindrical projection, which projects points of the sphere horizontally from the axis onto the circumscribed cylinder (see Fig. 5.25, left) is area preserving. In particular, the area of a sector cut from a sphere by two great circles under the angle α (measured in radians) is*

$$\mathcal{A}_S = 2R^2\alpha. \qquad (5.45)$$

Proof. We consider the two pictures on the left of Fig. 5.25. By Thales, we have $rH = Rh$. This implies that the two grey ribbons in the upper picture have the same area. Thus the sphere has the same area (slice by slice and piece by piece) as the circumscribing cylinder. □

Theorem 5.7 (A. Girard 1626, see also Euler (E514, 1781), *Opera* 26, p. 205). *The area of a spherical triangle with angles α, β and γ (measured in radians) is*

$$\mathcal{T} = R^2 \cdot (\alpha + \beta + \gamma - \pi). \qquad (5.46)$$

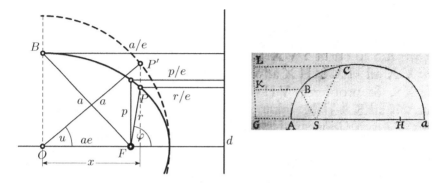

Fig. 5.26. A point P on an ellipse, F the focus (Sun), φ the true anomaly, u the eccentric anomaly, a the semi-major axis, and e the eccentricity (left); illustration from Newton's *Principia* (right)

Proof. The three great circles obtained by producing the three sides of the triangle divide the surface of the sphere into eight spherical triangles: T, $\mathcal{T}_a$, $\mathcal{T}_b$, $\mathcal{T}_c$ and four antipodal triangles having the same respective areas, see the picture on the right in Fig. 5.25. By (5.44), their areas satisfy

$$\mathcal{T} + \mathcal{T}_a + \mathcal{T}_b + \mathcal{T}_c = 2\pi \cdot R^2 \ . \tag{5.47}$$

But $T \cup T_a$, $T \cup T_b$ and $T \cup T_c$ are sectors with angles α, β and γ, respectively. Hence, by (5.45)

$$\mathcal{T} + \mathcal{T}_a = 2\alpha \cdot R^2 \ , \qquad \mathcal{T} + \mathcal{T}_b = 2\beta \cdot R^2 \ , \qquad \mathcal{T} + \mathcal{T}_c = 2\gamma \cdot R^2 \ .$$

Adding the three identities and subtracting (5.47) gives the desired result. $\square$

5.9 Trigonometric Formulas for the Conics

The trigonometric functions allow one to derive distance and area formulas for the conics, which will soon be very important.

Distances from the focus. Given a point P on an ellipse (see Fig. 5.26), we want to derive formulas for its distance PF to a focus F. We denote this distance by r. Astronomers call the angle φ the *true anomaly* and the angle u, after embedding the ellipse in the circle of radius a, the *eccentric anomaly*.[9] We then have the formulas

$$r = \frac{p}{1 + e \cos \varphi} \qquad \text{(formula for the true anomaly)},$$
$$r = a - ex = a - ae \cos u \qquad \text{(formula for the eccentric anomaly)}. \tag{5.48}$$

---

[9]To readers, who are surprised to see the angle at the *centre* of the ellipse called eccentric ("away from the centre"), we recall that the real centre of the solar system is the Sun, i.e. the focus.

Proof. The distance BF is equal to a by (3.5). The lengths r/e, p/e and a/e are determined by our first definition of the ellipse (see Fig. 3.4). Then we see in Fig. 5.26 the relations $r\cos\varphi + \frac{r}{e} = \frac{p}{e}$ and $a\cos u + \frac{r}{e} = \frac{a}{e}$. These equations, when solved for r, lead to the formulas (5.48). □

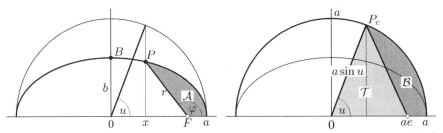

Fig. 5.27. Computation of the area $\mathcal{A}$ swept out by the radius vector

Formula for the area. The area $\mathcal{A}$ swept out by the line joining F and P (see Fig. 5.27, left) plays an important role in astronomy, as we will soon see. We stretch the ellipse into a circle (see Fig. 5.27, right) to get $\mathcal{B} = \frac{a}{b}\mathcal{A}$. Since $\mathcal{B}$ is the difference of a circular sector (of area $\frac{a^2}{2} \cdot u$) and a triangle (whose area $\mathcal{T}$ we get from Eucl. I.41 and from the fact that OF is equal to ae),[10] we obtain

$$\mathcal{B} = \frac{a^2}{2}(u - e\sin u) \qquad \text{and} \qquad \mathcal{A} = \frac{ab}{2}(u - e\sin u). \qquad (5.49)$$

5.10 The Great Discoveries of Kepler and Newton

> "Astronomy is older than physics. In fact, it got physics started by showing the beautiful simplicity of the motion of the stars and planets, the understanding of which was the *beginning* of physics."
>
> (R. Feynman, 1964, Chap. 3.4)

> " ... i libri di Apollonio, ... delle quali sole siamo bisogni nel presente trattato. [The books of Apollonius, the only ones which we require in the present treatise.]"
>
> (Galilei 1638, giornata quarta)

Three great works marked the emergence of modern science (see the first quotation): Kepler's *Astronomia Nova* (1609), Galilei's *Discorsi* (1638) and Newton's *Principia* (1687). The discoveries of all three works were based mainly on tools from elementary geometry (Thales, Euclid and Apollonius, see the second quotation), however in a highly ingenious way. So they fit well into our book, but don't expect easy bedtime reading here.

---

[10]This can be seen either from $a^2 - b^2 = a^2e^2$ (see (3.8)) and Pythagoras, or from the second formula in (5.48) with $u = 0$.

Kepler's laws.

> "... itaque futilum fuisse meum de Marte triumphum; forte fortuito incido in secantem anguli 5°.18'. quæ est mensura æquationis Opticæ maximæ. Quem cum viderem esse 100429, hic quasi e somno expergefactus, & novam lucem intuitus ... [When my triumph over Mars appeared to be futile, I fell by chance on the observation that the secant of the angle 5°18' is 1.00429, which was the error of the measure of the maximal point. I awoke as if from sleep, & a new light broke on me.]"   (J. Kepler 1609, Cap. LVI, p. 267)

Before Kepler, the knowledge in astronomy was, after thousands of years of measurements and calculations (by the Babylonian priests, Greek philosophers, Ptolemy, Copernicus' *De revolutionibus* and Tycho Brahe) as follows: The planets move around the Sun on *eccentric* circles, i.e. the Sun is not precisely at the centre of these circles. This model was quite compatible with the innumerable measurements made with unequalled precision by Tycho Brahe for all the planets known at that time, *with the exception of the planet Mars*.

After years of "pertinaci studio elaborata Pragæ", Kepler finally discovered the following laws (the first two in Kepler 1609, the last one in Kepler 1619):

Kepler 1. *Planets move on elliptic orbits with the Sun at one of the foci.*

Kepler 2. *The planets orbiting the Sun sweep out equal areas in equal time.*

Kepler 3. *The squares of the periods of revolution are proportional to the cubes of the semi-major axes.*

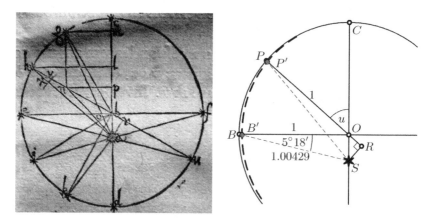

Fig. 5.28. The discovery of Kepler's first law (*Astronomia Nova*, Chap. 56); Kepler's drawing (left), modern drawing (right).

Kepler's calculations and meditations, which led to the discovery of his laws, fill hundreds of pages in his *Astronomia Nova* (1609). The decisive breakthrough occurs in Chap. 56 [11] and is explained in Fig. 5.28: The best possible circle for the orbit of Mars, which we take of radius 1, would have an eccentricity $e = OS$ such that the angle SBO is $5°18'$, where B is the point with the greatest distance from the axis SOC. But the true distance BS for Mars measured by Brahe was smaller by a factor $1/1.00429$ than the distance BS for the point on that circle. Luckily, Kepler remarked that this value is precisely $\cos 5°18'$ and "a new light broke on him" (see the quotation): we should move the point B to the point B', whose distance $B'S$ is the same as that of BO; in other words, *we have to replace the hypotenuse* (which is BS) *by the leg* (BO). Kepler tried the same recipe at other points: move the point P to the position P', such that the length $P'S$ is equal to that of the leg PR. This becomes

$$P'S = PR = 1 + e \cos u , \qquad (5.50)$$

because the angle u, called the *eccentric anomaly*, reappears as angle SOR, so that $OR = e \cos u$. These distances (5.50), which "are confirmed by very numerous and very sure measurements" (end of Chap. 56), are precisely those of the second formula of (5.48) and thus the points *describe an ellipse*.

Newton's proof of Kepler 2. Once Kepler's laws were discovered, one wanted to understand them in the light of the foundations of mechanics, which Galilei (1638, Giornata terza) had laid down and which Newton had turned into the following two crystal clear laws:

Lex 1. Without force a body remains in uniform motion on a straight line.

Lex 2. The change of motion is proportional to the motive force impressed.

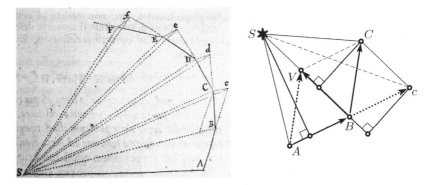

Fig. 5.29. Newton's proof for Kepler 2; reproduction from Newton's *Principia* (left); the triangles ABS, BcS and BCS having the same area (right)

---

[11] For more details on the first parts of the book, which culminate in the discovery of Kepler 2 (Chap. 40), we refer to Wilson (1968), Thorvaldsen (2010) and Wanner (2010).

Theorem 5.8 (Theorem 1 of the *Principia*, Engl. transl. 1729). *"The areas, which revolving bodies describe by radii drawn to an immoveable centre of force, do lie in the same immoveable planes, and are proportional to the times in which they are described."*

Proof. We imagine a celestial body moving on an orbit $ABCDE \ldots$ under the influence of a force f acting from the Sun (see Fig. 5.29). The crucial idea is to let it advance for a certain time interval Δt *without* force from A to B (under Lex 1 uniformly on a straight line) and to compensate the missing force by one giant kick with force

$$f \cdot \Delta t \tag{5.51}$$

at the point B. Without this kick, the body would continue during the second time step Δt in uniform motion to the point c. The two triangles ABS and BcS, having the same base and the same altitude, have the same area by Eucl. I.41. Now the kick at B is in direction of the Sun, hence the velocity vector AB is transformed into a velocity AV such that BVS are aligned (by Lex 2). As a consequence, the movement for the second time interval leads from B to C in such a way that cC is parallel to BS. We conclude, again by Eucl. I.41, that the triangles BcS and BCS also have the same area.

Continuing in the same way, we find that all triangles ABS, BCS, CDS, etc., which correspond to equal time steps Δt, have equal areas. Thus Kepler's second law is proved, at least for discrete force impulses. For the case that we "now let the numbers of those triangles be augmented, and their breadth diminished *in infinitum*", Newton had prepared a "cor. 4. lem. 3." to conclude that the law will also be true in the case of a force acting "continually".[12] □

This "Theorema 1" of the *Principia* did away with the first 40 chapters of Kepler's *Astronomia Nova* and its proof, more than 300 years later, has lost none of its beauty and elegance.

The discovery of the law of gravitation from Kepler 1 & 2

> "And it is the glory of Geometry that from those few principles, fetched from without, it is able to produce so many things."
> (I. Newton, from the Preface of the *Principia*, Engl. transl. 1729)

> "... one of the most dramatic moments of the *real beginnings* was when Newton suddenly understood *so* much from *so* little ..."
> (R. Feynman, lecture of March 13, 1964)

Theorem 5.9 (Prop. 11 of Newton's *Principia*). *A body P, orbiting according to Kepler 1 and 2,[13] moves under the effect of a centripetal force, directed to the centre S, satisfying the law*

---

[12] Today we would interpret the above procedure as a numerical method for differential equations (more precisely, the *symplectic Euler method*, cf. e.g. Hairer, Lubich and Wanner, 2006, p. 3), and rely on convergence results for such methods. The same argumentation applies to all subsequent proofs of this chapter.

[13] Not the original wording; Newton did not mention Kepler in the *Principia*.

$$f = \frac{Const}{r^2} \, , \qquad \text{where } r \text{ is the distance } SP. \qquad (5.52)$$

For the *proof*, we first establish a relation between the physical force and a geometrical quantity. For this we look at Newton's drawing in his manuscript from 1684 reproduced in Fig. 5.30, left: We imagine a body moving with initial velocity in direction AB attracted by a centre of force situated far away in direction AC. This force will deviate the body during a certain time interval Δt to a curved orbit AD. If there were no initial velocity, the body would move to C, so that $ACDB$ would be a parallelogram. But the distance AC, for a fixed time interval Δt, is proportional to the force (Lex 2). We conclude that

$$\begin{matrix} \textit{the acting force is proportional to the distance } BD \\ \textit{between the point on the tangent and the point on the orbit.} \end{matrix} \qquad (5.53)$$

For this distance, denoted by RQ in the sequel, Newton discovered a nice property:

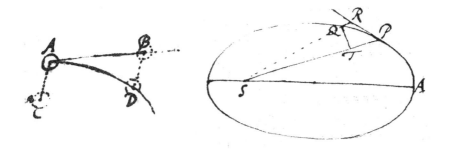

Fig. 5.30. Reproductions from Newton's autograph (1684), manuscript Cambridge Univ. Lib. Add. 3965[6]; the force acting on a moving body (left); picture for Newton's lemma (right). Reproduced by kind permission of the Syndics of Cambridge University Library

Newton's lemma. *Let APQ be an ellipse with focus S and suppose P to be the position of the planet moving towards Q, while the point R moves on the tangent with S, Q, R collinear. Let T be the orthogonal projection of Q onto PS (see Fig. 5.30, right). Then, if the distance PQ tends to zero, we have*

$$RQ \approx Const \cdot QT^2 \, , \qquad (5.54)$$

where the constant is independent of the position of P on the ellipse.

Proof. The proof is displayed in Fig. 5.31. We begin by collecting what we know from Apollonius (see Chap. 3): we know that the tangent PR is parallel to the diameter DCK, conjugate to GCP (Apoll. II.6). We denote the lengths of these diameters by $2d$ and $2c$ respectively. Through H, the other focus, and Q we draw parallels to DK which yield the points I and X on SP and V on

CP.[14] We further know that the normal PF of length h is the angle bisector of SPH (Apoll. III.48), i.e. the triangle IPH is isosceles, hence $IP = PH$ (Eucl. I.6). We next have $SE = EI$ by Thales since $SC = CH$ (Apoll. III.45). Therefore, since $SE + EI + IP + PH = 2a$ (Apoll. III.52), we obtain our first interesting result,

$$EP = EI + IP = a. \tag{5.55}$$

The *key idea* of the proof is now the following one: if our ellipse were a circle, we would know by Eucl. III.35 (or Eucl. II.14) that $GV \cdot VP = QV^2$. But in the case of the ellipse, we have to divide these values by the lengths of the corresponding conjugate diameters and obtain

$$\frac{GV \cdot VP}{c^2} = \frac{QV^2}{d^2} \quad \text{i.e.} \quad (3): \quad VP = \frac{c^2}{GV} \cdot \frac{QV^2}{d^2}.$$

To complete the proof, we have to express VP in terms of RQ and QV in terms of QT. Note that the triangle XVP is similar to ECP and QTX is similar to PFE (orthogonal angles), whence by (5.55)

$$(2): \quad XP = VP \cdot \frac{a}{c}, \qquad (6): \quad QX = QT \cdot \frac{a}{h}.$$

In order to make more progress, we now leave the path of exemplary Greek rigour and suppose PQ very (infinitely) small, i.e. we identify

$$(1): \quad RQ \approx XP, \qquad (4): \quad GV \approx GP = 2c, \qquad (5): \quad QV \approx QX.$$

A simple calculation now gives by using, in this order, (1), (2), (3), (4), (5), (6),

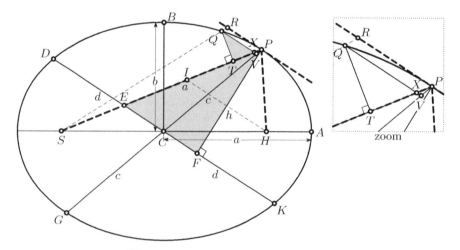

Fig. 5.31. Newton's proof of his lemma

---

[14] All upper case letters of this proof are the original ones of Newton, but not the lower case letters a, b, c, d and h, which we use to simplify the formulas.

$$RQ \approx \frac{a^3}{2h^2d^2} \cdot QT^2 \,.$$

From Apoll. VII.31 (Exercise 2 on page 73) we finally have $hd = ab$ ($\frac{1}{4}$ area of the circumscribed parallelogram), which turns the above formula into

$$RQ \approx \frac{a}{2b^2} \cdot QT^2 \tag{5.56}$$

where, as stated, the constant[15] is independent of the position of P. □

Proof of Theorem 5.9. The main theorem is finally obtained by combining the above three results:

(a) The force f is proportional to RQ (equation (5.53));

(b) RQ is proportional to QT^2 (Newton's lemma (5.54));

(c) QT is inversely proportional to SP, because $\frac{QT \cdot SP}{2}$ (the area of the triangle SPQ) is constant (Δt fixed, Kepler 2);

hence f is inversely proportional to SP^2. □

R. Feynman and the reciprocal problem

> "Pour voir présentement que cette courbe $ABC \ldots$ est toûjours une Section Conique, ainsi que Mr. Newton l'a supposé, *pag. 55. Coroll. I.* sans le démontrer; il y faut bien plus d'adresse. [To see now that this curve $ABC \ldots$ is always a conic section, as Mr. Newton has assumed without proof on *p. 55, Coroll. I*, requires considerably more ability.]" (Joh. Bernoulli, 1710)

> "... no calculus required, no differential equations, no conservation laws, no dynamics, no angular momentum, no constants of integration. This is Feynman at his best: reducing something seemingly big, complicated, and difficult to something small, simple, and easy." (B. Beckman, 2006)

The reciprocal result, that a body orbiting under the influence of a central force obeying the inverse-square law always follows an elliptic, parabolic or hyperbolic arc, was much harder to prove. Joh. Bernoulli, who gave a proof for the problem in 1710 using differential calculus, stated proudly that answering this question "requires considerably more ability" (see the quotation). A *geometric* explanation, as elegant as the proofs above, had to wait for another three centuries and was presented by R. Feynman in his lecture of March 13, 1964 at Caltech (see Feynman, Goodstein and Goodstein 1996, also Beckman 2006).[16]

---

[15] Newton remarked that this constant is the reciprocal of the latus rectum.

[16] The authors are grateful to Christian Aebi and Bernard Gisin, Geneva, for valuable references to the literature.

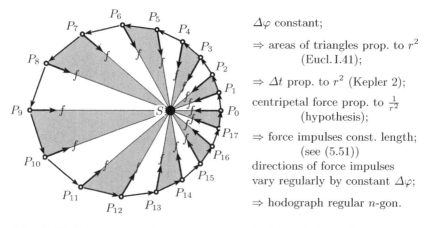

$\Delta\varphi$ constant;

$\Rightarrow$ areas of triangles prop. to r^2
  (Eucl. I.41);

$\Rightarrow \Delta t$ prop. to r^2 (Kepler 2);

centripetal force prop. to $\frac{1}{r^2}$
  (hypothesis);

$\Rightarrow$ force impulses const. length;
  (see (5.51))

directions of force impulses
vary regularly by constant $\Delta\varphi$;

$\Rightarrow$ hodograph regular n-gon.

Fig. 5.32. Feynman's variant with equal angles instead of equal time steps

Equal angles instead of equal time steps. We suppose that we have a force acting according to the inverse-square law. As we observed in (5.51) and (5.52), the force impulses, for constant time steps, decrease like $\frac{1}{r^2}$ with increasing r. We now choose *equal angles* at the Sun. By Eucl. VI.19, the areas of the triangles SP_iP_{i+1} are proportional to r^2. Therefore, by Kepler 2, the time steps Δt (which multiply the force) are proportional to r^2 as well and

$$\begin{array}{c} \textit{the force impulses will all have the same length.} \\ \text{Moreover, } \textit{their directions form a regular star.} \end{array} \qquad (5.57)$$

The situation is summarised in Fig. 5.32.

The hodograph. We now draw the *velocities* as points in a space with origin O (see Fig. 5.33, left). The velocity $\dot{P}_0$ at the perihelion P_0 is fastest and directed upwards. Then the impulses f push the velocities $\dot{P}_1, \dot{P}_2, \ldots$ first to the left, then downwards, until at the aphelion (here P_9) the velocity is slowest and directed exactly downwards. All impulses f have, by (5.57), constant length and their directions increase regularly by the same amount $\Delta\varphi$. We therefore get a regular n-gon and, for $\Delta\varphi \to 0$, we obtain the surprising result:

$$\begin{array}{c} \textit{The velocity } \dot{P} \textit{ of a planet orbiting under the effect of} \\ \textit{a central force inversely proportional to } r^2 \textit{ describes a circle.} \end{array} \qquad (5.58)$$

The centre C of the circle is not at the origin O, except for circular motion with constant speed. If the origin O were on or outside the circle, we would have parabolic or hyperbolic motion.

It is interesting that such an elegant result escaped the attention of Euler, Lagrange and Laplace. Only in the work of Hamilton did the velocities (momenta) acquire the same importance as the positions.

Conclusion. Now comes the most difficult step (Feynman: "I took a long time to find that"). We have to find a connection between the orbit in Fig. 5.32

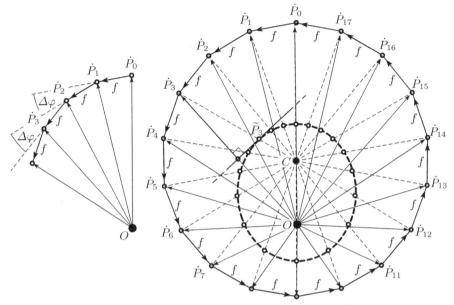

Fig. 5.33. The hodograph of Kepler motion, producing a circle (left); synthesis of the two pictures (right)

and the hodograph in Fig. 5.33. After some time for thought, we draw in the second picture the curve of points P having the same distance from O as from the circle and obtain Fig. 5.33 (right). We know from Chap. 3 (see in particular Fig. 12.8 (b)) that this curve is an ellipse with O and C as foci. We denote by $\widetilde{P}_i$ the points of this ellipse situated on the rays $C\dot{P}_i$. These points are located under the same angles with respect to C as are the corresponding points P_i with respect to S in Figure 5.32.

We next consider Fig. 3.4 (right): the tangent at P is orthogonal to FB (in the notation of that figure). Applied to Fig. 5.33, this means that the tangent at $\widetilde{P}_i$ to the ellipse is *orthogonal* to $O\dot{P}_i$. On the other hand, the tangent at P_i to the orbit in Fig. 5.32 (left) is *parallel* to $O\dot{P}_i$. We conclude that *the two ovals are identical, just rotated by* 90° *and, perhaps, scaled differently.* Since we know that the "oval" in Fig. 5.33 is an ellipse, with C as focus, we have that the orbit in Fig. 5.32 is also an ellipse with S as focus.   □

This was "Feynman at his unique best" (see the quotation); later D.L. and J.R. Goodstein discovered that precisely the same proof had been published in 1877 by another great physicist, James Clerk Maxwell.

> "It is not easy to use the *geometric* method to discover things, it is very difficult, but the elegance of the demonstrations after the discoveries are made, is really very great. The power of the *analytic* method is that it is much easier to discover and to prove things, but not in any degree of elegance. There is a lot of dirty

paper with x-es and y-s and crossed out cancellations and so on ...
(laughter)."

(R. Feynman, lecture of March 13, 1964, 35th minute)

This "dirty paper with x-es and y-s" leads us to the next chapters ...

5.11 Exercises

1. Prove, for the circular quadrilateral with sides a, b, c, d of Ptolemy's
   Lemma 5.1 and Fig. 5.4, the formulas

   $$\delta_1 : \delta_2 = (ab + cd) : (ad + bc), \quad \delta_1^2 = (ac + bd)(ab + cd) : (ad + bc) \quad (5.59)$$

   which can be found in Förstemann (1835).

2. Multiply the values of $\cos \alpha$ for $\alpha = 0, \frac{\pi}{6}, \frac{2\pi}{6}, \frac{3\pi}{6}, \frac{4\pi}{6}, \frac{5\pi}{6}, \frac{6\pi}{6}$ by 6 and design
   a simple rule for French fisherman to find the tidal height as the sea level
   falls, hour per hour, during approximately 6 hours from high water to low
   water.

3. (Exercise suggested by P. Henry (2009)) Reconstruct Viète's proof of the
   addition formulas (5.6) — which in Viète were not "formulas", but half a
   page of Latin text — by supposing $BC = \sin \alpha$, $AC = \cos \alpha$, $BD = \sin \beta$,
   $AD = \cos \beta$ to be known (see Fig. 5.34) and by computing, using Thales,
   Pythagoras and Eucl. III.20, $BE = \sin(\alpha + \beta)$ and $AE = \cos(\alpha + \beta)$.

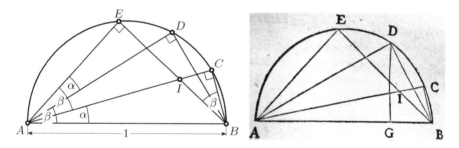

Fig. 5.34. Proof of Viète; right: illustration from van Schooten's edition 1646

4. Verify the values of sine, cosine and tangent given in Table 5.2.

5. Consider an arbitrary triangle with sides a, b, c. Prove the following beautiful expressions for the half-angles:

   $$\sin \frac{\alpha}{2} = \sqrt{\frac{(s - c)(s - b)}{bc}}, \qquad \cos \frac{\alpha}{2} = \sqrt{\frac{(s - a)s}{bc}},$$

   $$\tan \frac{\alpha}{2} = \sqrt{\frac{(s - c)(s - b)}{(s - a)s}}, \qquad \text{where} \quad s = \frac{a + b + c}{2} \qquad (5.60)$$

Table 5.2. Values of sine, cosine and tangent obtained from regular polygons

α	radians	$\sin\alpha$	$\cos\alpha$	$\tan\alpha$
$0°$	0	0	1	0
$15°$	$\dfrac{\pi}{12}$	$\dfrac{\sqrt{2}}{4}\left(\sqrt{3}-1\right)$	$\dfrac{\sqrt{2}}{4}\left(\sqrt{3}+1\right)$	$2-\sqrt{3}$
$18°$	$\dfrac{\pi}{10}$	$\dfrac{\sqrt{5}-1}{4}$	$\dfrac{1}{2}\sqrt{\dfrac{5+\sqrt{5}}{2}}$	$\sqrt{1-\dfrac{2}{5}\sqrt{5}}$
$30°$	$\dfrac{\pi}{6}$	$\dfrac{1}{2}$	$\dfrac{\sqrt{3}}{2}$	$\dfrac{\sqrt{3}}{3}$
$36°$	$\dfrac{\pi}{5}$	$\dfrac{1}{2}\sqrt{\dfrac{5-\sqrt{5}}{2}}$	$\dfrac{\sqrt{5}+1}{4}$	$\sqrt{5-2\sqrt{5}}$
$45°$	$\dfrac{\pi}{4}$	$\dfrac{\sqrt{2}}{2}$	$\dfrac{\sqrt{2}}{2}$	1
$60°$	$\dfrac{\pi}{3}$	$\dfrac{\sqrt{3}}{2}$	$\dfrac{1}{2}$	$\sqrt{3}$
$75°$	$\dfrac{5\pi}{12}$	$\dfrac{\sqrt{2}}{4}\left(\sqrt{3}+1\right)$	$\dfrac{\sqrt{2}}{4}\left(\sqrt{3}-1\right)$	$2+\sqrt{3}$
$90°$	$\dfrac{\pi}{2}$	1	0	∞

is the semi-perimeter of the triangle.

6. Use Exercise 5 to derive the identity

$$\sin^2\frac{\alpha}{2} + \sin^2\frac{\beta}{2} + \sin^2\frac{\gamma}{2} + 2\sin\frac{\alpha}{2}\sin\frac{\beta}{2}\sin\frac{\gamma}{2} = 1 \qquad (5.61)$$

for the angles α, β, γ of a triangle.

7. Derive the product formulas

$$2\cdot\sin\frac{u+v}{2}\cdot\cos\frac{u-v}{2} = \sin u + \sin v\,,$$
$$2\cdot\cos\frac{u+v}{2}\cdot\cos\frac{u-v}{2} = \cos u + \cos v\,, \qquad (5.62)$$
$$2\cdot\sin\frac{u+v}{2}\cdot\sin\frac{u-v}{2} = \cos v - \cos u\,.$$

8. Derive, once by analytic calculations and once by a geometric argument, the *law of tangents* due to Viète (1593b) for two angles α, β and the opposite sides a, b of a triangle:

$$\frac{a-b}{a+b} = \frac{\tan\frac{\alpha-\beta}{2}}{\tan\frac{\alpha+\beta}{2}}\,. \qquad (5.63)$$

9. Extend the formulas (5.8) by proving the following product formulas for the sine function (Euler E562, 1783, §8)

$$\begin{aligned}
\sin 1\alpha &= 1 \cdot \sin \alpha, \\
\sin 2\alpha &= 2 \cdot \sin \alpha \cdot \sin\left(\tfrac{\pi}{2} + \alpha\right), \\
\sin 3\alpha &= 4 \cdot \sin \alpha \cdot \sin\left(\tfrac{\pi}{3} + \alpha\right) \cdot \sin\left(\tfrac{2\pi}{3} + \alpha\right), \\
\sin 4\alpha &= 8 \cdot \sin \alpha \cdot \sin\left(\tfrac{\pi}{4} + \alpha\right) \cdot \sin\left(\tfrac{2\pi}{4} + \alpha\right) \cdot \sin\left(\tfrac{3\pi}{4} + \alpha\right),
\end{aligned} \tag{5.64}$$

and for the cosine function (Euler, 1783, §5)

$$\begin{aligned}
\cos 1\alpha &= 1 \cdot \sin\left(\tfrac{\pi}{2} + \alpha\right), \\
\cos 2\alpha &= 2 \cdot \sin\left(\tfrac{\pi}{4} + \alpha\right) \cdot \sin\left(\tfrac{3\pi}{4} + \alpha\right), \\
\cos 3\alpha &= 4 \cdot \sin\left(\tfrac{\pi}{6} + \alpha\right) \cdot \sin\left(\tfrac{3\pi}{6} + \alpha\right) \cdot \sin\left(\tfrac{5\pi}{6} + \alpha\right), \\
\cos 4\alpha &= 8 \cdot \sin\left(\tfrac{\pi}{8} + \alpha\right) \cdot \sin\left(\tfrac{3\pi}{8} + \alpha\right) \cdot \sin\left(\tfrac{5\pi}{8} + \alpha\right) \cdot \sin\left(\tfrac{7\pi}{8} + \alpha\right).
\end{aligned} \tag{5.65}$$

10. Discover, by iterating the formulas (5.8) and (5.9), the beautiful analytic expressions of Viète (1593b) for the perimeters of the square, regular octagon, regular 16-gon, regular 32-gon etc., inscribed in the circle of radius 1. This will finally lead to the famous product

$$\frac{2}{\pi} = \sqrt{\frac{1}{2}} \cdot \sqrt{\frac{1}{2} + \frac{1}{2}\sqrt{\frac{1}{2}}} \cdot \sqrt{\frac{1}{2} + \frac{1}{2}\sqrt{\frac{1}{2} + \frac{1}{2}\sqrt{\frac{1}{2}}}} \cdots \tag{5.66}$$

11. Check the sine values of van Schooten (1683) in Fig. 5.35 (left) for the angles 38°20′ and 51°40′ and check if they are really "accuratissimo".

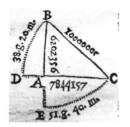
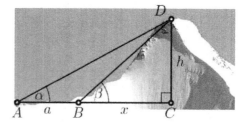

Fig. 5.35. Left: van Schooten's sine values; right: measuring the altitude of a famous mountain (photo by Marco Borello, Milano)

12. (An exercise from practical geodesy.) To measure the height of an oak tree by Thales' theorem as at the beginning of Chap. 1, one must have access to the base of the tree. But this is not always possible if one wants, for example, to determine the altitude of mountain (see Fig. 5.35 (right)). In this case, one measures *two* angles α and β from two points A and B with distance a. Compute the height h from these data.

13. Prove that if the Euler line of a triangle is parallel to a side, say to AB, then $\tan\alpha \cdot \tan\beta = 3$.

14. Find the side lengths of the orthic triangle in Fig. 4.9 (b), page 86, as

$$EF = a\cos\alpha\,, \qquad FD = b\cos\beta\,, \qquad DE = c\cos\gamma\,. \qquad (5.67)$$

15. Give a trigonometric proof of Morley's theorem in Sect. 4.10, i.e. suppose that the radius r of the circumcircle is given and, with the notation of Fig. 4.27, the angles α, β, γ. Then compute the distances AB, AR, and finally QR, which will lead to a symmetric formula.

16. Prove the cosine rule and the sine rule for arbitrary spherical triangles by adapting the ideas of Sect. 5.3, i.e. suppose that the triangle of Fig. 5.8 (right) is a spherical triangle and apply the formulas of Sect. 5.6 in precisely the same way as was done in Sect. 5.3 for planar triangles. In particular replace Pythagoras' theorem by the cosine rule (5.23) and the formulas used from (5.5) by (5.26) (respectively (5.25) in case of the sine rule).

17. Let A and B be points on the unit sphere lying on the same circle of latitude φ, with γ as difference of their longitudes. Let N be the North Pole.

    (a) Find the area of the triangle ABN bounded by the great circles NA and NB and the circle of latitude AB.

    (b) Find the area of the spherical triangle ABN bounded by *three* great circles.

    (c) Compare the results, in particular in the case where A approaches B.

18. In 1932, the *International Astronomical Union* defined the regions in the sky belonging to the various stellar constellations. Among the easy ones, *Corvus* was defined by the following boundary:
    circle of latitude $-11°$ from $12^{\mathrm{h}}\,50^{\mathrm{m}}$ to $11^{\mathrm{h}}\,50^{\mathrm{m}}$;
    meridian $11^{\mathrm{h}}\,50^{\mathrm{m}}$ from $-11°$ to $-24°\,30'$;
    circle of latitude $-24°\,30'$ from $11^{\mathrm{h}}\,50^{\mathrm{m}}$ to $12^{\mathrm{h}}\,35^{\mathrm{m}}$;
    meridian $12^{\mathrm{h}}\,35^{\mathrm{m}}$ from $-24°\,30'$ to $-22°$;
    circle of latitude $-22°$ from $12^{\mathrm{h}}\,35^{\mathrm{m}}$ to $12^{\mathrm{h}}\,50^{\mathrm{m}}$;
    meridian $12^{\mathrm{h}}\,50^{\mathrm{m}}$ from $-22°$ to $-11°$.
    Find the area of this region for the sphere of radius 1, i.e. with *steradians* (sr) as unit.

19. The shortest connection between two points on a sphere is along a great circle. For an airplane, flying from Paris ($49°$ northern latitude, $3°$ eastern longitude) to Vancouver ($49°$ northern latitude, $123°$ western longitude), determine the angle β between this great circle and the east-west direction at the point of departure.

20. A Norwegian fishing boat sends an SOS distress call from an unknown position in the Norwegian Sea. The signal is received in Trondheim

$(63°\,26'\,\text{N},\ 10°\,24'\,\text{E})$ from the direction N $74°\,13'$ W (i.e. $74°\,13'$ to the west of north) and in Tromsø $(69°\,39'\,\text{N},\ 18°\,59'\,\text{E})$ from the direction N $107°\,17'$ W. Where should the rescue team be sent?

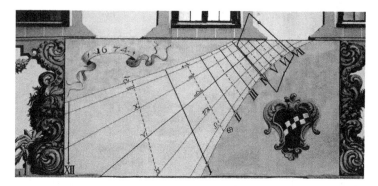

Fig. 5.36. A sundial at the Cistercian monastery Stams; photo by K. Galehr-Nadler

21. Construct a sundial for the Cistercian monastery at Stams, Tyrol. The monastery has the coordinates $47°\,17'$ N and $10°\,59'$ E, the wall is directed roughly to the west (its normal is $11°$ to the south of west). The large deviation from the east-west direction results in big distortions. The sundial should eventually look like that in Fig. 5.36.

22. Prove the following half-angle formulas, which nicely extend the formulas of Exercise 5 to a *spherical* triangle with sides a, b, c:

$$\sin\frac{\alpha}{2} = \sqrt{\frac{\sin(s-b)\sin(s-c)}{\sin b \sin c}}, \qquad \cos\frac{\alpha}{2} = \sqrt{\frac{\sin(s-a)\sin s}{\sin b \sin c}},$$

$$\tan\frac{\alpha}{2} = \sqrt{\frac{\sin(s-b)\sin(s-c)}{\sin(s-a)\sin s}}, \qquad \text{where} \quad s = \frac{a+b+c}{2} \tag{5.68}$$

is the semi-perimeter (arclength) of the triangle.
Show also the dual formulas for the side lengths (half-side formulas):

$$\sin\frac{a}{2} = \sqrt{-\frac{\cos(\sigma-\alpha)\cos\sigma}{\sin\beta\sin\gamma}}, \qquad \cos\frac{a}{2} = \sqrt{\frac{\cos(\sigma-\beta)\cos(\sigma-\gamma)}{\sin\beta\sin\gamma}},$$

$$\tan\frac{a}{2} = \sqrt{-\frac{\cos(\sigma-\alpha)\cos\sigma}{\cos(\sigma-\beta)\cos(\sigma-\gamma)}}, \qquad \text{where} \quad \sigma = \frac{\alpha+\beta+\gamma}{2}. \tag{5.69}$$

Hint. In complete analogy to the solution of Exercise 5, insert the spherical law of cosines (5.35) into (5.9). You'll also need the last formula of (5.62).

23. Show that the three angle bisectors (i.e. great circles that bisect the angles) of a spherical triangle meet at a single point, the incentre I. Further show that the radius ρ of the incircle is given by

$$\tan \rho = \sqrt{\frac{\sin(s-a)\sin(s-b)\sin(s-c)}{\sin s}} \, , \quad \text{where } s = \frac{a+b+c}{2} \, .$$

24. The *perpendicular bisector of a spherical segment* is the great circle that bisects the segment and is perpendicular to it. Show that the three perpendicular bisectors of the sides of a spherical triangle meet at a single point, the circumcentre O. Further show that the radius r of the circumcircle is given by

$$\cot r = \sqrt{-\frac{\cos(\sigma-\alpha)\cos(\sigma-\beta)\cos(\sigma-\gamma)}{\cos \sigma}} \, , \quad \text{where } \sigma = \frac{\alpha+\beta+\gamma}{2} \, .$$

25. Prove the *altitude theorem* (Eucl. II.14) for a right-angled spherical triangle

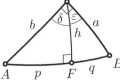

$$\sin^2 h = \tan p \cdot \tan q \, . \tag{5.70}$$

26. Derive, for *circular* planetary motion, the inverse-square law of gravitation from Kepler's third law; i.e. from $T^2 = Const \cdot a^3$, where T is the period of revolution. Choose a fixed time interval Δt during which the planet moves from P to Q (see the figure). Use the fact that, similarly to (5.53) and for mass 1,

$$f \approx \frac{RQ}{\Delta t^2} \, . \tag{5.71}$$

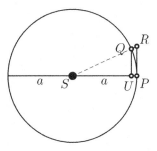

Part II

Analytic Geometry

"Ce que les Anciens avoient démontré sur les courbes, quelque important, quelque subtil qu'il fut, n'étoit pourtant qu'un amas de Propositions particulières ... jusq'à l'invention de l'Algébre ; moyen ingénieux de réduire les Problèmes au Calcul le plus simple & le plus facile que la Question proposée puisse admettre. Cette clef universelle des Mathématiques ... a produit une véritable révolution dans les Sciences ... [What the Ancients had proved about curves, however important or subtle it might have been, was nevertheless just a collection of particular propositions ... until the invention of algebra; an ingenious means of reducing the problems to the simplest and easiest calculations which the proposed question will allow. This universal key to mathematics ... caused a genuine revolution in science ...]"   (G. Cramer, 1750)

Algebra,[1] with its identities and equations, grew out of geometrical figures in Euclid's Book II and al-Khwārizmī's book (see the first row of Fig. II.1 and the explanations in Hairer and Wanner, 1997, pp. 2–4). During the following centuries, mainly at the hands of Stifel, Cardano, Viète (see Fig. II.2 below) and Descartes, this science became an ever more powerful instrument in its own right. The stages of this development, at first only figures and Arabic text, then better and better symbols for algebraic operations, finally the introduction of letters for known and unknown numerical values, is documented in Fig. II.1. The use of this tool by Viète and Descartes to solve geometrical problems then led to a great revolution in geometry. This resulted, under the influence of Euler's *Introductio in analysin infinitorum*, vol. II, in the creation of what is today called *analytic geometry*.

---

[1] From Ḥisāb al-jabr w'al-muqābala [The compendious book on calculation by completion and balancing], Muḥammad ibn Mūsā al-Khwārizmī, Baghdad (830)

The first victories of this new science are described in Chaps. 6 and 7. Gauss' role in renewing interest for one of the problems of ancient Euclidean geometry, the question of constructibility with ruler and compass, is treated in Chap. 8. Analytic geometry in higher dimensions, vector spaces and linear maps are discussed in Chaps. 9 and 10.

In these chapters, we will frequently touch on neighbouring subjects, such as calculus or linear algebra, from a geometric point of view. The "European Mathematics Subject Area Group" has ensured that students are offered complementary courses on these subjects for more profound discussions.

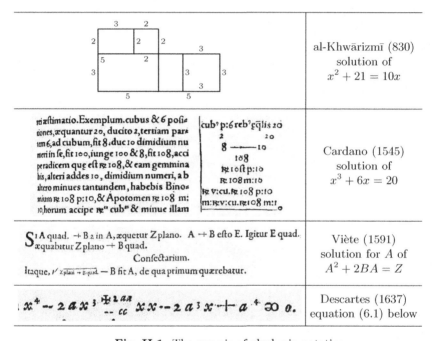

Fig. II.1. The genesis of algebraic notation

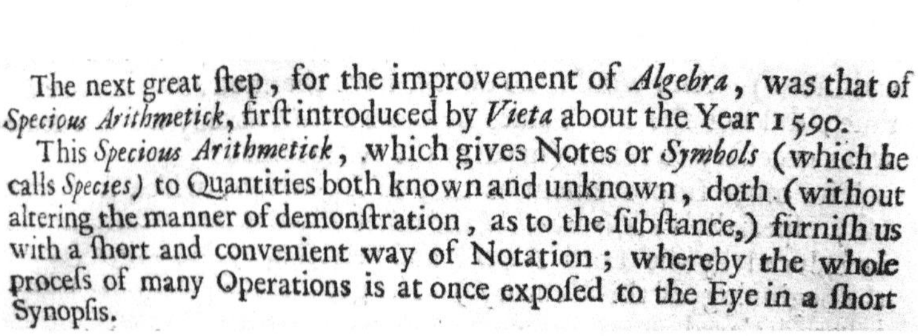

The next great ſtep, for the improvement of *Algebra*, was that of *Specious Arithmetick*, firſt introduced by *Vieta* about the Year 1590.

This *Specious Arithmetick*, which gives Notes or *Symbols* (which he calls *Species*) to Quantities both known and unknown, doth (without altering the manner of demonſtration, as to the ſubſtance,) furniſh us with a ſhort and convenient way of Notation ; whereby the whole proceſs of many Operations is at once expoſed to the Eye in a ſhort Synopſis.

Fig. II.2. Wallis on the algebra of Viète (reproduced from Wallis (1685))

6

Descartes' Geometry

6.1 The Principles of Descartes' Geometry

> "... affin de faire voir qu'on peut construire tous les Problesmes
> de la Geometrie ordinaire, sans faire autre chose que le peu qui
> est compris dans les quatre figures que i'ay expliquées. Ce que
> ie ne croy pas que les anciens ayent remarqué, car autrement ils
> n'eussent pas pris la peine d'en escrire tant de gros liures, ou le
> seul ordre de leurs propositions nous fait connoistre qu'ils n'ont
> point eu la vraye methode pour les trouuer toutes, mais qu'ils ont
> seulement ramassé celles qu'ils ont rencontrées. [... to show that
> it is possible to construct all the problems of ordinary geometry
> by doing no more than the little covered in the four figures that I
> have explained. This is one thing which I believe the ancients did
> not observe, for otherwise they would not have put so much labor
> into writing so many thick books in which the very sequence of the
> propositions shows that they did not have a sure method for find-
> ing all, but rather gathered together those propositions on which
> they had happened by accident.]"   (R. Descartes, *La Geome-
> trie*, 1637, p. 304; English translation by Smith and Latham, 1925)

Descartes' *Geometrie*, published in 1637 as an appendix to his *Discours de la
méthode* (from p. 297 onwards; first separate publication Paris 1664), is one
of the most influential scientific works of the 17th century. For example, it
was one of the only two books[1] that the young Isaac Newton owned – he read
them very carefully.

Descartes begins by noting that for "tous les Problesmes de Geometrie" it
is sufficient to "connoistre la longeur de quelques lignes droites" [Any problem
in geometry can easily be reduced to such terms that a knowledge of the
lengths of certain straight lines is sufficient for its construction[2]] and that
"souuent on n'a pas besoin de tracer ainsi ces lignes sur le papier, & il suffist

---

[1] The second was John Wallis' *Arithmetica Infinitorum*.

[2] This and the following English translations by Smith and Latham (1925).

A. Ostermann and G. Wanner, *Geometry by Its History*,
Undergraduate Texts in Mathematics, DOI: 10.1007/978-3-642-29163-0_6,
© Springer-Verlag Berlin Heidelberg 2012

de les designer par quelques lettres, chascune par vne seule. Comme pour adiouster la ligne *BD* a *GH*, ie nomme l'vne *a* & l'autre *b*, & escris *a + b*" [often it is not necessary thus to draw the lines on paper, but it is sufficient to designate each by a single letter. Thus, to add the lines *BD* and *GH*, I call one *a* and the other *b*, and write *a + b*]. From this historical moment dates the use of *lower case letters* to denote (geometric) magnitudes. Two pages further Descartes writes "C'est a dire, *z*, que ie prens pour la quantité inconnuë ..." [That is, *z*, which I take for the unknown quantity ...] which is the origin of using the last letters of the alphabet to denote unknowns.

Geometry	Algebra
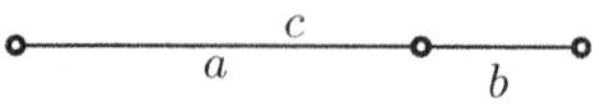	sum $c = a + b$
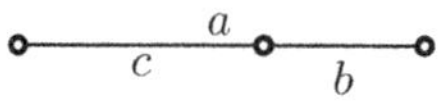	difference $c = a - b$
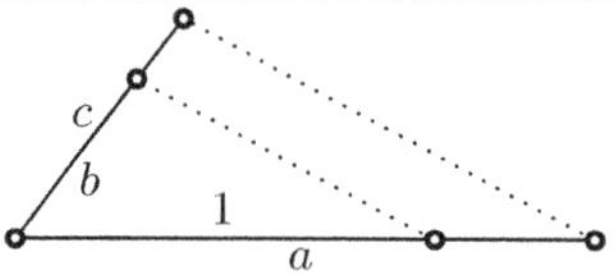	product $c = a \cdot b$
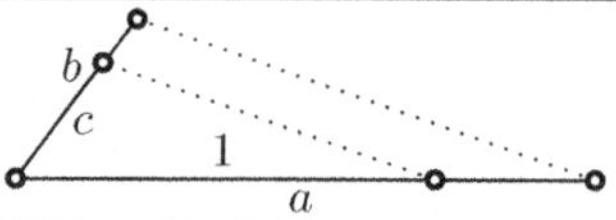	quotient $c = \dfrac{b}{a}$
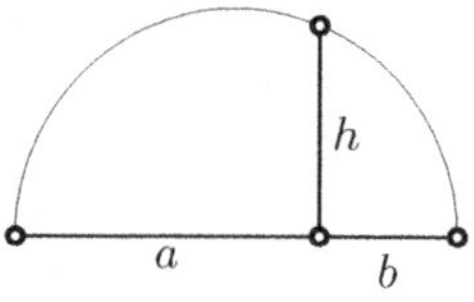	root $h = \sqrt{a \cdot b}$ (Eucl. II.14)

Fig. 6.1. Descartes' dictionary between geometric figures and algebraic identities

What Descartes describes is a dictionary between geometric and algebraic operations, see Figs. 6.1 and 6.2. This dictionary allows one to translate a geometric problem into an algebraic one and vice-versa. By passing from one formulation to the other, the solution may become simpler. We shall demon-

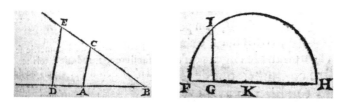

Fig. 6.2. Facsimile of Descartes' drawings from the 1664 edition

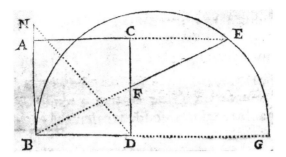

Fig. 6.3. Descartes' example; reproduced from the 1664 edition

strate below by a couple of examples the advantages of this new method and discuss a number of new theorems which it allows one to obtain.

Example. We start with an example of Descartes himself (see Fig. 6.3): Let a square AD be given and a segment BN. We want to find a point E on the produced side AC with the property that the segment EF on the line EB has the same length as BN. The problem is treated in Pappus (*Collection* Book VII, Prop. 72, see also Heath, 1921, vol. II, p. 412) and is attributed by Pappus to Heraclitus. Descartes admits that Pappus gave a geometric solution by producing BD to G so that $DG = DN$ and then drawing the circle with diameter BG. However, says Descartes, "those not familiar with this construction would not be likely to discover it". On the other hand, the algebraic method is straightforward: We denote the given lengths $AB = BD$ and BN by a and c respectively, and by x one of the unknown lengths, say $DF = x$, so that $BF = \sqrt{a^2 + x^2}$. Thales' theorem for the similar triangles BDF and ECF gives

$$\frac{x}{\sqrt{a^2 + x^2}} = \frac{a - x}{c},$$

since EF is required to have length c. Multiplying out we obtain[3]

$$x^4 - 2ax^3 + (2a^2 - c^2)x^2 - 2a^3x + a^4 = 0. \tag{6.1}$$

For such an equation, Euler devised an elegant idea which uses its *symmetry*: divide the equation by $a^2 x^2$ and set

$$\frac{x}{a} + \frac{a}{x} = y \qquad \text{to obtain} \qquad y^2 - 2y - \frac{c^2}{a^2} = 0. \tag{6.2}$$

We thus have to solve successively two quadratic equations, which correspond to the two circles of Pappus' construction (see Exercise 1).

---

[3]The same equation as in Fig. II.1; we see that the notation has not changed much since Descartes' work. Descartes' habit of writing xx for x^2 remained standard until the times of Gauss.

6.2 The Regular Heptagon and Enneagon

The theory of regular polygons with a large number of vertices, left in a somewhat chaotic state by the Greeks (see Chap. 4), finally became accessible in a satisfactory way through progress in algebra. Abū'l-Jūd Muḥammad ibn al-Layth, 11th century and François Viète, see Viète (1593a), found the equations for solving the regular heptagon (7-gon) and the regular enneagon (9-gon). Kepler saw in the regular polygons the principal reason for the harmonies in the world and dedicated the entire first book of his *Harmonices Mundi* (1619) to these figures, *quæ proportiones harmonicas pariunt*.

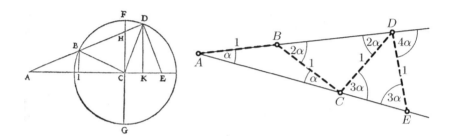

Fig. 6.4. Illustration from Viète (1593a, left); the "Viète ladder" (right)

Idea for the solution. We apply an idea which an Arabic text attributes to Archimedes (see Heath, 1921, p. 240), and which was rediscovered by Abū'l-Jūd (see Fig. 6.7) and Viète (see Fig. 6.4, left). We modify Fig. 1.9 (a) on page 9 by moving the point A outside the circle (see Fig. 6.4, left) so that $AB = BC$. We thus obtain a broken line $ABCDE$ ... with segments all of the same length (see Fig. 6.4, right). We conclude that the angle to the left of C is α (by Eucl. I.5), the angle at B is 2α (exterior angle, see (1.2) on page 8), which repeats at D (again Eucl. I.5), the angle to the right of C is 3α (exterior angle of ACD), then at D we find 4α, and so on. We call this figure, which can be inscribed in any angle α, a *Viète ladder*.

The regular heptagon. We want to compute the diagonals x and y of the regular heptagon with side length 1 (see Fig.6.5, left).

Solution. The angles at A are all $\frac{\pi}{7}$ by Eucl. III.20. We take $\alpha = \frac{\pi}{7}$ in the Viète ladder and obtain Fig. 6.5, right. The triangles ABC, BCD and CDE are then respectively similar to the triangles BAG, CAF and DAE of the figure on the left. Thus Thales determines the distances $AC = x$, $BD = \frac{y}{x}$ and $CE = \frac{1}{y}$ in the ladder. But this figure can also be interpreted as the heptagon of the left figure *folded together as a Geisha fan* along the dashed lines through A. Therefore $AC = x$ (once again) and $AD = AE = y$. So we obtain

$$1 + \frac{y}{x} = y \qquad \text{and} \qquad x + \frac{1}{y} = y. \tag{6.3}$$

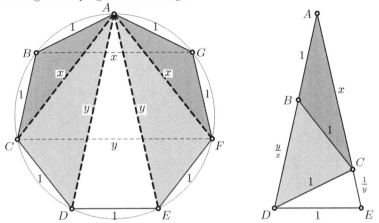

Fig. 6.5. The regular heptagon as a Geisha fan (left) and folded (right)

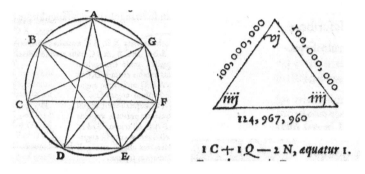

Fig. 6.6. The regular heptagon; on the left: drawing of the heptagon from Viète (1593a); on the right: the value of $y - 1$ to 9 digits (with one misprint, above); the equation for $z = y - 1$, which is $z^3 + z^2 - 2z = 1$, in the notation of Viète: $C =$ Cubos, $Q =$ Quadrato, $N =$ Numero (below)

We can eliminate x, or y, between the two equations. This gives

$$y^3 - 2y^2 - y + 1 = 0 \qquad \text{and} \qquad x^3 - x^2 - 2x + 1 = 0, \qquad (6.4)$$

two equations of degree 3. We will see in Chap. 8 that the (positive) solution of these equations cannot be constructed with Euclid's instruments, but we can compute it numerically to any desired precision:

$$\begin{aligned} y &= 2.24697960371746706105\dots \\ x &= 1.80193773580483825247\dots \end{aligned} \qquad (6.5)$$

The regular enneagon. We denote the lengths of the diagonals which we require by x, y, z and now take $\alpha = \frac{\pi}{9}$ in the Viète ladder. This time, after folding, Thales' theorem gives us $AC = x$, $BD = \frac{z}{x}$, $CE = 1$, $DF = \frac{1}{z}$. The distances $AD = y$, $AE = AF = z$ are unchanged under folding, thus

$$1 + \frac{z}{x} = y, \qquad x + 1 = z, \qquad y + \frac{1}{z} = z. \qquad (6.6)$$

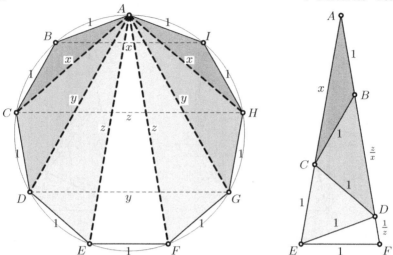

Fig. 6.7. The regular enneagon as a Geisha fan (left) and folded (right)

Algebraic manipulations (express y in terms of z from the last equation, x from the second, and insert everything into the first) yield

$$z^3 - 3z^2 + 1 = 0 \,. \tag{6.7}$$

This equation, found by Abū'l-Jūd in the 11th century (see e.g. M. Cantor, 1894, p. 715), was one of the reasons for the eager search for a closed-form solution of cubic equations, a dream which became reality only some five centuries later (Tartaglia, Cardano, Viète; see below). As for the heptagon, the equations allow one to compute the solution numerically:

$$z = 2.879385241571816768108218555\ldots$$
$$y = 2.532088886237956070404785305\ldots$$
$$x = 1.879385241571816768108218555\ldots$$

6.3 The Trisection of an Angle and Cubic Equations

> "Quid igitur quærit à Geometris Adrianus Romanus? Datum angulum trifariam secare ... Quid ab Analystis? Datum solidum sub latere & dato coëfficiente plano adfectum, multa cubi, resolvere ... Quare quærenti Adriano licet sive in Geometricis sive in Arithmetricis satisfacere. [What does Adrianus Romanus therefore ask the geometers? To trisect a given angle ... And what the analysts? Given a solid figure obtained by multiplying one side and a coefficient assigned base, to find the value of the cube ... Therefore one has the choice to satisfy the inquiring Adrianus either in geometry or analysis.]"
> (F. Viète, 1595, pp. 312/313)

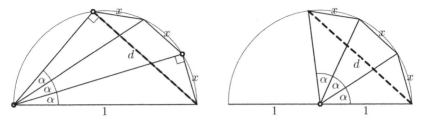

Fig. 6.8. The trisection of an angle (left: as sine; right: as chords)

Problem. Trisect an angle with given sine d (see Fig. 6.8), i.e. find α or $x = \sin\alpha$ such that $\sin 3\alpha = d$.

Solution (Viète, 1593a). Using the formulas (5.6), we have

$$\sin 3\alpha = \sin(2\alpha + \alpha) = 3\sin\alpha\,\cos^2\alpha - \sin^3\alpha = 3\sin\alpha - 4\sin^3\alpha \qquad (6.8)$$

hence

$$\sin^3\alpha - \frac{3}{4}\sin\alpha + \frac{\sin 3\alpha}{4} = 0 \qquad \text{or} \qquad x^3 - \frac{3}{4}x + \frac{d}{4} = 0. \qquad (6.9)$$

Any method for solving all such equations thus allows one to trisect any angle. One can also use a similar formula for the cosine (see Viète's example in Fig. 6.9).

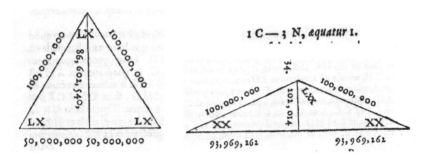

Fig. 6.9. Trisection of the 60° angle by Viète (1593a) with the help of the equation $y^3 - 3y = 1$ for $y = 2\cos 20°$; all given digits correct

Solving cubic equations. Algebra helps geometry, but geometry can also help algebra. If we assume that we *are able* to trisect any angle, say with the help of trigonometric and inverse trigonometric functions, then we can use this tool the other way round to solve certain cubic equations. This idea, discovered by Viète (1595, see the quotation) works as follows.

Let a cubic equation

$$z^3 + az^2 + bz + c = 0 \qquad (6.10)$$

be given. It has been known since Cardano that the substitution $z + \frac{a}{3} = y$ leads to an equation without the term y^2:

$$y^3 - py + q = 0, \quad \text{where} \quad p = \frac{a^2}{3} - b, \quad q = \frac{2a^3}{27} - \frac{ab}{3} + c. \quad (6.11)$$

This equation is quite similar to (6.9). In order to make the two equations identical, we set $y = \mu \sin \alpha$ and insert $x = \frac{y}{\mu}$ into (6.9). Comparing the equations we get

$$p = \frac{3\mu^2}{4} \quad \text{and} \quad q = \frac{\mu^3 \sin 3\alpha}{4}. \quad (6.12)$$

The first condition determines μ (this is possible if $p \geq 0$), and the second then determines α (this is possible if $|\frac{27q^2}{4p^3}| \leq 1$). We finally obtain

$$z = -\frac{a}{3} + 2\sqrt{\frac{p}{3}} \sin\left(\frac{1}{3} \arcsin\left(\frac{q}{2}\left(\frac{3}{p}\right)^{\frac{3}{2}} \right) + \frac{2k\pi}{3} \right), \quad k = 0, 1, 2 \quad (6.13)$$

for the solutions of (6.10).

6.4 Regular Polygons in the Unit Circle

Problem. Construct regular polygons inscribed in a given circle, which we take to be the unit circle.

Solution. In order to extend formula (6.8) to an expression for $\sin(n\alpha)$ for arbitrary (odd) values of n, we start from

$$\sin((n+2)\alpha) + \sin((n-2)\alpha) = 2\sin(n\alpha)\cos(2\alpha), \quad (6.14)$$

obtained by adding the expressions given by (5.6) for $\sin((n \pm 2)\alpha)$. Here we insert $\cos(2\alpha) = \cos^2 \alpha - \sin^2 \alpha = 1 - 2\sin^2 \alpha$ and find

$$\sin((n+2)\alpha) = (2 - 4\sin^2 \alpha)\sin(n\alpha) - \sin((n-2)\alpha). \quad (6.15)$$

Since we are interested in polygons inscribed in the unit circle, we go over to the chords (5.4)

$$x = \text{cord}\, \alpha = 2\sin\frac{\alpha}{2}, \quad d_n = \text{cord}(n\alpha) = 2\sin\frac{n\alpha}{2}, \quad (6.16)$$

and obtain from (6.15) an even simpler formula,

$$d_{n+2} = (2 - x^2) \cdot d_n - d_{n-2} \quad (n = 1, 3, 5, \ldots). \quad (6.17)$$

This equation allows us, starting from $d_{-1} = -x$, $d_1 = x$, to compute the chords $d_n = \text{cord}(n\alpha)$ recursively for all odd positive values of n, with the result:

$$d_1 = +x^1$$
$$d_3 = -x^3 \; +3x^1$$
$$d_5 = +x^5 \; -5x^3 \; +5x^1$$
$$d_7 = -x^7 \; +7x^5 \; -14x^3 \; +7x^1$$
$$d_9 = +x^9 \; -9x^7 \; +27x^5 \; -30x^3 \; +9x^1$$
$$d_{11} = -x^{11} \; +11x^9 \; -44x^7 \; +77x^5 \; -55x^3 \; +11x^1$$
$$d_{13} = +x^{13} -13x^{11} \; +65x^9 \; -156x^7 +182x^5 \; -91x^3 \; +13x^1$$
$$d_{15} = -x^{15} +15x^{13} \; -90x^{11} \; +275x^9 -450x^7 \; +378x^5 -140x^3 \; +15x^1$$
$$d_{17} = +x^{17} -17x^{15} +119x^{13} -442x^{11} +935x^9 -1122x^7 +714x^5 -204x^3 +17x^1$$

$$(6.18)$$

et eo infinitum continuando ordine (see Fig. 6.10).

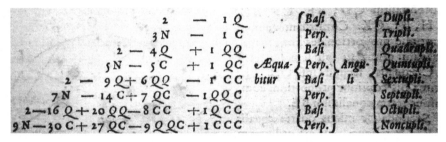

Fig. 6.10. Table of chords for multiple angles from Viète's *Responsum* (1595), reproduced from van Schooten's edition (1646), p. 319; Viète also gives chords for even n; these contain the cosine, i.e. "Perp." is replaced by "Basi"

Remark. If n is not a prime number, say $n = m \cdot k$, then d_n can also be obtained by inserting d_k into d_m, i.e. the angle α is multiplied first by k then by m. For example, $d_9 = -(d_3)^3 + 3d_3 = -(-x^3 + 3x)^3 + 3(-x^3 + 3x)$.

$\boldsymbol{n = 3}$. If $\alpha = \frac{2\pi}{3}$, then $d_3 = 0$, hence $x^3 - 3x = 0$. We divide this equation by x and set $x^2 = y$, so that $y = 3$; thus *the sides of the equilateral triangle inscribed in the unit circle have length* $\sqrt{3}$, a result in accordance with the formula $R = \frac{\sqrt{3}}{3}$ of Table 1.1 on page 18.

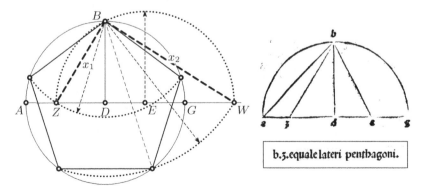

Fig. 6.11. Ptolemy's construction of inscribed pentagon; right: reproduced from Ptolemy–Regiomontanus (1496)

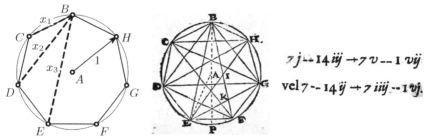

Fig. 6.12. Diagonals of the heptagon inscribed in the unit circle; right: facsimile of Kepler's drawing and equation (Kepler 1619)

$n = 5$. Here, we solve $d_5 = x^5 - 5x^3 + 5x = 0$ and obtain for the squares $y = x^2$ of the diagonals the equation $y^2 - 5y + 5 = 0$. Hence, *the squares of the lengths of the diagonals of the regular pentagon inscribed in the unit circle are* $(5 \pm \sqrt{5})/2$. In particular, the side length is $\sqrt{\frac{5-\sqrt{5}}{2}} = \sqrt{3 - \Phi}$, again in accordance with Table 1.1. This result can be turned into the construction displayed in Fig. 6.11, which was already known to Ptolemy: let ADG be a diameter, E the midpoint between D and G and B vertically above D. Draw the circle with centre E passing through B, cutting the diameter AG in the points Z and W. Then BZ and BW are the diagonals. We also see, from the formula $R = \Phi$ in Table 1.1, that ZD *is the side of the inscribed decagon.*

$n = 7$. The equation $d_7 = -x^7 + 7x^5 - 14x^3 + 7x = 0$ for the lengths of the three diagonals of the regular heptagon inscribed in the unit circle was, independently of Viète, also published by Kepler (1619) and is attributed by him to Jost Bürgi.[4] See Fig. 6.12. The squares of the lengths of the diagonals thus are the three roots of $y^3 - 7y^2 + 14y - 7 = 0$.

6.5 Van Roomen's Famous Challenge

In his book *Methodus polygonorum* (1593), the Flemish mathematician Adriaan van Roomen (in Latin, Adrianus Romanus) challenged "all mathematicians from all over the world" to solve the equation

$$
\begin{aligned}
x^{45} &- 45x^{43} + 945x^{41} - 12300x^{39} + 111150x^{37} - 740259x^{35} + 3764565x^{33} \\
&- 14945040x^{31} + 46955700x^{29} - 117679100x^{27} + 236030652x^{25} \\
&- 378658800x^{23} + 483841800x^{21} - 488494125x^{19} + 384942375x^{17} \\
&- 232676280x^{15} + 105306075x^{13} - 34512075x^{11} + 7811375x^9 \\
&- 1138500x^7 + 95634x^5 - 3795x^3 + 45x = \sqrt{\frac{7}{4} - \sqrt{\frac{5}{16}} - \sqrt{\frac{15}{8} - \sqrt{\frac{45}{64}}}}
\end{aligned}
$$

$$(6.19)$$

---

[4] "... sic procedit *Justus Byrgius*, Mechanicus Caesaris et Landgravij Hassiae; qui in hoc genere ingeniosissima et inopinabilia multa est commentus."

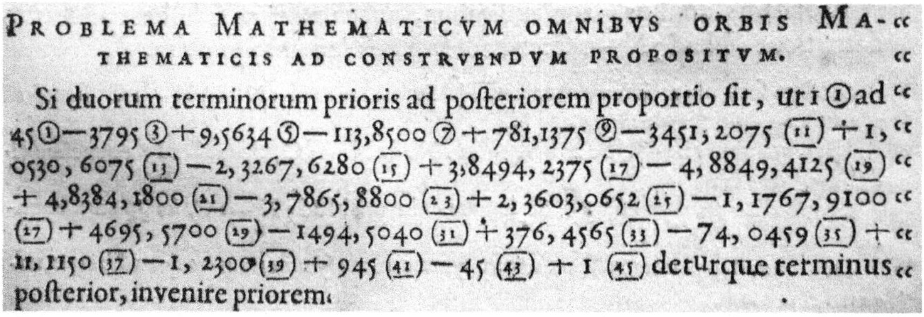

PROBLEMA MATHEMATICVM OMNiBVS ORBIS MA- cc
THEMATICIS AD CONSTRVENDVM PROPOSITVM.   cc

Si duorum terminorum prioris ad posteriorem proportio sit, ut 1 ① ad cc
45 ① — 3795 ③ + 9,5634 ⑤ — 113,8500 ⑦ + 781,1375 ⑨ — 3451, 2075 ⑪ + 1, cc
0530, 6075 ⑬ — 2, 3267, 6280 ⑮ + 3,8494, 2375 ⑰ — 4, 8849, 4125 ⑲ cc
+ 4,8384, 1800 ㉑ — 3, 7865, 8800 ㉓ + 2, 3603,0652 ㉕ — 1, 1767, 9100 cc
㉗ + 4695, 5700 ㉙ — 1494, 5040 ㉛ + 376, 4565 ㉝ — 74, 0459 ㉟ + cc
11, 1150 ㊲ — 1, 2300 ㊴ + 945 ㊶ — 45 ㊸ + 1 ㊺ deturque terminus cc
posterior, invenire priorem.

Fig. 6.13. Van Roomen's challenge; reproduced from van Schooten's edition of Viète's Opera (1646), p. 305

(see also Fig. 6.13). He even added a list of ten outstanding scientists who could be considered capable of solving the problem — three Germans, two Italians, three Dutch, one Danish, one Flemish, but *no* French mathematician (see the paper by Henry (2009) for an detailed account of this problem). The French king (Henri IV) was not pleased to hear this and ordered Viète to come and solve the problem. Three hours later Viète presented a first solution to the king.

Those among us who can barely solve equations of the second or third degree are left speechless by Viète's audacity in attacking such a problem. The above equation, however, is not just *any* equation of degree 45, but a very particular one:

(a) the left hand side of (6.19) is d_{45} from (6.18);

(b) the right hand side is $\text{cord}(24°)$; this can be verified by computing $\sin(12°) = \sin(30° - 18°)$ from the values of Table 5.2.

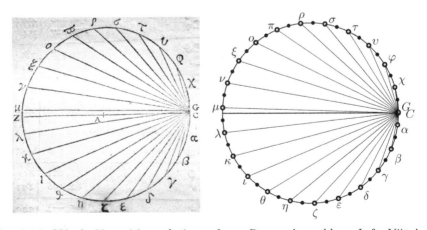

Fig. 6.14. Viète's 23 positive solutions of van Roomen's problem. Left: Viète's drawing from van Schooten's edition (p. 313); right: a modern computer plot. The points C and G, which are distant by only $32'$, cannot be distinguished

So the first solution presented to King Henri IV was $x = \text{cord}(\frac{24}{45}°) = \text{cord}(32')$, "quæsita fit $\frac{930,839}{100,000,000}$", see Viète (1595), p. 213. To find the remaining 22 positive solutions of the problem, Viète required some additional hours of thought: if we increase $\alpha = 32'$ by $\frac{2 \cdot 360}{45}° = 16°$, then $\sin \frac{45\alpha}{2}$, and with it the corresponding chord, will again have the same value. Therefore a second solution will be $x = \text{cord}(32' + 16°)$, a third one $x = \text{cord}(32' + 32°)$ and so on, until $x = \text{cord}(32' + 352°)$, every other vertex of a regular 45-gon (see Fig. 6.14).

6.6 A Geometric Theorem of Fermat

The following theorem was included, as usual without proof, in a letter of Fermat (June 1658) to Digby and addressed to the "Illustrissimos Viros Vicecomitum Brouncker et Johannem Wallisium" in order to demonstrate proudly to these Englishmen ("quae Angliam invisere non erubescent") his ability, not only in "numeros integros", but also in "Geometria".[5]

Theorem 6.1. *Let AMB be a semicircle of radius 1 drawn on the horizontal side of a rectangle $EFBA$ of height $\sqrt{2}$. For an arbitrary point M on this semicircle, let R and S be the intersection points of the lines ME and MF with the diameter AB (see Fig. 6.15, left). Then*

$$AS^2 + RB^2 = AB^2 \ . \tag{6.20}$$

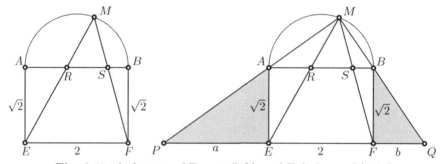

Fig. 6.15. A theorem of Fermat (left); and Euler's proof (right)

Proof. We do not know how Fermat found his result. An elegant idea of Euler (see Euler, 1750) is as follows: we produce the segments MA, MB and EF (Eucl. Post. 2; we have not forgotten) and obtain two triangles EAP and FQB, both similar to MBA (see Fig. 6.15, right). Hence by Thales

$$\frac{a}{\sqrt{2}} = \frac{\sqrt{2}}{b} \qquad \text{or} \qquad ab = 2 \ . \tag{6.21}$$

---

[5]He could not have guessed that at the same time, a 15-year-old English boy was preparing to become, some years later, the greatest scientist since Archimedes.

Since, again by Thales, the ratios AS/PF, RB/EQ and AB/PQ are equal, formula (6.20) is equivalent[6] to $PF^2 + EQ^2 = PQ^2$, or, by Descartes' dictionary, to

$$(a+2)^2 + (2+b)^2 = (a+2+b)^2,$$

which, when multiplied out, is the same as (6.21). □

6.7 Heron's Formula for the Area of a Triangle

Problem. Given the three sides a, b, c of a triangle, find its area $\mathcal{A}$. The answer is given by the famous formula of Heron of Alexandria (approx. A.D. 10–70).

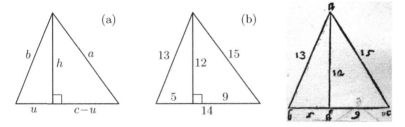

Fig. 6.16. Medieval solution of Heron's problem; right: reproduced from Tartaglia 1560

Theorem 6.2 (Heron's formula). *The area $\mathcal{A}$ of a triangle with sides a, b, c is given by*

$$\mathcal{A} = \sqrt{s(s-a)(s-b)(s-c)}, \qquad (6.22)$$

where $s = \dfrac{a+b+c}{2}$ is the semi-perimeter. Equivalently,

$$4\mathcal{A} = \sqrt{(a+b+c)(-a+b+c)(a-b+c)(a+b-c)}. \qquad (6.23)$$

Proof. We will give Heron's original proof below. A much easier approach was discovered by Arab mathematicians, and in Europe during the Renaissance; see in particular Leonardo Pisano's *Practica geometriae* (1220, p. 35) or Tartaglia's huge treatise on "numbers and measurements" (1560, in the "quarta parte").

By Eucl. I.41, we can obtain the area if we know the altitude h (see Fig. 6.16 (a)). It is easier to begin by calculating the quantity u, which is given by Eucl. II.13 (formula (2.2) on page 38) as $2uc = b^2 + c^2 - a^2$. Then we find h by Pythagoras as $h^2 = b^2 - u^2$.

---

[6]The authors are grateful to Bernard Gisin for suggesting this simplification of Euler's proof.

The standard example treated by all ancient writers is the triangle with sides 13, 14, 15 (see Fig.6.16 (b)). We choose 14 for c (which leads to the easiest calculations) and obtain (compare with the reproduction from Tartaglia)

$$
\begin{aligned}
u &= \tfrac{169+196-225}{28} \\
&= \tfrac{365-225}{28} \\
&= \tfrac{140}{28} = 5
\end{aligned}
$$

169. & l'altro ⋅ 96. la cui fumma fara 365.

(che fara 225) reftara 140.

partendo adunque 140 per 28. ne venira 5.

(6.24)

and by Pythagoras $h = 12$. Our triangle consists of two right-angled triangles juxtaposed along their sides of length 12. It has area $\mathcal{A} = 14 \cdot 6 = 84$.

The same algorithm, written in "modern" algebraic notation, reduces to a sequence of propositions from the first two books of Euclid:[7]

$$
\begin{aligned}
16\mathcal{A}^2 &= 4h^2c^2 & &\text{(Eucl. I.41)} \\
&= 4b^2c^2 - 4u^2c^2 & &\text{(Eucl. I.47)} \\
&= 4b^2c^2 - (b^2 + c^2 - a^2)^2 & &\text{(Eucl. II.13)} \\
&= (2bc + a^2 - b^2 - c^2)(2bc - a^2 + b^2 + c^2) & &\text{(Eucl. II.5)} \\
&= (a^2 - (b-c)^2)((b+c)^2 - a^2) & &\text{(Eucl. II.4)} \\
&= (a - b + c)(a + b - c)(b + c - a)(b + c + a). & &\text{(Eucl. II.5)} \qquad \square
\end{aligned}
$$

(6.25)

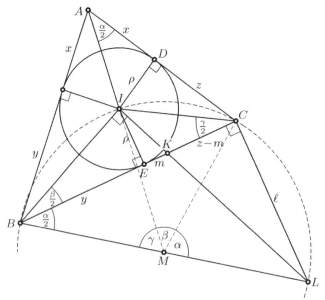

Fig. 6.17. Heron's proof of his formula

[7]The authors wish to acknowledge helpful discussions with Christian Aebi about this presentation of the proof.

Heron's original proof. This proof, as given by Heath (1926, vol. II, pp. 87–88), is as follows.

Using (5.18) we see that (6.22) is proved, once we know that

$$\rho = \sqrt{\frac{(s-a)(s-b)(s-c)}{s}} \tag{6.26}$$

for the radius of the incircle. Let x, y and z be the respective distances of A, B and C from the points of tangency of the incircle (see Fig. 6.17). Since $y+z = a$, $x+z = b$ and $x+y = c$, we have $x+y+z = s$ and hence $x = s-a$, $y = s-b$, $z = s-c$, cf. Fig. 4.6 on page 83.

The main idea is now to draw at C the perpendicular to CB and at I the perpendicular to BI. Let L be the intersection of these two perpendiculars. We then draw the circle with centre M and diameter BL; it passes through I and C.

We see that $\frac{\beta}{2}$ and $\frac{\gamma}{2}$ are inscribed angles, hence the corresponding central angles at M are β and γ (Eucl. III.20). By Eucl. I.32, the third angle at the centre is α. Again by Eucl. III.20, the corresponding inscribed angle at B is $\frac{\alpha}{2}$. Hence the triangles BCL and ADI are similar. The triangles IEK and LCK are also similar. Applying Thales twice, we have

$$\frac{y+z}{x} = \frac{\ell}{\rho} = \frac{z-m}{m} .$$

Adding 1 to each side gives

$$\frac{x+y+z}{x} = \frac{s}{x} = \frac{z}{m} \qquad \Rightarrow \qquad m = \frac{xz}{s} .$$

Finally, ρ is the *altitude* of the right-angled triangle BIK. By (1.10), we have

$$\rho^2 = ym = \frac{xyz}{s} ,$$

which is the required relation (6.26). $\square$

Corollary 6.3 (Lhuilier, 1810/11). *The area of a triangle is given by*

$$\mathcal{A} = \sqrt{\rho \cdot \rho_a \cdot \rho_b \cdot \rho_c} , \tag{6.27}$$

where ρ_a, ρ_b, ρ_c are the radii of the excircles.

Proof. By using the formulas (4.10) for $\rho_a \cdot \rho_b \cdot \rho_c$, (6.26) for ρ and multiplying out, we get the expression under the root of (6.22). $\square$

Remarks. (i) A similar proof of Heron's formula, with a different auxiliary triangle, was given by Euler (E135, 1750, §8).

(ii) For a proof based on trigonometric identities, use

$$\rho = x \cdot \tan \frac{\alpha}{2} = (s - a) \cdot \tan \frac{\alpha}{2}$$

(see Fig. 6.17) and insert the third formula of (5.60) in Exercise 5 of Chap. 5, page 150. Another possibility is to multiply the first two formulas of (5.60) and obtain, with (5.8),

$$\sqrt{s(s-a)(s-b)(s-c)} = bc \sin \frac{\alpha}{2} \cos \frac{\alpha}{2} = \frac{bc}{2} \sin \alpha = \frac{hc}{2} .$$

(iii) For a proof using matrices ("Gram's matrix") see (10.15) on page 297.

(iv) Two elegant proofs of Thébault for enthusiasts of Euclid and Apollonius are given in Exercises 10 and 11 of this chapter, page 181.

6.8 The Euler–Brahmagupta Formula for a Cyclic Quadrilateral

A quadrilateral that is inscribed in a circle is called *cyclic*. Heron's formula has a beautiful analogue for cyclic quadrilaterals.

Theorem 6.4 (Euler E135, 1750, §12; Brahmagupta). *The area $\mathcal{A}_q$ of a cyclic quadrilateral $ABCD$ with sides a, b, c and d is given by*

$$\mathcal{A}_q = \sqrt{(s-a)(s-b)(s-c)(s-d)}, \qquad (6.28)$$

where $s = \dfrac{a+b+c+d}{2}$ is the semi-perimeter.

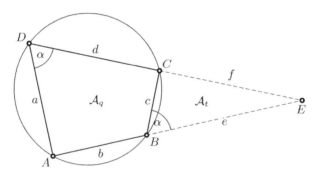

Fig. 6.18. Area of a cyclic quadrilateral

Proof. The proof given here follows the ideas of Euler's paper. For another proof, which does not use Heron's formula, see Exercise 13 below.

We produce AB and DC to obtain the point E (see Fig. 6.18).[8] By Eucl. III.22, see Fig. 2.15 (a) on page 39, the triangles ADE and CBE are similar, hence

---

[8]If AB is parallel to DC, and AD to BC, then a point such as E does not exist. In this case, however, the theorem is trivial since the quadrilateral is then a rectangle.

$$\frac{e}{c} = \frac{f + d}{a}, \qquad \frac{f}{c} = \frac{e + b}{a}.$$

This is a linear system for e and f. If we subtract, respectively add the two equations, they yield simple formulas for $e - f$ and $e + f$. The result is, after simplification,

$$\frac{e - f}{c} = \frac{d - b}{a + c}, \qquad \frac{e + f}{c} = \frac{d + b}{a - c}. \qquad (6.29)$$

The union of our quadrilateral with the triangle CBE is the triangle ADE, similar to CBE with similarity factor $\frac{a}{c}$. By Eucl. VI.19, we thus have

$$\mathcal{A}_q + \mathcal{A}_t = \mathcal{A}_t \cdot \frac{a^2}{c^2} \quad \Rightarrow \quad \mathcal{A}_q = \mathcal{A}_t \cdot \left(\frac{a^2 - c^2}{c^2}\right) = \mathcal{A}_t \cdot \frac{a + c}{c} \cdot \frac{a - c}{c}, \quad (6.30)$$

where $\mathcal{A}_t$ is the area of the triangle CBE. For the area of this triangle with sides c, e and f, we insert the penultimate expression of (6.25) and obtain

$$16\mathcal{A}_q^2 = 16\mathcal{A}_t^2 \cdot \frac{(a + c)^2}{c^2} \cdot \frac{(a - c)^2}{c^2} \qquad \text{(from (6.30))}$$

$$= \frac{(a + c)^2}{c^2} \cdot (c^2 - (e - f)^2) \cdot \frac{(a - c)^2}{c^2} \cdot ((e + f)^2 - c^2) \quad \text{(from (6.25))}$$

$$= ((a + c)^2 - (d - b)^2) \cdot ((d + b)^2 - (a - c)^2). \qquad \text{(from (6.29))}$$

The last formula simplifies with Eucl. II.5 to

$$16\mathcal{A}_q^2 = (a + c + d - b)(a + c - d + b)(d + b + a - c)(d + b - a + c),$$

which is (6.28). Already the penultimate formula is a nice result.   □

6.9 The Cramer–Castillon Problem

> "Dans ma jeunesse ... un vieux Géometre, pour essayer mes forces en ce genre, me proposa le Problème que je vous proposai, tentez de le résoudre et vous verrez, combien il est difficile."
>
> (G. Cramer in 1742; cited in Euler's *Opera*, vol. 26, p. xxv)

> "Sur un problème de géométrie plane qu'on regarde comme fort difficile"   (J. Castillon, 1776; title of his publication)

> "Le lendemain du jour dans lequel je lus à l'Académie ma solution du Problème concernant le cercle et le triangle à inscrire dans ce cercle, en sorte que chaque côté passe par un de trois points donnés, M. de la Grange m'en envoya la solution algébrique suivante."
>
> (J. Castillon, 1776; see *Oeuvres de Lagrange*, vol. 4, p. 335)

Given a circle and n points $A_1, A_2, \ldots, A_n$ in its plane (see Fig. 6.19, left), the problem consists in finding a *polygon* with n vertices $B_1, B_2, \ldots, B_n$ inscribed in the circle, and such that for each i, the side $B_i B_{i+1}$, possibly produced, and with $B_{n+1} = B_1$, passes through A_i (see Fig. 6.19, right).

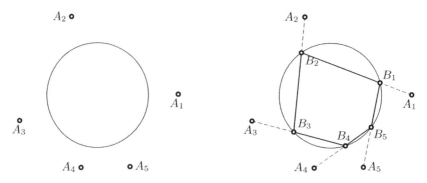

Fig. 6.19. The Cramer–Castillon problem (left) and a solution (right)

This problem has a long history. For a detailed account, see Wanner (2006). For $n = 3$, a special case goes back to Pappus (*Collection*, Prop. VII.117). The general case was proposed to Cramer by an unknown "vieux Géometre [old geometer]". Cramer himself showed the problem to the young Castillon in 1742 ("you'll see how difficult it is", see the first quotation). It took more than 30 years before Castillon found a *geometric* solution in 1776, and the problem kept the reputation of being very difficult (second quotation).

After Castillon's presentation to the Berlin Academy, Lagrange found an analytic solution *in one evening* (last quotation). This is a really striking example of the power of the analytic method. Lagrange's solution was later simplified and generalised to arbitrary n by Carnot (1803).

The Möbius transformation

> "Wenn man den schlichten, stillen Mann [Möbius] vor Augen hat, muss es einen einigermassen in Erstaunen setzen, dass sein Vater ... den Beruf eines Tanzlehrers ausübte. Um die Verschiedenheit der Generationen vollends vor Augen zu führen, erwähne ich, dass ein Sohn des Mathematikers der bekannte Neurologe ist, der Verfasser des vielbesprochenen Buches 'Vom physiologischen Schwachsinn des Weibes'. [If we imagine this quiet and modest man [Möbius], we might be somewhat surprised to hear that his father was a dance teacher. But in order to show even more strikingly the contrast between generations, I mention that one of the sons of the mathematician is the well-known neurologist, author of the much discussed book 'On the physiological imbecility of woman'.]"
>
> (F. Klein, 1926, p. 117)

The main tool for Carnot's solution of the Cramer–Castillon problem is the so-called *Möbius transformation*

$$u \mapsto v = \frac{pu + q}{ru + s}, \tag{6.31}$$

where p, q, r, s are given quantities.

Carnot discovered that the *composition* of two such transformations

$$u_2 = \frac{p_1 u_1 + q_1}{r_1 u_1 + s_1}, \qquad u_3 = \frac{p_2 u_2 + q_2}{r_2 u_2 + s_2} \tag{6.32a}$$

is again a Möbius transformation:

$$u_3 = \frac{p u_1 + q}{r u_1 + s}, \tag{6.32b}$$

where

$$p = p_2 p_1 + q_2 r_1, \qquad q = p_2 q_1 + q_2 s_1,$$
$$r = r_2 p_1 + s_2 r_1, \qquad s = r_2 q_1 + s_2 s_1. \tag{6.32c}$$

In the language of matrices (see Chap. 10), these relations can be expressed by

$$\begin{bmatrix} p & q \\ r & s \end{bmatrix} = \begin{bmatrix} p_2 & q_2 \\ r_2 & s_2 \end{bmatrix} \begin{bmatrix} p_1 & q_1 \\ r_1 & s_1 \end{bmatrix}, \tag{6.33}$$

which is the *product* of the two coefficient matrices. Similarly, the *inverse* map is again a Möbius transformation with the *inverse* matrix as coefficient scheme. Therefore, the transformations with $ps - qr \neq 0$ form a *group* with respect to composition.

An analytic solution of the Cramer–Castillon problem. By a scaling argument, the radius of the given circle may be taken as 1. We slightly simplify Carnot's proof by using a suggestion of Gauss (see *Werke*, vol. 4, p. 393, "Zusatz V").

Auxiliary Problem. We consider a point A given by its distance a from the centre and an angle α above the horizontal line (see Fig. 6.20). Then, for a given point B_1 on the circle, determined by an angle φ above the horizontal, let B_2 denote the (other) intersection of the circle and the line through A and B_1. We want to determine ψ, the angle which corresponds to B_2.

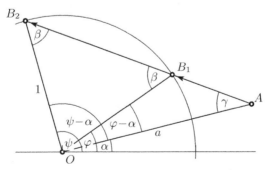

Fig. 6.20. Geometric proof of (6.34)

Solution. The two angles marked β are equal by Eucl. I.5. We apply the law of tangents (5.63) to the triangle AB_2O to obtain

$$\frac{a-1}{a+1} = \frac{\tan\dfrac{\beta-\gamma}{2}}{\tan\dfrac{\beta+\gamma}{2}}.$$

From (1.2) (β is an exterior angle of AB_1O) and Eucl. I.32 applied to OAB_2, we have

$$\frac{\beta-\gamma}{2} = \frac{\varphi-\alpha}{2} \quad \text{and} \quad \frac{\beta+\gamma}{2} = 90° - \frac{\psi-\alpha}{2}.$$

Finally, since $\tan(90° - \delta) = 1/\tan\delta$, the above formula becomes

$$\frac{1}{\tan\dfrac{\psi-\alpha}{2}} \cdot \frac{a-1}{a+1} = \tan\frac{\varphi-\alpha}{2}, \tag{6.34}$$

where $\frac{a-1}{a+1}$ is a given constant. We now apply the addition theorem (5.6) for tan to each side of (6.34). For α fixed, both sides represent Möbius transformations for $u_1 = \tan\frac{\varphi}{2}$ and $u_2 = \tan\frac{\psi}{2}$. Thus, by the group property, there exists a relation (6.32a) with given constants p_1, q_1, r_1, s_1.

Solution of the Cramer–Castillon problem. We start from an arbitrary point B_1 with an unknown $u_1 = \tan\frac{\varphi_1}{2}$ and determine the point B_2 as explained above. We then compute $B_3, B_4, \ldots$ in a similar manner and finally have to satisfy the condition $B_{n+1} = B_1$. Applying (6.32c) repeatedly we will have

$$u_{n+1} = u_1 = \frac{pu_1 + q}{ru_1 + s} \tag{6.35}$$

with the new coefficient matrix

$$\begin{bmatrix} p & q \\ r & s \end{bmatrix} = \begin{bmatrix} p_n & q_n \\ r_n & s_n \end{bmatrix} \cdots \begin{bmatrix} p_2 & q_2 \\ r_2 & s_2 \end{bmatrix} \begin{bmatrix} p_1 & q_1 \\ r_1 & s_1 \end{bmatrix}. \tag{6.36}$$

Relation (6.35) is a quadratic equation for u_1 with two solutions, in general.

Another method for obtaining the above Möbius transformation is given in Exercise 14 below, and a particular example in Exercise 15.

6.10 Exercises

1. Verify Pappus' solution in Fig. 6.3 by showing the equivalence of his construction with the equations in (6.2).

2. Nicomedes' construction for doubling the cube in the form of Pappus' Prop. IV.24 of his *Collection* states the following: *Let $AB\Gamma\Lambda$ be a rectangle, let Δ bisect AB and E bisect $B\Gamma$. Draw the line $\Lambda\Delta H$ and the perpendicular EZ such that $\Gamma Z = A\Delta$. Then construct Θ and K such that $\Gamma\Theta$ is parallel to HZ and $\Theta K = A\Delta$ (see Fig. 6.21 (a); this last construction requires the use of the conchoid). Then*

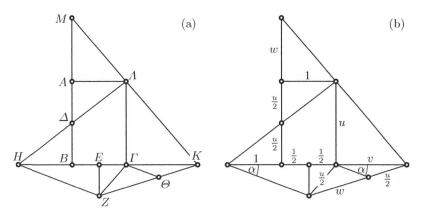

Fig. 6.21. Nicomedes' construction for doubling the cube: Pappus' Proposition IV.24 (a) and it's proof (b)

$$MA^3 : A\Lambda^3 = \Lambda\Gamma : A\Lambda. \tag{6.37}$$

In Ver Eecke's French edition, the proof consists of three pages of text and explanations in footnotes. Find, using modern algebraic notations (see Fig. 6.21 (b)), a proof in five lines.

3. 1000 years after Abū'l-Jūd, it is time to apply the Geisha fan method to a new challenge, the computation of the diagonals w, x, y, z of the regular 11-gon (or hendecagon) of side length 1 (see the picture).

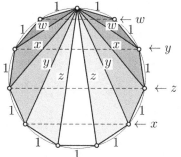

4. Prove the following identity from Ramanujan's notebooks (see Ramanujan, 1957, vol. II, p. 263): If ABC is a right-angled triangle and if we draw a circle with centre A and radius AC and another with centre B and radius BC (see the picture), then

$$RS^2 = 2 \cdot AR \cdot SB.$$

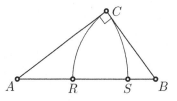

5. Solve the following problem from Einstein's *Maturitätsexamen* of Sept. 1896 at the *Aargauische Kantonsschule* (see Hunziker, 2001): Given the distances $AI = 1$, $BI = \frac{1}{2}$ and $CI = \frac{1}{3}$ of the vertices of a triangle from the incentre I, find the radius ρ of the incircle. *Hint.* Use formula (5.61).

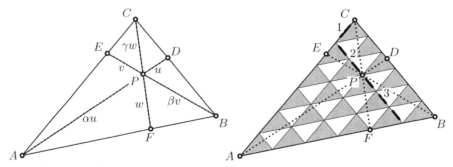

Fig. 6.22. The identities of Euler (left) and their Stone Age proof (right)

6. Prove the following "Porisma tertium" of Fermat (1629c, *Oeuvres*, vol. I, p. 79) (see the picture): Let AB be a fixed diameter of a circle and NM a fixed parallel secant. Then the ratio $(AR \cdot SB)/(AS \cdot RB)$ is the same for every choice of C on the semicircle above AB (see the figure).

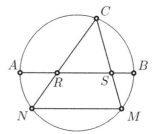

7. Prove three identities in Euler's last paper on Euclidean geometry (published in 1815, written in 1780): in a triangle ABC let the lines AD, BE and CF intersect in point P (see Fig. 6.22, left); then

$$\frac{PD}{AD} + \frac{PE}{BE} + \frac{PF}{CF} = 1\,, \qquad \frac{1}{\alpha+1} + \frac{1}{\beta+1} + \frac{1}{\gamma+1} = 1\,,$$

$$\frac{AP}{AD} + \frac{BP}{BE} + \frac{CP}{CF} = 2\,, \qquad \frac{\alpha}{\alpha+1} + \frac{\beta}{\beta+1} + \frac{\gamma}{\gamma+1} = 2\,,$$

$$\frac{PA}{PD}\frac{PB}{PE}\frac{PC}{PF} = \frac{PA}{PD} + \frac{PB}{PE} + \frac{PC}{PF} + 2\,, \qquad \alpha\beta\gamma = \alpha + \beta + \gamma + 2\,.$$

$$(6.38)$$

Hint. Begin by proving the first relation. Then set $\alpha = \frac{PA}{PD}$, $\beta = \frac{PB}{PE}$, $\gamma = \frac{PC}{PF}$ and transform the formulas in the left column into the algebraic expressions involving α, β, γ in the right column and show that these are equivalent.

8. Solve a problem from the famous *Liber Abaci* (1202) of Leonardo Pisano, called Fibonacci, and three variants by C. Aebi:[9] Two towers, AC and BD, of respective height 30 and 40 feet, are 50 feet apart (see Fig. 6.23). There is a fountain F between these towers. Two birds start flying at the same time, one from the top of each tower, on a straight line and at the

---

[9]Private communication

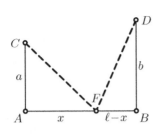 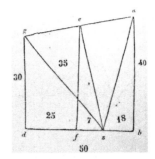

Fig. 6.23. The problem of Leonardo Pisano; right: facsimile from *Liber Abaci*, 1857 edition

same speed, and reach the fountain simultaneously. How far is the centre of the fountain from each tower? [10]

Var. 1. The birds fly from the towers to the fountain and, always at the same speed, back to the foot of their corresponding tower. They arrive *there* at the same time. Where is the fountain now?

Var. 2. The birds fly from the towers to the fountain, back to the foot of their corresponding tower, and finally back to the top, arriving *there* at the same time. In other words: the triangles CFA and DFB have the same perimeter. Where is the fountain now?

Var. 3. Where is the fountain if the triangles CFA and DFB have the same area?

9. Inspired by Fig. 6.16, find a small obtuse-angled triangle with integer side lengths and integer area.

10. Victor Thébault (1882–1960), the inventor of thousands of original problems in number theory and geometry, one of which became particularly famous (see Sect. 7.11 below), also found two very elegant proofs of Heron's formula. His first proof from Thébault (1931) is illustrated in Fig. 6.24, left: draw the circles with diameters AB and AC, with respective centres C_1 and B_1, radii $\frac{c}{2}$ and $\frac{b}{2}$, and radical axis AKA'. Then use Eucl. I.41 to get $\mathcal{A} = AK \cdot 2C_1B_1$ and square the formula. The key of the proof is the use of Eucl. III.35 and the computation of the power of D with respect to the right circle.

11. Reconstruct Thébault's second proof (1945) of Heron's formula (Fig. 6.24, right), which is for admirers of Apollonius: project the points B and C orthogonally onto the inner and outer angle bisectors at A (which are orthogonal). Then show that $\mathcal{A}$ is equal to the area of the rectangle with sides

---

[10] "In quodam plano sunt due turres, quarum una est alta passibus 30, altera 40, et distant in solo passibus 50; infra quas est fons, ad cuius centrum uolitant due aues pari uolatu, descendentes pariter ex altitudine ipsarum; queritur distantia centri ab utraque turri." (Liber Abaci, 1857 edition, vol. 1, p. 398)

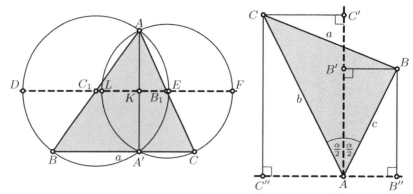

Fig. 6.24. Thébault's proofs of Heron's formula

AB'' and AC', and also to the area of the rectangle with sides AB' and AC''. The key of the proof is then the use of the corollary to Apoll. III.42 (formula (3.18)).

12. Show that the three angle bisectors of a spherical triangle meet in a single point, the incentre I. Show that the radius ρ of the incircle is given by

$$\tan \rho = \sqrt{\frac{\sin(s-a)\sin(s-b)\sin(s-c)}{\sin s}}, \quad \text{where} \quad s = \frac{a+b+c}{2}.$$

13. Give a *direct* proof of the Euler–Brahmagupta formula (6.28) without using Heron's formula.
    Hint. Draw the diagonal AC in Fig. 6.18 and compute its length twice by applying the law of cosines to the triangles ACB and ACD. Express the area of the quadrilateral by that of the two triangles.

14. Derive a second method of obtaining the Möbius transformation for the map $B_1 \mapsto B_2$ of Fig. 6.20 by using Cartesian coordinates: suppose that the point A has Cartesian coordinates (a_1, b_1) and insert for B_1 and B_2 the coordinates of (1.13), i.e. $(\frac{1-u_1^2}{1+u_1^2}, \frac{2u_1}{1+u_1^2})$ and $(\frac{1-u_2^2}{1+u_2^2}, \frac{2u_2}{1+u_2^2})$. Express the collinearity of A, B_1 and B_2 by Thales' theorem and show that

$$u_2 = \frac{b_1 u_1 + a_1 - 1}{(a_1+1)u_1 - b_1} \quad \text{with matrix} \quad \begin{bmatrix} b_1 & a_1 - 1 \\ a_1 + 1 & -b_1 \end{bmatrix}. \tag{6.39}$$

15. Given the four points $A_1, \ldots, A_4$ with coordinates

$$(1.8, 0.8), \quad (-1.4, 1.7), \quad (-0.4, -0.2), \quad (-1.8, -0.4),$$

find the solutions of the corresponding Cramer–Castillon problem.[11]

---

[11] Example suggested by F. Sigrist (Neuchâtel).

16. Problem proposed by Armenia/Australia for the 35th Int. Math. Olympiad (held in Hong Kong, July 12–19, 1994). ABC is an isosceles triangle with $AB = AC$. Suppose that (i) M is the midpoint of BC and O is the point on the line AM such that OB is perpendicular to AB; (ii) Q is an arbitrary point on the segment BC, different from B and C; and (iii) E lies on the line AB and F on the line AC in such a manner that E, Q and F are distinct and collinear. Prove that OQ is perpendicular to EF if and only if $EQ = QF$.

17. This exercise is one of the *Math Challenges* from π *in the Sky*, Issue 9, Dec. 2005 (Pacific Inst. Math. Sciences): Let AA', BB', CC' be the angle bisectors of the triangle ABC. If the angle $B'A'C' = 90°$, find the angle BAC.

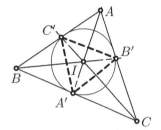

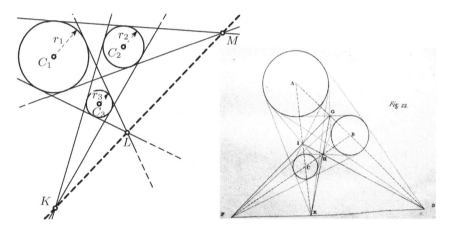

Fig. 6.25. A theorem of Monge (right: original illustration from Monge 1795)

18. (A theorem of Monge 1795.) Let three circles be given and let K, L and M be the intersections of the common tangents of these circles, taken in pairs. Then these points are collinear (see Fig. 6.25). Prove this result.

Remark. Puissant gave an analytical proof in his book (1801), consisting of several pages of formula-jungle. Monge found the theorem by considering a sphere in space. A simple proof was given by Steiner (1826c).

The following eight exercises are enjoyable geometric problems concerning the circle ("problematum ad circulum pertinentium"). They were invented by Euler for Caput XXII of his *Introductio* (1748, vol. II). All these questions are very natural, but lead to equations that are difficult to solve. Of course, there is no hope of a geometric treatment.

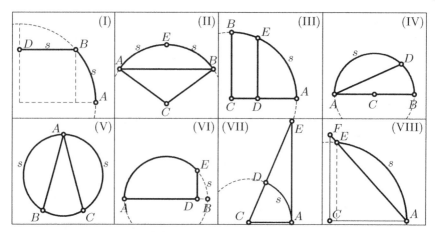

Fig. 6.26. Euler's problems

19. Find the point B on a quarter-circle, such that the arc AB has the same length s as BD (see Fig. 6.26 (I)).

20. Find a sector $ACBE$ of a circle which is divided by the chord AB into two pieces of equal area (see Fig. 6.26 (II)).

21. Divide the quarter-circle ABC by a vertical line ED into two pieces of equal area (see Fig. 6.26 (III)).

22. Divide the semicircle $ABDA$ by a line AD into two pieces of equal area (see Fig. 6.26 (IV)).

23. Divide a circle by two lines AB and AC (where A is on the circle) into three pieces of equal area (see Fig. 6.26 (V)).

24. Find for a semicircle the arc s, such that the arc AE has the same length as the broken line ADE (see Fig. 6.26 (VI)).

25. Find a right-angled triangle CAE which the circle CDA divides into two pieces of equal area (see Fig. 6.26 (VII)).

26. Find for a quarter-circle the arc AE, whose length s is equal to the distance AF (see Fig. 6.26 (VIII)).

7

Cartesian Coordinates

"C'est à l'aide de ce secret que Descartes, à l'âge de vingt ans,
parcourant l'Europe dans le simple appareil d'un jeune soldat
volontaire, résolvait d'un coup d'œil, et comme en se jouant, tous
les problèmes géométriques que les mathématiciens de divers pays
s'envoyaient mutuellement ... [It was with this secret method that
20-year-old Descartes, travelling through Europe dressed simply
as a young volunteer soldier, solved at a glance, as if playing a
game, all the geometrical problems with which mathematicians of
various countries challenged one another ...]"

(J.-B. Biot, *Essai de Géométrie analytique*, 1823, p. 75)

The so-called *Cartesian coordinates*,[1] used to determine the position of a point
in the plane (see Fig. 7.1, left), first appeared (in a somewhat hidden form) in
Descartes' solution of a problem of Pappus (see below). They came into general
use only a few decades later. We owe important simplifications to Newton
(1668), who freely used *negative* values for coefficients and coordinates. A
clear exposition is given in Euler's *Introductio* (1748), vol. II, §1–4.

7.1 Equations of Lines and Circles

The equation of a circle. Euclid's Postulate 3 has the following algebraic
counterpart: the coordinates x, y of a point P on the circle with centre $C =
(x_0, y_0)$, passing through a given point $P_1 = (x_1, y_1)$ satisfy, by Pythagoras'
theorem (see Fig. 7.1, right),

$$(x - x_0)^2 + (y - y_0)^2 = r^2, \quad \text{where} \quad r^2 = (x_1 - x_0)^2 + (y_1 - y_0)^2. \quad (7.1)$$

Equations of lines. The equations of a line are an algebraic incarnation of
Thales's theorem. We call p the *slope* of the line. The four useful relations are
(see Fig. 7.2):

---

[1]The Latinised name of René Descartes is *Renatus Cartesius*.

A. Ostermann and G. Wanner, *Geometry by Its History*,
Undergraduate Texts in Mathematics, DOI: 10.1007/978-3-642-29163-0_7,
© Springer-Verlag Berlin Heidelberg 2012

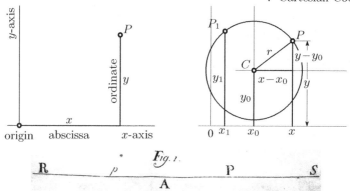

Fig. 7.1. Cartesian coordinates and the equation of a circle; below: first publication of a coordinate axis (Euler 1748; A is the origin, R, S are the endpoints of the line, P is a point with a positive coordinate, p is a point with a negative one)

$$y = px + q \qquad \text{(given ordinate } q \text{ at origin and slope } p\text{),} \qquad (7.2a)$$

$$y = y_0 + p(x - x_0) \qquad \text{(given point } P_0 \text{ and slope } p\text{),} \qquad (7.2b)$$

$$y = y_0 - \frac{1}{p}(x - x_0) \qquad \text{(perpendicular to slope } p \text{ through } P_0\text{),} \qquad (7.2c)$$

$$y = y_0 + \frac{y_1 - y_0}{x_1 - x_0}(x - x_0) \qquad \text{(two given points } P_0, P_1\text{).} \qquad (7.2d)$$

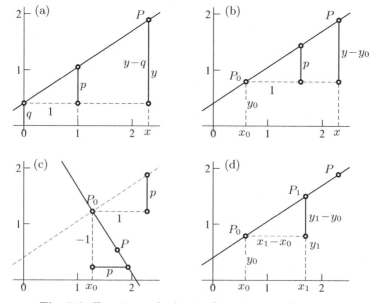

Fig. 7.2. Equations of a line in Cartesian coordinates

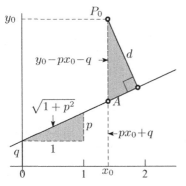

The distance of a point from a line. Let P_0 be a point with coordinates x_0, y_0. We wish to calculate the distance d from this point to the line $y = px + q$. Since the two grey triangles in the figure are similar (orthogonal angles), we get by Thales

$$d = \frac{|px_0 + q - y_0|}{\sqrt{1 + p^2}}. \qquad (7.3)$$

Constructions with ruler and compass. We can now characterise constructions with ruler and compass.

Lemma 7.1. *Every construction with ruler and compass in Euclidean geometry corresponds to a composition of rational operations and square roots in Cartesian geometry, and vice versa.*

Proof. Any such construction can provide only

1. The line through two points (Eucl. Post. 1):

$$y = y_0 + \frac{y_1 - y_0}{x_1 - x_0}(x - x_0) \qquad \Rightarrow \qquad y = px + q.$$

2. The intersection of two lines:

$$\begin{aligned} y &= p_1 x + q_1 \\ y &= p_2 x + q_2 \end{aligned} \qquad \Rightarrow \qquad x = \frac{q_2 - q_1}{p_1 - p_2}, \qquad y = p_1 x + q_1.$$

3. The intersection of a line with a circle:

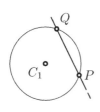

$$\begin{aligned} (x - x_0)^2 + (y - y_0)^2 &= r^2 \\ y &= px + q \\ \Rightarrow \quad (x - x_0)^2 + (px + q - y_0)^2 &= r^2 \\ \Rightarrow \quad Ax^2 + 2Bx + C &= 0, \\ \Rightarrow \quad x = \frac{-B \pm \sqrt{B^2 - AC}}{A}. \end{aligned}$$

4. The intersection of two circles:

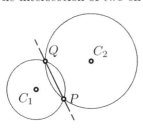

$$\begin{aligned} (x - x_0)^2 + (y - y_0)^2 &= r_0^2 \\ (x - x_1)^2 + (y - y_1)^2 &= r_1^2 \end{aligned}$$

(subtract to get)

$$\begin{aligned} -2x(x_0 - x_1) + (x_0^2 - x_1^2) \\ -2y(y_0 - y_1) + (y_0^2 - y_1^2) = r_0^2 - r_1^2 \end{aligned}$$

(equation of a line, return to 3).

Conversely, the above algebraic operations correspond to constructions with ruler and compass, as can be seen by referring to Descartes' dictionary (see Fig. 6.1) from right to left.   □

Remark. The line obtained in point 4 of the above proof is the *radical axis* of the two circles (see Exercise 8 on page 55).

Briefly, Euclidean geometry, in the strict sense, corresponds to all calculations that involve linear or quadratic equations. On the other hand, all curves of higher degree or algebraic curves belong to a new type of geometry of which Descartes was very proud (in the introduction of his "second book" *De la nature des lignes courbes*).

7.2 The Problem of Pappus

Here is *the* historical problem that absorbed Descartes' interest for five or six weeks; it is the origin of *his* geometry: "La question donc qui auoit esté commencée a resoudre par Euclide, & poursuiuie par Apollonius, sans auoir esté acheuée par personne, estoit telle" (R. Descartes, 1637, p. 306):

Statement of the problem.[2] Let three (or four) lines a, b, c (and d) be given. For a point P, we denote by PA, PB, PC (and PD) the distances of P to these lines. We wish to determine the locus of all points P for which

$$PA \cdot PB = (PC)^2 \qquad \text{or} \qquad PA \cdot PB = PC \cdot PD. \qquad (7.4)$$

One can generalise this problem to five, six, seven or more lines. Pappus (in the introduction to Book VII of his *Collection*) claimed that for three or four lines the curve "est unam ex tribus conicis sectionibus"; for more than four lines, however, the curves were "non adhuc cognitos".

In order to solve the problem of Pappus, Descartes proposes to fix the position of the point P by two values,[3] choosing one as $x = OA$, and the other as

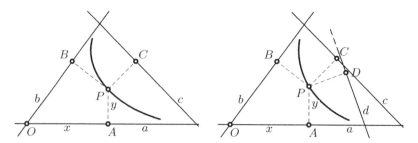

Fig. 7.3. Pappus' problem for three lines (at left) and four lines (at right)

---

[2]The original problem has been slightly modified without changing its essence.

[3]Since we are dealing with unknowns, we take letters from the end of the alphabet; the letter z has already been used above, so we take x and y.

$y = AP$ ("Que le segment de la ligne AB, qui est entre les poins A & B, soit nommé x. & que BC soit nommé y"). This is how *Cartesian coordinates* were born.

With (7.3), each of the factors PA, PB, PC, etc. in (7.4) is of the form $a_i x + b_i y + c_i$. Multiplying out the products, condition (7.4) becomes, in the case of three or four lines, an equation of the form

$$ax^2 + 2bxy + cy^2 + 2dx + 2ey + f = 0 \tag{7.5}$$

where a, b, c, d, e and f are known constants. Descartes now proceeds as follows: for any fixed value of y he obtains a quadratic equation

$$\alpha x^2 + 2\beta x + \gamma = 0 \quad \Rightarrow \quad x = \frac{-\beta \pm \sqrt{\beta^2 - \alpha\gamma}}{\alpha} \tag{7.6}$$

that determines two points (or one point, or none) on the curve. In this way, by separating innumerable different cases, Descartes claims to show that the curves are indeed conics. The proofs were reconsidered by Cramer (1750)[4] and by Euler (1748, vol. II, Caput V).

The last and most important simplification of the theory dates back to the 18th and 19th centuries with the systematic use of *eigenvalues and eigenvectors* (Lagrange, Cayley). We will return to this problem in Chap. 10.

7.3 Conic Sections: Poles, Polars and Tangents

We have already used Cartesian coordinates in Chap. 3 in our treatment of the conic sections (parabola, ellipse and hyperbola):

$$y^2 = 2px\,, \qquad \frac{x^2}{a^2} + \frac{y^2}{b^2} = 1\,, \qquad \frac{x^2}{a^2} - \frac{y^2}{b^2} = 1\,, \tag{7.7}$$

see equations (3.2), (3.9) and (3.13), respectively. We will obtain here new proofs and results with this method. We start with an analytic proof of Apoll. III.52, i.e. $\ell_1 + \ell_2 = 2a$ with

$$\ell_1 = \sqrt{(x+c)^2 + y^2}\,, \quad \ell_2 = \sqrt{(x-c)^2 + y^2}\,, \quad c^2 = a^2 - b^2\,.$$

A direct verification is cumbersome. However, Euler (1748, vol. II, §128) discovered the existence of the simple formulas[5]

$$\ell_1 = a + ex\,, \quad \ell_2 = a - ex\,, \qquad e^2 = 1 - \frac{b^2}{a^2}\,, \tag{7.8}$$

---

[4]Cramer desperately admits, after some 20 pages of calculation: "On voit par cet échantillon, quelle varieté de Cas se présenteroit dans les Lignes des Ordres supérieurs ..."

[5]Which we already encountered in Chap. 5, formula (5.48) on page 140.

where e is the eccentricity from (3.8) on page 66. The verification of these formulas, for instance $(x+c)^2+y^2 = (a+ex)^2$, is a straightforward calculation using $c = ae$ from (3.8).

Poles and polars. The concept of poles and polars is developed in the second book of Apollonius. This theory becomes particularly elegant in an analytical setting.

Definition 7.2. Let $P_0 = (x_0, y_0)$ be a given point. Replace the three equations of a conic in (7.7) by

$$y_0 \cdot y = px_0 + px \,, \qquad \frac{x_0 \cdot x}{a^2} + \frac{y_0 \cdot y}{b^2} = 1 \,, \qquad \frac{x_0 \cdot x}{a^2} - \frac{y_0 \cdot y}{b^2} = 1 \,, \qquad (7.9)$$

respectively, according to the following rule: each quadratic term is split into two factors, one for P and one for P_0; each linear term is divided into two halves, one for P and one for P_0. The resulting equations define straight lines. These lines are called *polars* of the given *pole* P_0, with respect to the given conic.

Theorem 7.3. *If the point P_0 lies on the conic, i.e. if*

$$y_0^2 = 2px_0 \qquad or \qquad \frac{x_0^2}{a^2} + \frac{y_0^2}{b^2} = 1 \qquad or \qquad \frac{x_0^2}{a^2} - \frac{y_0^2}{b^2} = 1 \,, \qquad (7.10)$$

then the corresponding polar is the tangent to the conic at P_0.

Proof. Let $P = (x, y)$ be another point on the polar of P_0 (see Fig. 7.4, left). Then, in the case of the parabola, we compute $y^2 - 2px$ by using $(y - y_0)^2 = y^2 - 2yy_0 + y_0^2$, insert (7.9) and (7.10), and obtain after simplification

$$y^2 - 2px = (y - y_0)^2 \,.$$

Since this square is positive for $y \neq y_0$, *all* other points of the polar lie on the same side of the curve, hence the polar must be a tangent. In the case of the ellipse, a similar calculation gives

$$\frac{x^2}{a^2} + \frac{y^2}{b^2} - 1 = \frac{(x - x_0)^2}{a^2} + \frac{(y - y_0)^2}{b^2} > 0 \,,$$

with the same conclusion. The argument for the hyperbola is slightly more complicated, because here we obtain

$$\frac{x^2}{a^2} - \frac{y^2}{b^2} - 1 = \frac{(x-x_0)^2}{a^2} - \frac{(y-y_0)^2}{b^2} = \left(\frac{x-x_0}{a} - \frac{y-y_0}{b}\right) \cdot \left(\frac{x-x_0}{a} + \frac{y-y_0}{b}\right).$$

This last expression is always negative, because a polar of a point on the hyperbola is steeper than both asymptotes. Therefore the two factors in the last expression have different signs.   □

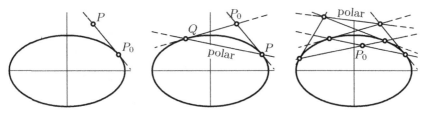

Fig. 7.4. Pole and polar

Theorem 7.4. *If a point P_1 lies on the polar of P_0, then the point P_0 lies on the polar of P_1.*

Proof. Both properties are equivalent to one of the conditions

$$y_0 \cdot y_1 = px_0 + px_1, \quad \frac{x_0 \cdot x_1}{a^2} + \frac{y_0 \cdot y_1}{b^2} = 1, \quad \frac{x_0 \cdot x_1}{a^2} - \frac{y_0 \cdot y_1}{b^2} = 1$$

which are perfectly symmetric. □

This result, together with Theorem 7.3, leads to the next two theorems (see Fig. 7.4, middle and right):

Theorem 7.5. *If the point P_0 lies outside the conic, then the polar of P_0 is the line through the two points of contact of the tangents to the conic from P_0.*

Theorem 7.6. *If the point P_0 lies inside the conic, then the polar of P_0 is the set of all points whose polar passes through P_0.*

Condition of contact for a line. The problem is to find a condition for a line $y = px + q$ to be a tangent to the ellipse (7.7). In order to solve this problem, we write the equation in the form $-\frac{p}{q} x + \frac{1}{q} y = 1$ and compare it to (7.9). We see that we must have

$$\frac{x_0}{a^2} = -\frac{p}{q} \quad \text{and} \quad \frac{y_0}{b^2} = \frac{1}{q}.$$

Taking squares we get $\frac{x_0^2}{a^2} = \frac{p^2}{q^2} \cdot a^2$ and $\frac{y_0^2}{b^2} = \frac{1}{q^2} \cdot b^2$. The point (x_0, y_0) lies on the ellipse, if the sum of these two terms is 1 (see (7.10)) and we have:

Theorem 7.7. *The line with equation $y = px + q$ is tangent to the ellipse (7.7) if and only if*

$$a^2 p^2 + b^2 = q^2. \tag{7.11}$$

If we denote by $h = q - pa$ and $h' = q + pa$ the values of $y = px + q$ above the endpoints of the major axis, then condition (7.11) becomes the elegant formula

$$h \cdot h' = b^2; \tag{7.12}$$

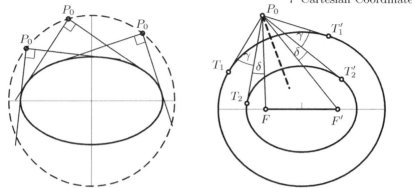

Fig. 7.5. Monge's circle (left); Poncelet's "second theorem" (right)

this is an analytic proof of Apoll. III.42 (see Exercise 11 of Chap. 3, page 76). Euler, who rediscovered this result in (1748, vol. II, §121), called it an "egregia proprietas".

Tangents from a given point. Let P_0 with coordinates x_0, y_0 be a fixed point located outside the ellipse. When a line through P_0 is tangent to the ellipse, what is its slope? By (7.2b), this line has an equation of the form $y = y_0 + p(x - x_0)$, i.e. $y = px + q$ with $q = y_0 - px_0$. Inserting this into (7.11) leads to the next theorem.

Theorem 7.8. *The slopes p_1 and p_2 of the two tangents to the ellipse through a given point $P_0 = (x_0, y_0)$ are the solutions of the quadratic equation*

$$(a^2 - x_0^2)\, p^2 + 2x_0 y_0\, p + (b^2 - y_0^2) = 0 \,;\tag{7.13}$$

hence, by a result in algebra called "Viète's formulas", they satisfy

$$p_1 + p_2 = -\frac{2x_0 y_0}{a^2 - x_0^2}, \qquad p_1 p_2 = \frac{b^2 - y_0^2}{a^2 - x_0^2}\,.\tag{7.14}$$

Monge's circle.[6] *The set of points, from which an ellipse is seen under a right angle, is the circle with radius*

$$R = \sqrt{a^2 + b^2}\tag{7.15}$$

and with the same centre as the ellipse, called "Monge's circle" (see Fig. 7.5, left).

Proof. The tangents are orthogonal if $p_1 p_2 = -1$ (see Fig. 7.2 (c)). Inserting this into the second expression of (7.14) leads to $x_0^2 + y_0^2 = a^2 + b^2$. □

---

[6]G. Monge 1746–1818, famous French "géomètre", with L. Carnot founder of the École Polytechnique in Paris (1794).

Poncelet's "second theorem". We let α_1 and α_2 with $\tan \alpha_1 = p_1$ and $\tan \alpha_2 = p_2$ be the angles which the two tangents through P_0 make with the y-axis. The angle $\frac{\alpha_1 + \alpha_2}{2}$ gives the direction orthogonal to the angle bisector of $T P_0 T'$ (see Fig. 7.5, right). We compute $\tan(\alpha_1 + \alpha_2)$ using (5.6) and obtain with (7.14)

$$\tan(\alpha_1 + \alpha_2) = \frac{p_1 + p_2}{1 - p_1 p_2} = \frac{-\frac{2x_0 y_0}{a^2 - x_0^2}}{1 - \frac{b^2 - y_0^2}{a^2 - x_0^2}} = \frac{-2x_0 y_0}{a^2 - b^2 - x_0^2 + y_0^2}, \tag{7.16}$$

an expression which depends only on $c^2 = a^2 - b^2$, i.e. on the position of the foci, and not on the individual values of a and b. We conclude that *the tangents from a given point P_0 to two confocal ellipses have the same angle bisector.* In the limiting case, when the eccentricity tends to zero, the ellipse tends to the segment FF' and we have:

Theorem 7.9 (Poncelet's second theorem). *The angles between two tangents from a point P_0 to an ellipse and between the lines joining P_0 to the foci have the same bisector.*

Remark. Some authors call this "Poncelet's first theorem". It is not only a beautiful theorem of geometry, but was recently used in an elegant proof in operator theory (see Crouzeix, 2004, p. 473, after formula (4.2)).

7.4 Problems of Minimum and Maximum

The principal initiators of problems, in which the minimum or maximum of a quantity is required, were Apollonius (see the solution of Exercise 4 in Chap. 8) and Pierre de Fermat (1601–1665).[7] Fermat's own writings were published only posthumously by his son in 1679; others (Newton and Leibniz) had in the meantime developed the same ideas into a powerful calculus. This is another subject in mathematics which evolved from geometric problems. Later, in the hands of the Bernoulli brothers, Euler and Lagrange, this calculus turned into one of the pillars of modern science. Jakob Steiner devoted an entire course 1838–1839 in Berlin to *Maxima u. Minima*[8] and two long articles in *Crelle Journal* (1842), claiming that complicated analytical calculations should not hide the original geometric ideas and the immediate insight which they provide into the true nature of the problem.

Triangles between fixed and moving lines (attributed to Fermat in *Elem. Math.* 11 (1956), p. 114). Given a point P at distance c from O, a line parallel

---

[7]M. Cantor (see Cantor, 1900, p. 239, or Probl. 94 of the English edition of Dörrie, 1933) discovered in the correspondence of Regiomontanus (letter to Christian Roder from 1471) the statement of the probably oldest maximum problem of the post-Greek era. A variant of this problem is treated in Exercise 5 on page 233.

[8]A copy of this course is preserved in the *Burgerbibliothek* in Bern.

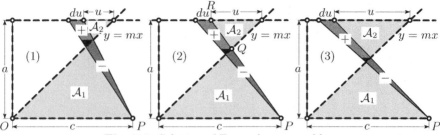

Fig. 7.6. Solution of Fermat's area problem

to OP at distance a and line of slope m through O (see Fig. 7.6), find for which distance u the sum of the areas $\mathcal{A}_1 + \mathcal{A}_2$ is smallest.

Solution. We increase u by a small amount du.[9] When u is small (Fig. 7.6 (1)), a relatively large area $\boxminus$ is removed from $\mathcal{A}_1$ and a small area $\boxplus$ is added to $\mathcal{A}_2$. So the total area decreases. On the other hand, if u is large (Fig. 7.6 (3)), the area $\boxplus$ is larger than $\boxminus$ and the total area increases. The *smallest* value for the sum of the areas is thus obtained precisely where gain and loss, $\boxplus$ and $\boxminus$, *are equal* (Fig. 7.6 (2)). We replace this by the condition that the sum of the areas $\boxplus$ and $\boxminus$ be twice the area $\boxminus$. These are two similar triangles and by Eucl. VI.19 we conclude that $RP = RQ + QP = \sqrt{2} \cdot QP$, hence by Thales,

$$u + c = \sqrt{2} \cdot c \qquad \text{or} \qquad u = (\sqrt{2} - 1) \cdot c . \tag{7.17}$$

Remark. We were slightly sloppy in the above proof. In fact we neglected the little black triangle. The secret in understanding calculus lies in knowing which quantities may be neglected (*"Elisis deinde superfluis ..."*) and which may not. After a long dispute, Fermat explains this in a letter to Mersenne and Descartes (June 1638, Fermat, *Oeuvres*, vol. 2, p. 157): "Divisons le reste par E et ôtons ensuite tout ce qui se trouvera mêlé avec E; ... [divide the rest by E and then neglect everything which is connected to E; ...]", after which Descartes replied: "que si vous l'eussiez expliqué au commancement en cette façon, je n'y eusse point du tout contredit, ... [if you had explained it in this manner from the beginning, I would not have contested it at all, ...]". The following examples should help to understand better and better these ideas; we hope that we need not "alia exempla addere", but that "hæc sufficiunt".

Fermat's Principle of Refraction

> "... et trouver la raison de la réfraction dans notre principe commun, qui est que la nature agit toujours par les voies les plus courtes et les plus aisées."
>
> (Fermat to Cureau de la Chambre, 1657)

The law of refraction of light was discovered independently by Willebrord Snel van Royen, latinised Snellius, and Descartes, latinised Cartesius, but only in

---

[9]This is Leibniz' ingenious notation; Fermat used E or e to denote a small quantity.

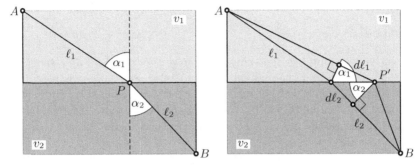

Fig. 7.7. Explanation of Fermat's principle of refraction

"rationem experientiæ"; Fermat thus proposed in several articles (numbers VIII and IX in "Methodus ad Disquirendam Maximam et Minimam" in Fermat, *Oeuvres*, vol. I, pp. 170 and 173; sent to Cureau de la Chambre in 1662) to find a mathematical "Synthesis" from "naturam operari per modos et vias faciliores et expeditiores" (nature acts in the simplest and most direct manner).

Let two fixed points A and B be given and two media, with speed of light v_1 in the medium above the x-axis, and v_2 below it. We wish to determine the point P, for which the time taken for light to travel from A to B is shortest (see Fig. 7.7, left). If we move the point P to P' by a small distance dx, this creates two small right-angled triangles (see Fig. 7.7, right) in which the angles α_1 and α_2 reappear as orthogonal angles. The distance ℓ_1 increases by $d\ell_1 = \sin\alpha_1 \cdot dx$, while ℓ_2 decreases by $d\ell_2 = \sin\alpha_2 \cdot dx$. So the time we lose in the upper region is $dx \cdot \sin\alpha_1/v_1$, and the time we gain in the lower region is $dx \cdot \sin\alpha_2/v_2$. As in the above example, P is in the optimal position if gain and loss are balanced, i.e. if

$$\frac{\sin\alpha_1}{v_1} = \frac{\sin\alpha_2}{v_2}, \qquad (7.18)$$

which is precisely the law observed experimentally.

Fermat's original proof, in manuscript IX mentioned above, is similar but, lacking elegant notation, extends over some 5 pages; the analytical calculation was proudly performed "in tribus lineis" (in three lines) by Leibniz in 1684 (see Hairer and Wanner, 1997, p. 93).

Cylinder with maximal surface area in a sphere. In more difficult examples Fermat explains the use of "triplicatas" variables. In this case Leibniz' notation exhibits all its power. We demonstrate this on the following problem, sent by Fermat to Mersenne on Nov. 10, 1642 (article VII, *Oeuvres*, vol. I, p. 167).

Problem. Inscribe in a given sphere of radius 1 a cylinder with radius y and height $2x$, of *maximal surface area* (see Fig. 7.8, left). In other words, dividing the surface area $2y^2\pi + 2x \cdot 2y\pi$ (see (2.11)) by 2π, we have to maximize $y^2 + 2xy$,

condition $x^2 + y^2 = 1$.

Solution.

$$2y\,dy + 2y\,dx + 2x\,dy = 0 \,,$$
$$2x\,dx + 2y\,dy = 0 \,. \tag{7.19}$$

The solution was found by replacing x and y by $x+dx$ and $y+dy$ respectively, using the principle that the maximised quantity, as well as the quantity defining the condition, must neither increase nor decrease, expanding and neglecting "negligiblios" small terms. In the equations thus obtained we eliminate dx from the first with the help of the second, divide by $2x\,dy$ and get

$$\left(\frac{y}{x}\right)^2 - \frac{y}{x} - 1 = 0 \,, \tag{7.20}$$

which is equation (1.3) for the *golden ratio*.

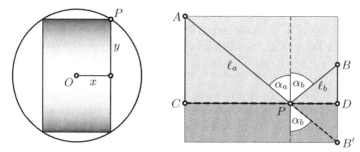

Fig. 7.8. Cylinder in sphere with maximal surface area (left); minimal sum of distances from two points (right)

Minimal sum of distances from two points. This is the problem with which Steiner began his course in 1838: let two fixed points A and B be given, on the same side of a fixed line CD. Find the point P on this line for which the sum of the distances $\ell_a + \ell_b = \min$ (see Fig. 7.8, right). We can answer the question in the same way as we derived the principle of refraction above (with $v_1 = v_2 = 1$). This gives, as for (7.18), the condition

$$\sin\alpha_a = \sin\alpha_b \,, \qquad \text{i.e.} \qquad \alpha_a = \alpha_b \,. \tag{7.21}$$

The optimal solution behaves like a reflected light ray; we admire once again the "intelligence" of light in obeying Fermat's principle about the action of nature. A second idea for treating the problem is to reflect the point B in CD to obtain B'; the shortest path is then a straight line (Eucl. I.20), and the angles are equal by Eucl. I.15.

The Fermat point of a triangle. *"Datis tribus punctis, quartum reperire, a quo si ducantur tres rectæ ad data puncta, summa trium harum rectarum sit minima quantitas"*; or in translation: *"Given three points A, B, C, find a fourth one F, such that for the three lines drawn from F to the given points, the sum of their lengths*

$$\ell_a + \ell_b + \ell_c = \min \tag{7.22}$$

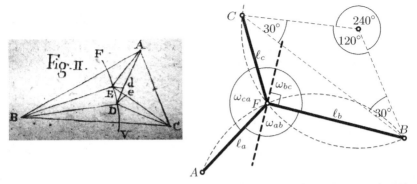

Fig. 7.9. The Fermat point of a triangle; left: drawing from Fagnano (1779)

is as small as possible". This is all that Fermat ever wrote about the "Fermat point" of a triangle, in the last two lines of a long article addressed around 1640 to Mersenne (number IV in *Oeuvres*, vol. I, p. 147–153). It was intended as a challenge for E. Torricelli, whose solution was the subject of Exercise 9 in Chap. 4. The first (lengthy) proof using the new principles was published by Fagnano (1779, see Fig. 7.9, left). The shorter version below was presented in Steiner's 1838 course.

The difficulty here is that the point we require can move with two degrees of freedom, and that we must minimise a sum of *three* terms.

Idea. We move F in a direction *orthogonal* to one of the lines, say to FB (see Fig. 7.9, right); then only two terms of the sum change (for small movements) and we are back in the previous situation. We thus obtain from (7.21) the condition $\omega_{ab} = \omega_{bc}$. If we choose another movement, orthogonal either to FA or to FC (all other movements are compositions of two of these) we conclude finally that all three angles must be equal: $\omega_{ab} = \omega_{bc} = \omega_{ca}$.[10] We thus have:

Theorem 7.10. *The Fermat point F minimising the sum of the distances $\ell_a + \ell_b + \ell_c$ is characterised by $\omega_{ab} = \omega_{bc} = \omega_{ca} = 120°$.*

Since we know that the angles AFB and BFC are both $120°$, we find the position of F at the intersection of two circles with angle $240°$ at the centre (see Fig. 7.9, right, and Eucl. III.20).

Steiner's challenge. If q is any exponent, we now try to minimise $\ell_a^q + \ell_b^q + \ell_c^q$. From $(\ell + d\ell)^q = \ell^q + q\ell^{q-1} d\ell + \ldots$ (binomial theorem), we obtain in the same way: $\ell_a^{q-1} \sin \omega_{ab} = \ell_c^{q-1} \sin \omega_{bc}$ and $\ell_b^{q-1} \sin \omega_{bc} = \ell_a^{q-1} \sin \omega_{ca}$. Hence:

Theorem 7.11 (J. Steiner, 1835a). *The point P minimising the sum of the powers $\ell_a^q + \ell_b^q + \ell_c^q$ is characterised by*

$$\frac{\sin \omega_{bc}}{\ell_a^{q-1}} = \frac{\sin \omega_{ca}}{\ell_b^{q-1}} = \frac{\sin \omega_{ab}}{\ell_c^{q-1}} . \tag{7.23}$$

---

[10]Note that the argument fails if one angle of the triangle is $120°$ or larger. In that case the minimising point lies at the obtuse angle vertex.

Corollary 7.12 (Fagnano, 1779). *The point P minimising $\ell_a^2 + \ell_b^2 + \ell_c^2$ is the barycentre G.*

Proof. This result follows from the above theorem (putting $q = 2$) by the law of sines (5.11) applied to the six small triangles of Fig. 4.8 (a), using the fact that the angles left and right of the feet of the medians have the same sine value, and the sides left and right of these feet have the same length.   □

In the next item (5.) of Steiner (1835a) we find the challenge: determine what happens to the point P solving (7.23), if the exponent q varies continuously from $q = 1$ to $q = 2$ and then $q \to \infty$, leading to a curve connecting the Fermat point F via the barycentre G to the circumcentre O. Steiner gave no indication of what he meant by the "eigenthümliche Beziehung (peculiar relation)" of this curve. We have computed it numerically for the triangle of Fig. 7.10.

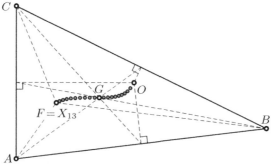

Fig. 7.10. Steiner's challenge for $q = \frac{1}{1-\lambda}$ with $0 < \lambda < 1$ in steps of $\frac{1}{20}$

Minimal sum of distances from four points. Given four points A, B, C and D forming a convex quadrilateral, find a point F such that the sum $\ell_a + \ell_b + \ell_c + \ell_d$ of its distances to the vertices is minimal. The answer is surprisingly simple.

Theorem 7.13 (Fagnano, 1779). *The point F which minimises the sum of the distances $\ell_a + \ell_b + \ell_c + \ell_d$ is the intersection point of the two diagonals AC and BD (see Fig. 7.11, left).*

Proof. We write the sum to be minimised as $(\ell_a + \ell_c) + (\ell_b + \ell_d)$. The first term is as small as possible if A, F, C are collinear, and the second term is minimal if B, F, D are collinear.   □

Minimal connecting graph. Gauss, in a letter to Schumacher dated March 21, 1836 (see *Werke*, vol. 10, p. 461), suggested a more interesting question ("kürzestes Verbindungssystem, ... eine ganz schickliche Preisfrage für unsere Studenten"): *find an itinerary of minimal total length, connecting a given set of points.* Gauss was led to this question by thinking about the shortest railway connection between Harburg, Bremen, Hannover and Braunschweig. The solution of a particular problem is drawn in Fig. 7.11, right. There are

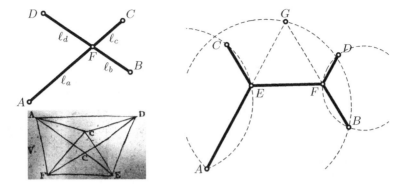

Fig. 7.11. The Fagnano problem for 4 points (left, with an original drawing of Fagnano 1779); the Schumacher–Gauss problem (right)

two branching points E and F which can move freely. Each of these, by Theorem 7.10, is in an optimal position if connected to the rest of the graph by three segments making angles of 120°. Several circles help to construct the solution by Eucl. III.20.

However, the solution is not always of this form: if you take a map of Germany and find the four towns mentioned above, the optimal solution consists of a straight line between Braunschweig and Hannover, together with the Fermat point for Hannover, Bremen and Harburg.

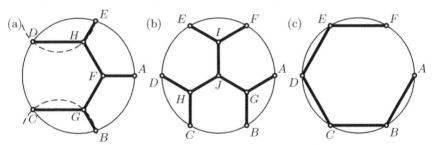

Fig. 7.12. The Schumacher–Gauss problem for the regular pentagon (a) and hexagon (b, only local minimum), (c, global minimum)

Pentagon and hexagon. The same algorithm leads to a nice solution for five points arranged to form a regular pentagon (see Fig. 7.12 (a)), but the "nice" solution for the hexagon (Fig. 7.12 (b)) of total length $3\sqrt{3} = 5.196\ldots$ is not the shortest connection, which has length 5 (Fig. 7.12 (c)).

7.5 Some Famous Curves and Their Tangents

> "L'un des plus féconds rapprochements que l'on ait faits dans les sciences est l'application de l'algèbre à la théorie des courbes ... La recherche de ces propriétés a conduit à l'analyse infinitésimale dont la découverte a changé la face des mathématiques."
>
> (Laplace, "Programme" à *l'École Normale*, 1er pluviôse, an III)

"Differentials are as slippery as ever."
(From a book review, Amer. Math. Monthly 65 (1958), p. 641)

Descartes, in his introduction to the second "Livre" of his Geometry, proudly discusses the wide field of new curves, which his method allows one to study and which the "Anciens" had been reticent to consider, by limiting themselves to ruler and compass and possibly to the conics. We now present the properties of some curves which owe their fame to their importance or their beauty. Comprehensive catalogues of interesting curves were published by G. Loria (1910/11) and by F.G. Teixeira (1905). The properties of some important plane curves can also be found in Lawrence (1972).

Folium of Descartes. The first concrete example was the curve defined by the equation

$$x^3 + y^3 - 3xy = 0 \,. \tag{7.24}$$

In a letter (dated Jan. 18, 1638) addressed to Fermat, but sent to Father Mersenne, Descartes challenged Fermat to find the tangents to this curve. Fermat, who already had his manuscript (Fermat, 1629b) in a drawer, remarked that what Descartes judged to be difficult could in fact be done easily and elegantly.[11] These calculations of Fermat differed from the differential calculus of Newton and Leibniz essentially only in notation and attempt at generality.

In Leibniz' notation, the solution "elegantissima" is as follows: we suppose that a point (x, y) satisfies (7.24) and ask under which condition a neighbouring point $(x + dx, y + dy)$, with dx and dy very small, also satisfies (7.24). We insert, multiply out, neglect terms dx^2, $dx\,dy$, dy^2, ... of higher order, subtract, and obtain

$$(x^2 - y)\,dx + (y^2 - x)\,dy = 0 \,. \tag{7.25}$$

The slope of the tangent is thus $\frac{dy}{dx} = -\frac{x^2-y}{y^2-x}$. The points at which the curves $x^3 + y^3 - 3xy = C$ are horizontal $(dy = 0)$ lie on the parabola $x^2 = y$, and the points at which these curves are vertical $(dx = 0)$ lie on the parabola $x = y^2$ (see the figure).

The conchoid. The conchoid was one of the first curves which Fermat considered. If we place the origin at the point A (see Fig. 7.13) and give the point C the coordinates (x, y), we have by Thales' theorem

---

[11] "quem difficilem judicabat D. DesCartes, cui nihil difficile, elegantissima ..."

$$\frac{y - c}{c} = \frac{b}{\sqrt{x^2 + y^2} - b},$$

which leads to

$$(y - c)\sqrt{x^2 + y^2} = by \quad \text{or} \quad (y - c)^2(x^2 + y^2) = b^2 y^2. \tag{7.26}$$

The curve is now "algebraised" and can be treated by the above method. The result, stated in the version of Joh. Bernoulli (1691), is as follows:

Theorem 7.14. *Let M be the intersection of the perpendicular to AC through A with the line through C parallel to the fixed line through D. Then the tangent at C to the conchoid of Nicomedes is parallel to MD.*

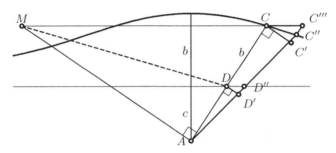

Fig. 7.13. The tangent to Nicomedes' conchoid

Proof. We give a geometrical proof, based on the idea of *composed movements* due to Roberval and Torricelli (1644): first rotate the line DC around A by a small angle, to the position $D'C'$ (see Fig. 7.13). But now D' does not lie on the horizontal line, where it should be. So by a second movement, we push the line upwards to become $D''C'''$ so that $D'D'' = C'C''$. Then the triangles $DD'D''$, $CC'C'''$ and CAM are similar. Further, since $C'''C'/C''C' = C'''C'/D''D' = CA/DA$, the triangles DAM and $C''C'C$ are also similar, which is the required result. □

The cycloid. This curve was invented by Galileo Galilei, and its geometrical properties were one of the major challenges in the disputes between the most eminent scientists of the early 17th century, Descartes, Fermat, Pascal, Roberval, Wallis and Torricelli, with Father Mersenne as go-between. In 1645 a 16-year-old Dutch boy with brilliant ideas, Christiaan Huygens, joined this circle.

Definition 7.15. Suppose that a circle (which we take of radius 1) rolls on a line DAE (see Fig. 7.14), and that P is a fixed point on this circle. The curve $DGPE$ described by P is called a *cycloid*. If we denote the distance AB by t, which is also the angle PCH (measured in radians), and if we take the point A as origin, we have for the coordinates of P

$$\begin{aligned} x &= t + \sin t \\ y &= 1 + \cos t \end{aligned} \quad -\pi \le t \le \pi. \tag{7.27}$$

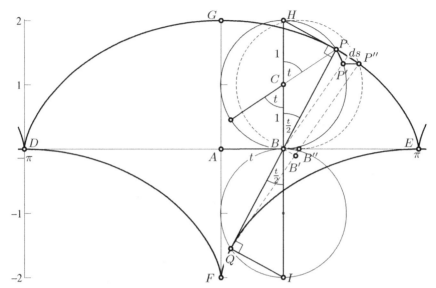

Fig. 7.14. The tangent to the cycloid

Equations (7.27) constitute a *parametric representation* of the curve, with t as parameter, and allow its geometrical properties to be established by analytic methods (see e.g. Hairer and Wanner, 1997, pp. 104, 117, 138, 142). C. Huygens (1673),[12] however, deduced these properties from geometrical results many years before the first publications of the differential calculus.

Theorem 7.16. *With the notation of Fig. 7.14 we have*
(a) the tangent to the cycloid at P is orthogonal to BP, i.e. passes through H and

$$\frac{dy}{dx} = -\tan\frac{t}{2} ; \qquad (7.28)$$

(b) the infinitesimal length of the arc PP'' is

$$ds = 2\cos\frac{t}{2}\, dt ; \qquad (7.29)$$

(c) the area $DAEGD$ is 3π, i.e. three times the area of the circle;
(d) the two neighbouring normals PB and $P''B''$ of the curve intersect at the point Q such that $QB = BP$. This is thus the centre of curvature and lies on a second cycloid FE;
(e) the arc length QE is equal to the distance QP (we say that the cycloid FE is the "evolute" of the curve GE);
(f) the total arc length DGE is 8.

---

[12]The historical importance of this book can hardly be exaggerated, since a year later the young Leibniz, during his stay in Paris, received his initiation to modern mathematics from Huygens. Two more years later Leibniz' differential calculus was born.

Proof.

(a) We again apply the idea of composed movements (see the previous proof): for a small increment dt, we move the point P to P' by rotating the circle through the angle dt, followed by a horizontal translation of P' to P'' and B to B'' of distance dt. Then the sides PP' and $P'P''$ are of equal length and orthogonal to PC and CB. The triangle $PP'P''$ is thus isosceles, smaller by the factor dt and orthogonal to PCB. Hence PP'' is orthogonal to PB.

(b) By Eucl. III.20, the angle PBC is $\frac{t}{2}$ and BPH is a right angled triangle. Hence $BP = 2\cos\frac{t}{2}$ and we have (7.29).

(c) If dt tends to zero, the line BP' tends to the altitude of $PP'P''$ and bisects PP'', therefore the area $BB'P''P$ is composed of a triangle and a rectangle, both of base $\frac{ds}{2}$ and altitude $2\cos\frac{t}{2}$. Therefore, by (1.6), the cycloid has an area three times that of the circle, which is composed only of the triangles.

(d) Since BB' is half of PP'', we also see that the two normals PB and $P''B''$ intersect at Q where $QB = BP = 2\cos\frac{t}{2}$. Because $QI = 2\sin\frac{t}{2} = 2\cos\frac{t-\pi}{2}$, we see that Q moves on a cycloid which is "dephased" by π. This cycloid has QP, which is orthogonal to QI, as tangent. This property means that Q is the centre of curvature, as we will see in Sect. 7.6 below.

(e) C. Huygens convinced himself of this by a picture as in Fig. 7.15 (left). Using integral calculus, you can simply integrate $2\sin\frac{t}{2}\,dt$ (see (7.29)) and obtain the same result.

(f) This follows because the arc length DGE is twice the arc length FE, which is $FG = 4$. □

With the above results, we can now verify directly the isochronal property (see e.g. Hairer and Wanner, 1997, equation (7.29), p. 142), which allowed Huygens to use the cycloid to design the most accurate pendulum clock of

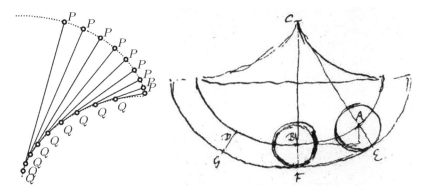

Fig. 7.15. Determination of arc length (left); autograph drawing of Huygens (1692/93, right)

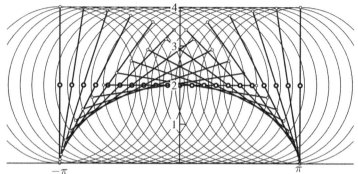

Fig. 7.16. The tangent to the cycloid

his time (see Fig. 7.15 right). One also obtains the brachistochrone property (Hairer and Wanner, 1997, p. 138), which initiated the development of the calculus of variations.

Theorem 7.17. *The cycloid can also be obtained as the envelope of a diameter of a circle of radius 2, rolling on a line with half the rotational speed, i.e. with the same horizontal speed, as the circle of Thm. 7.16 (see Fig. 7.16).*

Proof. The line PH of Fig. 7.14, if produced, is the diameter of a circle of radius 2, rotated through an angle of $\frac{t}{2}$. If this circle rolls along the line DE, it rotates at this moment around the base point B. This diameter touches the curve at the point where the diameter has no vertical speed, i.e. where the diameter is perpendicular to B. This is precisely the point P. The curve of Fig. 7.16 is thus the same as that of Fig. 7.14.   □

Steiner's deltoid. We will explain below the original motivation of Jacob Steiner's discovery (see Theorem 7.26). Let a point Q rotate on a circle and let a line be attached to it, *which rotates with half the speed in the opposite direction*, i.e. such that $\tau = \frac{t}{2}$. Thus, if this line is orthogonal to the circle for, say $t = 0$, it will again be orthogonal for $t = 120°$, and a third time for $t = 240°$. If these lines are extended, they produce a beautiful curve, which is *Steiner's deltoid* (1857).[13]

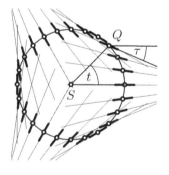

If we let a circle of radius 2 roll inside a circle of radius 3, then a diameter of the rolling circle will produce precisely the required movement and *the deltoid will be the envelope of the family of these diameters* (see Fig. 7.17 below and Fig. 7.28). Further, as for the cycloid (Theorem 7.17), we obtain the same curve *from the movement of a point on the circumference of a circle*

---

[13] Most references give the date 1856; this was the year of the presentation to the Academy, but the paper was published in the Crelle Journal of 1857.

of radius 1 rolling inside the big circle (see Fig. 7.17 above). The argument is the same as for the cycloid (see Fig. 7.17 left). Given the angle t, all the other angles are determined by the requirement that the arcs DP, DU and DA must be the same, i.e. the circles are rolling.

There is another surprise, also noted by Steiner: we observe that the end-points U and V of the rolling tangents, which are at constant distance 4, also both stay on the deltoid and move at half the speed in the opposite direction. To see this, we complete the broken line SQV to a parallelogram, i.e. inter-change the relative position of V with respect to Q, and of Q with respect to S. Then we have the same construction as for the point P, with t replaced by $-\frac{t}{2}$.

Equations. By looking at the positions of the points C and P in Fig. 7.17 (left) we obtain, as for (7.27),

$$\begin{aligned} x &= 2\cos t + \cos 2t \\ y &= 2\sin t - \sin 2t \end{aligned} \qquad 0 \le t \le 2\pi \qquad (7.30)$$

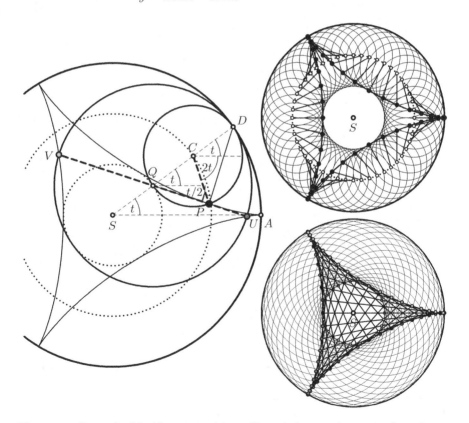

Fig. 7.17. Steiner's deltoid generated by rolling circles; as the trace of a point on the circumference (above); as the envelope of diagonals (below)

for the parametric representation of the deltoid in Cartesian coordinates. Steiner claims in the title of his paper that the curve is "of the fourth degree", without explaining why. Indeed, we obtain from (7.30) by squaring, adding and simplifying with (5.6) the relation

$$x^2 + y^2 = 5 + 4\cos 3t \tag{7.31}$$

which already shows the $120°$ symmetry of the curve. By setting $u = \cos t$, the first equation of (7.30) becomes

$$u^2 + u - \frac{x+1}{2} = 0 .$$

This is a quadratic equation for $u = \cos t$ and yields u as an algebraic expression in x:

$$u = -\frac{1}{2}\left(1 \mp \sqrt{2x+3}\right) .$$

Using $\cos 3t = \cos^3 t - 3\cos t$, the right-hand side of (7.31) can be written as

$$5 + 4u(4u^2 - 3) = -9 - 12x \pm 2(2x+3)\sqrt{2x+3} .$$

By inserting this into (7.31) we see that

$$(x^2 + y^2 + 9 + 12x)^2 = 4(2x+3)^3 , \tag{7.32}$$

which is, indeed, an algebraic equation of degree 4. By expanding (the first term with Eucl. II.4), we can transform this equation into

$$(x^2 + y^2 + 9)^2 = 8x^3 - 24xy^2 + 108 , \tag{7.33}$$

where the $120°$ symmetry of the curve is easier to see.

> "A l'egard des lignes de Mr. Bernoulli, vous avés raison, Monsieur, de ne pas approuver qu'on s'amuse à rechercher des lignes forgées à plaisir. [You are quite right, Monsieur, about the lines of Mr. Bernoulli, in not approving this investigation of lines created arbitrarily.]"
>
> (Letter of Leibniz to Huygens, Bronsvic $\frac{11}{21}$ Sept. 1691)

The lemniscate and the Cassini curves. Jac. Bernoulli introduced the lemniscate by giving the formula

$$x^2 + y^2 = a\sqrt{x^2 - y^2} , \tag{7.34}$$

where a is a (positive) parameter. Written in *polar coordinates*

$$x = r\cos\varphi \qquad y = r\sin\varphi , \tag{7.35}$$

this equation becomes

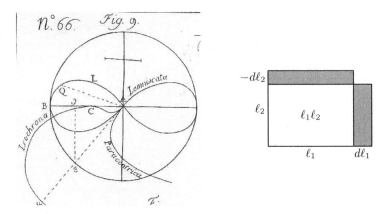

Fig. 7.18. Drawing of the lemniscate (together with other interesting curves) by Jac. Bernoulli 1695 (left); constant area of a rectangle (right)

$$r^2 = ar\sqrt{\cos^2\varphi - \sin^2\varphi} \qquad \text{or} \qquad r = a\sqrt{\cos 2\varphi} \qquad (7.36)$$

with (5.8).

The original motivation for this curve was not geometrical at all, but its arc length was used by Bernoulli ("Curvam Accessus & Recessus ...") to compute difficult integrals and solve differential equations for elastic curves, the *Isocrona Paracentrica* etc. ("... hujus Problematis omnium facillimam per rectificationem curvæ algebraicæ, quam *Lemniscatam* voco ...", see Exercise 22 below, Fig. 7.18, left, Jac. Bernoulli (1694) and (1695); see also J.E. Hofmann (1956) for more details).

Singularity. Squaring both sides in (7.34), we obtain

$$(x^2 + y^2)^2 + a^2 y^2 - a^2 x^2 = 0. \qquad (7.37)$$

If x, y tend to zero, the fourth-power term is negligible and the equation tends to $y^2 - x^2 = (y - x)(y + x) \approx 0$. This explains why the curve at the origin is asymptotically equal to the lines $y = \pm x$ (see the bold curve in Fig. 7.20). The shape of the curve thus "refert jacentis notæ octonarii ∞, ... sive lemnisci, *d'un noeud de ruban* Gallis".

Tangents to the lemniscate. The following remarkable property is attributed by Loria (1910/11, p. 217) to the mathematician Vechtmann (*Diss. inaug. phil. de curvis lemniscatae*, Göttingen 1843):

Theorem 7.18. *The perpendicular to the tangent to the lemniscate at a point P makes the angle*

$$3\varphi \qquad (7.38)$$

with the x-axis, where φ is the angle of the line PO with the x-axis (see Fig. 7.19 (a)).

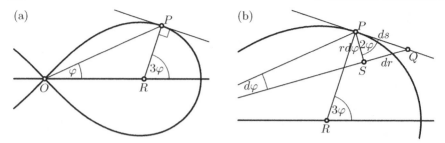

Fig. 7.19. Tangents to the lemniscate

Proof. We prove that the angle OPR is 2φ, hence the exterior angle of the triangle ORP is, by (1.2), equal to $\varphi + 2\varphi = 3\varphi$. By orthogonal angles (see Fig. 1.7 and Fig. 7.19 (b)), this means that $\tan 2\varphi = \frac{dr}{r\,d\varphi} = \frac{r\,dr}{r^2\,d\varphi}$. But this follows from $r^2 = a^2 \cos 2\varphi$ and, after differentiation, $r\,dr = a^2 \sin 2\varphi\,d\varphi$. $\square$

Cassini curves. We now choose $a^2 = 2$ in (7.37), add 1 to each side, and write $-2x^2 = 2x^2 - 4x^2$. This gives $(x^2 + y^2 + 1)^2 - 4x^2 = 1$, or, with Eucl. II.5,

$$((x+1)^2 + y^2)((x-1)^2 + y^2) = 1, \qquad \text{i.e.} \qquad \ell_1 \cdot \ell_2 = 1, \tag{7.39}$$

where ℓ_1 and ℓ_2 are the respective distances from the point (x, y) to the points $F_1 = (-1, 0)$ and $F_2 = (1, 0)$. This very nice property of the lemniscate was noticed neither by Jac. Bernoulli nor by Euler (1748, vol. II, Chap. XXI), despite the fact that a calculation such as the one above would have taken Euler not more than half a second. Generalising this characterisation of the lemniscate, we obtain, for an arbitrary positive constant C, the equation

$$\ell_1 \cdot \ell_2 = C \qquad \text{or} \qquad (x^2 + y^2 + 1)^2 - 4x^2 = C^2 \tag{7.40}$$

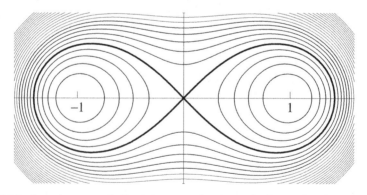

Fig. 7.20. Lemniscate and Cassini curves $\ell_1 \cdot \ell_2 = C$, drawn for the values $C = 0.2, 0.4, 0.6, 0.8, \mathbf{1}, 1.2, 1.4, \ldots$

for the so-called *Cassini curves*[14] (the other curves in Fig. 7.20).

In polar coordinates, the right-hand equation in (7.40) is $(r^2 + 1)^2 - 4r^2 \cos^2 \varphi = C^2$, or

$$r^4 - 2r^2 \cdot \cos 2\varphi + (1 - C^2) = 0, \qquad r = \sqrt{\cos 2\varphi \pm \sqrt{\cos^2 2\varphi + C^2 - 1}}. \quad (7.41)$$

This quadratic equation for r^2 has, depending on C and φ, either two, one, or no positive solutions.

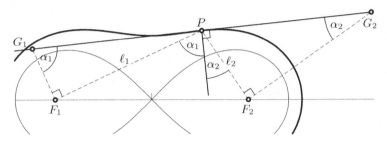

Fig. 7.21. Tangents to the Cassini curves

Theorem 7.19 (Steiner, 1835b). *The tangents to the Cassini curves have the following property: let α_1 and α_2 be the angles between the normal to the tangent at P and the lines joining P to F_1 and F_2 respectively (see Fig. 7.21); then*

$$\frac{\sin \alpha_1}{\sin \alpha_2} = \frac{\ell_1}{\ell_2}, \quad (7.42)$$

which is a sort of Snellius–Descartes' law for refraction with "speeds of light" ℓ_1 and ℓ_2. If G_1 is the point on the tangent such that $G_1 F_1$ is orthogonal to $F_1 P$ (and similarly for G_2), then $G_1 P = P G_2$.

Proof. We move the point P along the tangent by a small amount ds; then, precisely as in Fig. 7.7, ℓ_1 increases by $d\ell_1 = ds \sin \alpha_1$ and ℓ_2 decreases by $d\ell_2 = ds \sin \alpha_2$. However, since the area of the rectangle ℓ_1, ℓ_2 must remain constant (see Fig. 7.18, right), the two shaded rectangles have the same area, which means that $\frac{d\ell_1}{d\ell_2} = \frac{\ell_1}{\ell_2}$. This leads to (7.42).

The last statement follows from the fact that α_1 and α_2 repeat as orthogonal angles at G_1 and G_2 respectively, so that $G_1 P = \ell_1 / \sin \alpha_1$ and $P G_2 = \ell_2 / \sin \alpha_2$.   □

Remark. If we let $\alpha_1 = \alpha + \gamma$ and $\alpha_2 = \alpha - \gamma$ (i.e. 2α is the angle between the lines $P F_1$ and $P F_2$, while γ is the deviation of the normal from the angle bisector), formula (7.42) becomes, after inserting (5.6) and (5.7), the relation

---

[14]The famous astronomer Giovanni Domenico (or Jean-Dominique) Cassini (1625–1712) thought for a while that planets move on such curves. Thus the discovery of these beautiful curves is due to a false astronomical conjecture.

$$\tan \gamma = \tan \alpha \cdot \frac{\kappa - 1}{\kappa + 1}, \qquad \text{where} \qquad \kappa = \frac{\ell_1}{\ell_2}. \qquad (7.43)$$

In the case of the ellipse (see (3.5)) we have $\kappa = 1$ and we obtain $\gamma = 0$; another proof of Apoll. III.48.

7.6 Curvature

> "There are few Problems concerning Curves more elegant than this, or that give a greater Insight into their nature."
>
> (I. Newton 1671, Engl. pub. 1736, p. 59)

We agree with Newton (see the quotation) that the question of finding a measure for the curvature of a given curve at a given point P is very interesting. Newton treated this question in Problems V and VI of Newton (1671).

Solution. We draw the perpendicular through P to the tangent (see Fig. 7.22). Then we move P to a neighbouring point Q and again draw the perpendicular to the tangent. The point C where these perpendiculars meet will be, for a small displacement, the centre of the *osculating circle*, whose radius $r = PC$ is the *radius of curvature* at P. Its inverse $\kappa = \frac{1}{r}$ is the *curvature* at P.

Example: the ellipse. We write the coordinates of a point P on the ellipse with semi-axes a and b as (ac, bs) where $c = \cos u$ and $s = \sin u$. Then by (7.9) the tangent at P has slope $-\frac{bc}{as}$, so that by (7.2c) the perpendicular at P has the equation $y = bs + \frac{as}{bc}(x - ac)$. To simplify the calculations, we multiply this equation by the constant $\frac{b}{a}$ to obtain

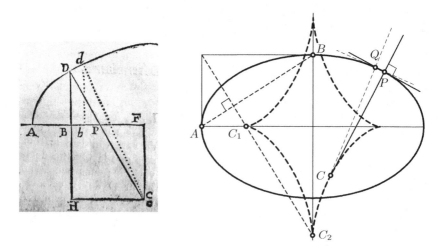

Fig. 7.22. Curvature of the ellipse; left: Newton's 1671 manuscript, reproduced from the 1740 French publication

$$\frac{b}{a}y = \left(\frac{b^2}{a} - a\right)s + tx, \qquad \text{where} \quad s = \sin u, \ t = \tan u. \tag{7.44}$$

If u now increases by a small quantity δ, then s will increase by δc (see formula (5.6) on page 117) and t will increase by $\frac{\delta}{c^2}$ (see formula (7.65) of Exercise 18 on page 237). The second perpendicular (at Q) is thus given by

$$\frac{b}{a}y = \left(\frac{b^2}{a} - a\right)(s + \delta c) + \left(t + \frac{\delta}{c^2}\right)x. \tag{7.45}$$

Subtracting (7.44) from (7.45) we obtain for the abscissa x_C of the intersection point C:

$$0 = \left(\frac{b^2}{a} - a\right)\delta c + \frac{\delta}{c^2}x_C \quad \Rightarrow \quad x_C = c^3\left(a - \frac{b^2}{a}\right), \ \ y_C = s^3\left(b - \frac{a^2}{b}\right) \tag{7.46}$$

(the result for y_C is found by symmetry). These formulas were discovered by Newton, in a different notation, with his calculus of "fluxions". If we allow u to vary in the range $0 \le u \le 2\pi$, then the point C describes a diamond-like curve with four cusps, drawn in Fig. 7.22. This curve is a (stretched) *astroid*.

The smallest and largest radii, $r_1 = \frac{b^2}{a}$ and $r_2 = \frac{a^2}{b}$ for the points A and B respectively, can be constructed by Thales' theorem using the perpendicular from the point $(-a, b)$ to AB (see Fig. 7.22, right). Notice the coincidence that $r_1 = p$, the latus rectum.

The tractrix and its curvature. Towards the end of the 17th century, mathematicians started to study a completely new type of curve. In contrast to Descartes and his followers who defined curves by algebraic formulas and investigated their properties, for  example their tangents, we now assume that we *know* certain properties of the tangents, and conversely try to find the equations of curves. Greater familiarity in dealing with moving points ("fluxions") or "infinitely small" quantities allowed mathematicians to master these new curves, which acquired the name "transcendental".[15] The tractrix was proposed around 1670 by the famous French architect and *medicus* Claude Perrault to many mathematicians in Paris and Toulouse as the solution of the following problem (see Fig. 7.23 (b)): *Find the curve such that on each tangent, the segment between the point of tangency and the intersection with the x-axis has the same length.* To illustrate this problem, he would take out a silver fob watch (horologio portabili suae thecae argenteae) and drag it across the table by pulling at the watchchain. The first published solutions were given in Huygens (1692) and Leibniz (1693). For more details about the analytical treatment, see Hairer and Wanner (1997, p. 135) and for a very rich and complete account of the whole theory, see Tournès (2009). We will obtain here many properties of this curve by simple geometric arguments.

---

[15]Unde *triplices* habemus *quantitates: rationales, Algebraicas, & transcendentes.* (Leibniz, 1693, p. 385)

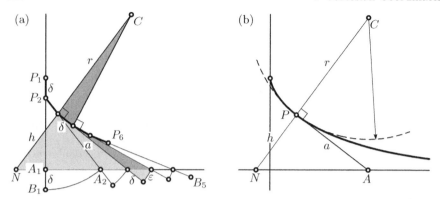

Fig. 7.23. The tractrix and its curvature; left: $\delta = 0.22a$, right: $\delta = 10^{-3}a$

Suppose that the watch-chain of length a is initially perpendicular to the x-axis, between P_1 and A_1 (see Fig. 7.23 (a)). We then pull A_1 to B_1 in the direction of the chain by a small distance δ, so that the watch moves by the same distance from P_1 to P_2. Then the point B_1 is brought back to the point A_2 on the x-axis along an arc centred at P_2. From here we continue to pull in the direction of the chain by another δ, return to the x-axis and so on. Thus the watch describes a broken line $P_1, P_2, P_3, \ldots$ If δ becomes smaller and smaller, this line tends to a curve, which is our tractrix (Fig. 7.23 (b)). See also Leibniz' original drawing in Fig. 7.24 (left). After this figure appeared in the September 1693 issue of the *Acta Eruditorum*, Leibniz was not satisfied and inserted in the November issue some *Corrigenda* which explained the above procedure more clearly.

Curvature of the tractrix. To find the radius of curvature r at a given point P, we draw two consecutive perpendiculars to the broken line (see Fig. 7.23 (a)). If we also add the perpendicular PN of length h, with N on the x-axis, we obtain two pairs of similar triangles (in grey), and have by Thales

$$\frac{r}{a} = \frac{\delta}{\varepsilon} \quad \text{and} \quad \frac{\delta}{\varepsilon} = \frac{a}{h} \qquad \Rightarrow \qquad r = \frac{a^2}{h}. \tag{7.47}$$

Thus the tractrix has the remarkable property that *the product rh has a constant value along the curve.*

Area under the tractrix. We see from Thales applied to the two triangles in Fig. 7.24 (right) that

$$\frac{dy}{dx} = -\frac{y}{\sqrt{a^2 - y^2}} \quad \text{or, neglecting the sign,} \quad \sqrt{a^2 - y^2} \cdot dy = y \cdot dx, \tag{7.48}$$

i.e. the two dark rectangles in this figure have the same area. Since this is true for all points x, y on the curve, we obtain that both surfaces have the same area, i.e. *the area under the tractrix is equal to $\frac{a^2\pi}{4}$* (Huygens, 1692, §IV).

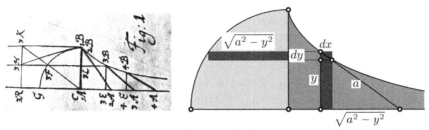

Fig. 7.24. Leibniz' drawing of the tractrix (left); area under the tractrix (right)

The tractroid. The next idea of Huygens was to rotate the picture on the right of Fig. 7.24 around the x-axis. The tractrix then generates a surface of revolution called the *tractroid*, while the circle on the left generates half of a sphere. The grey rectangle to the right yields a disc of thickness dx on a circle of radius y, i.e. a cylinder of volume $y^2\pi\,dx$. The rectangle to the left yields an annulus, which we can, for dy small, straighten to a prism of base $\sqrt{a^2 - y^2}\,dy$ and height $2\pi y$. By (7.48) this volume is precisely twice the volume of the disc. We conclude that *the volume of the tractroid is* $\frac{a^3\pi}{3}$, one fourth of the volume of the sphere of radius a (also found by Huygens, 1692, §V).

Curvature of the tractroid. The theory of curvature of surfaces, initiated by the classical papers Euler (E333, 1767b), Meusnier (1785) and Gauss (1828), requires much more calculus. The result is that, for a given point P, there is a different curvature in each direction, but there are two principal directions, perpendicular to each other, with a *minimal* and a *maximal curvature* κ_1 and κ_2. These determine the curvatures in all other directions. The so-called

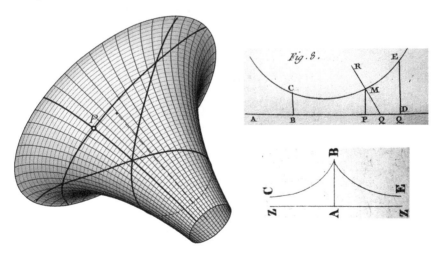

Fig. 7.25. Left: the tractroid, three geodesics and triangles without Eucl. I.32; right: drawing of Meusnier (1785) explaining the curvature of a surface of revolution (above), drawing of the tractroid (without calling it thus) by Minding (1839, below)

Gaussian curvature is defined by $\kappa = \kappa_1 \cdot \kappa_2$ and plays an important role in the study of surfaces which can be deformed into each other. Despite the fact that these theories are far beyond the scope of this book, we can see in Fig. 7.25 that for the tractroid $\kappa_2 = \frac{1}{r}$ (for the curvature of the generating tractrix) and $\kappa_1 = -\frac{1}{h}$ of the opposite sign (h is the radius of the sphere tangent at P and centred in N). We conclude from (7.47) that $\kappa = -\frac{1}{a^2}$ independently of the position of P, i.e. *the tractroid has constant negative curvature* (Minding, 1839). We finally mention the discovery of Beltrami (1868) that the "geodesics" on the tractroid (i.e. the curves of shortest distance, which correspond to straight lines in Euclidean geometry) can be used to create a non-Euclidean geometry.[16] This is demonstrated in Fig. 7.25, where three geodesics are drawn which form two triangles for which Eucl. I.32 is not satisfied.

7.7 The Euler Line by Euler

Euler (1767a) describes his remarkable discovery of the "Euler line", which we have already met in Chap. 4 (see page 91), in the paper *Solutio facilis problematum quorundam geometricorum difficillimorum*. The curious title "Easy solution of certain very difficult geometrical problems" expresses Euler's enthusiasm for the power of the analytical method. We now follow these calculations, which will lead to some further surprises.

Theorem 7.20. *Let a triangle ABC with sides a, b and c be placed in such a way that A is at the origin and B at distance c from A on the positive x-axis. Then the coordinates of the four remarkable points of the triangle are*

O *(circumcentre):* $\quad x_O = \dfrac{c}{2}$ $y_O = \dfrac{c}{8\mathcal{A}}\left(a^2 + b^2 - c^2\right)$

I *(incentre):* $\quad x_I = \dfrac{c + b - a}{2}$ $y_I = \dfrac{2\mathcal{A}}{a + b + c}$

G *(barycentre):* $\quad x_G = \dfrac{b^2 + 3c^2 - a^2}{6c}$ $y_G = \dfrac{2\mathcal{A}}{3c}$

H *(orthocentre):* $\quad x_H = \dfrac{b^2 + c^2 - a^2}{2c}$ $y_H = \dfrac{(b^2 + c^2 - a^2)(a^2 + c^2 - b^2)}{8\mathcal{A}c}$

where $\mathcal{A}$ is the area of the triangle, which can be computed by Heron's formula (6.22) on page 171.

Proof. Orthocentre H: We use the notation of Fig. 4.9. We obtain the formula for $x_H = AF = b\cos\alpha$ with the law of cosines (5.10). We have a similar formula for BD and obtain from the area formula (1.6) that

---

[16] "... nous appellerons *pseudospheriques* les surfaces de courbure constante négative [... we call *pseudospheres* the surfaces of constant negative curvature]", Beltrami, French transl., p. 259.

$$AD = \frac{2\mathcal{A}}{a}.\tag{7.49}$$

Finally, Thales' theorem gives $y_H = FH = \frac{AF \cdot BD}{AD}$, which leads to the stated expression.

Barycentre G: As for (7.49) we compute the coordinates of C as $x_C = x_H$ and $y_C = \frac{2\mathcal{A}}{c}$. The coordinates of B are $(c, 0)$ and those of A are $(0, 0)$. The coordinates of G are the corresponding arithmetic means.

Circumcentre O: We recall the two important formulas (5.16) and (5.18) for the radii of the circumcircle and the incircle,

$$R = \frac{abc}{4\mathcal{A}}, \qquad \rho = \frac{2\mathcal{A}}{a+b+c}.\tag{7.50}$$

Inspired by Fig. 5.8 (right), we then have $x_O = \frac{c}{2}$ and $y_O = R\cos\gamma$. The stated result follows by inserting R and using the cosine rule.

Incentre I: The value $x_I = s - a$ is given in Fig. 4.6, and $y_I = \rho$.   □

Distances. Once Euler had the coordinates of these points, he was able to compute their distances in a straightforward way by Pythagoras' theorem, as in (7.1). For example, the distance between I and O is obtained as

$$IO^2 = (x_I - x_O)^2 + (y_I - y_O)^2 = \left(\frac{b-a}{2}\right)^2 + \left(\frac{2\mathcal{A}}{a+b+c} - \frac{c}{8\mathcal{A}}(a^2 + b^2 - c^2)\right)^2.$$

This must now be multiplied out and simplified with considerable computational skill. In all these calculations, $\mathcal{A}$ appears only as $\mathcal{A}^2$, therefore the square roots don't appear and all the algebra we need involves rational expressions in a, b, c. Without the use of a modern computer algebra system, Euler arrived, in §19, after six pages of "facilis" calculations, at the following results, which we adapt to our notation.

Theorem 7.21. *In the triangle ABC of Theorem 7.20 we have the distances*

$$IO^2 = R^2 - 2R\rho,$$
$$IH^2 = 4R^2 - \tfrac{1}{4}P + 4R\rho + 3\rho^2,\tag{7.51}$$
$$HO^2 = 9R^2 - \tfrac{1}{2}P + 8R\rho + 2\rho^2,$$

where $P = (a + b + c)^2$, and also $HG = \frac{2}{3}HO$ and $GO = \frac{1}{3}HO$.

The last two results led Euler to the conclusion that H, G, O are collinear.

We shall give here two other consequences of the above formulas. Firstly, we observe that the first formula of (7.51), which was found independently by W. Chapple (1746), depends only on R and ρ, but is independent of the individual values of a, b, c. This has as interesting consequence the following porism, first noticed by Lhuilier (1810/11), which is a special case of Poncelet's closure theorem (1822), see Theorem 11.7.

Theorem 7.22. *Let two circles $\mathcal{C}_1$, $\mathcal{C}_2$ be given, $\mathcal{C}_2$ inside $\mathcal{C}_1$, of radius R and ρ respectively. Let the distance d of their centres satisfy $d^2 = R^2 - 2R\rho$. Then, if we construct, starting from any point A on $\mathcal{C}_1$, the points B, C and D on $\mathcal{C}_1$ by drawing successive tangents to $\mathcal{C}_2$, we will always have $D = A$.*

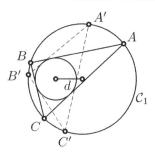

Our next application is a discovery of Feuerbach (1822), which we have already noted in Fig. 4.13:

Theorem 7.23. *The incircle and the Feuerbach circle (nine-point circle) of a triangle are tangent to each other.*

Proof. The nine-point centre N of the Feuerbach circle is the midpoint of the segment HO (see Fig. 4.13) and its radius is $R/2$. We compute the distance IN by using Pappus' formula (4.5) on page 90, which gives $IN^2 = \frac{1}{2}IO^2 + \frac{1}{2}IH^2 - \frac{1}{4}HO^2$. By inserting the values of (7.51) and multiplying out, we obtain $IN = \frac{R}{2} - \rho$. This proves the result.   □

Remark. In fact Feuerbach proved much more: *The nine-point circle is also tangent to all three excircles.*

The Nagel line. Euler's formulas of Theorem 7.20 provide us with an easy proof of an interesting discovery of Nagel (see Baptist, 1992, p. 77 ff.). We recall that the Nagel point X_8 is the intersection of the lines joining the vertices to the points of contact of the excircles with the opposite sides (see Thm. 4.12). Let F_c on AB be one of these points and F the midpoint of AB. We compute the slope of the line connecting the incentre I with F and that of the line connecting C with F_c:

$$\frac{y_I}{x_I - \frac{c}{2}} = \frac{y_C}{x_C - x_{F_c}} = \frac{4\mathcal{A}}{(a+b+c)(b-a)}.$$

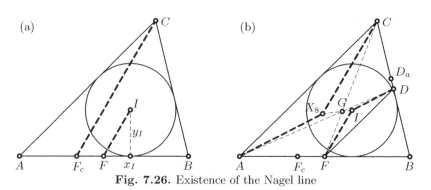

Fig. 7.26. Existence of the Nagel line

Here we have used $x_{F_c} = s - b = \frac{c-b+a}{2}$ from (4.9), inserted all other values from Theorem 7.20 and simplified the formulas. We see that *these lines are parallel* (see Fig. 7.26 (a)). If we now draw two of these pairs of lines (see Fig. 7.26 (b)), we see that the triangles AX_8C and DIF are similar, with ratio $2:1$, and with the centroid G as centre of similarity. We have thus proved:

Theorem 7.24 (Nagel 1836). *The Nagel point X_8, the barycentre G and the incentre I are collinear; G divides the segment X_8I in the ratio $2:1$.*

7.8 The Simson Line and Sturm's Circles

The *Simson Line* was discovered by William Wallace in 1797 (see Coxeter and Greitzer, 1967, p. 41). It reappeared in an article on "Practical Geometry" by F.-J. Servois (1813/14), who wrongly attributed it to Robert Simson (1687–1768) ("le théorème suivant, qui est, je crois, de Simson"). Servois was one of the many brilliant mathematicians of that period enrolled in the French military. The "practical" problem was to locate the position of a straight line (perhaps the trajectory of a cannon ball) behind an inaccessible obstacle.

Theorem 7.25. *Let P be a point on the circumcircle of a triangle ABC, and let D, E, F be the feet of the perpendiculars from P to the (possibly produced) sides of the triangle (see Fig. 7.27 (a)). Then these three points are collinear. The line through these three points is called the Simson line of P with respect to ABC. Furthermore, if one of these perpendiculars, say PF (perpendicular to AB), is produced to meet the circumcircle in the point R, then the Simson line of P is parallel to CR.*

Proof. Draw circles with diameters PA, PB and PC; two of the right angles at D, E and F will lie on each of these circles (see Fig. 7.27 (b)). The two angles δ at P are equal by Eucl. III.22, since they are opposite to the same

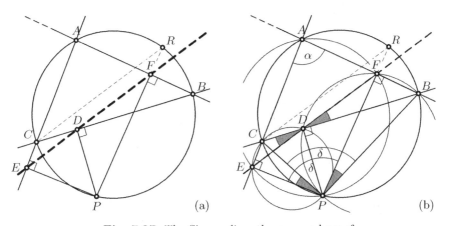

Fig. 7.27. The Simson line: theorem and proof

angle α at A, once in the cyclic quadrilateral $PEAF$, and once in $PCAB$. By subtracting the angle FPC, the two grey angles at P are also equal. Applying Eucl. III.21 twice, the two grey angles appear at D (and are equal there as well). Since CDB is a straight line by assumption, EDF is also a straight line (by the converse of Eucl. I.15).[17]

Finally, by Eucl. III.20, the grey angle BPR reappears a fifth time as BCR at C. Hence, by Eucl. I.27, the line CR is parallel to the Simson line.   □

Theorem 7.26 (J. Steiner, 1857). *If the point P describes the circumcircle of ABC, the corresponding Simson lines create a Steiner deltoid, which is tangent to the nine-point circle. The points of contact trisect those arcs of this circle which are outside the triangle (see Fig. 7.28).*

Remark. The directions of the Steiner deltoid of ABC are related to those of Morley's triangle of the same triangle.

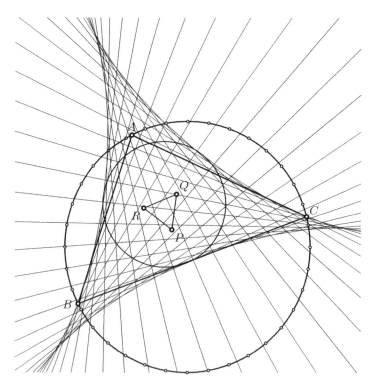

Fig. 7.28. The Simson lines of a point P rotating on the circumcircle together with Morley's triangle of ABC

---

[17]This is precisely the elegant proof of Servois. A simpler proof for this first part of the theorem is as follows (private communication of John Steinig): Show by three applications of Eucl. III.21 that angle PED = angle PCD = angle PCB = angle PAB = angle PAF = angle PEF.

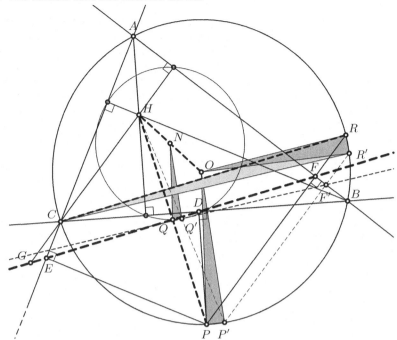

Fig. 7.29. Proof of Steiner's theorem about the Simson line and the deltoid

Proof. Steiner's original paper consists of 7 pages, nearly all in boldface, of detailed descriptions of this curve, without any hint of a proof. Only a few words let us guess what he intended. The nine-point circle and the circumcircle are in similarity position with ratio $1:2$ and with the orthocentre H as similarity centre (see Fig. 4.13). In particular, the nine-point circle bisects the segments HA, HB and HC.

The crucial fact is that the midpoint Q between H and P, which by similarity lies on the nine-point circle, also lies on the Simson line. This is true in the case where P is one of the vertices A, B or C of the triangle, because then the Simson line coincides with one of the altitudes (the property is also true when P is a point *opposite* one of the vertices, when the Simson line coincides with one of the *sides* of the triangle). For an arbitrary point P, we choose any of the three directions of the altitudes, for example HC and its parallel PF. Then what we wish to prove is that $PF = HG$ (see Fig. 7.29 and Exercise 10 of Chap. 2, page 55). By the last assertion of Theorem 7.25, $FR = GC$. We now move P to P'. Then $P'F'$ is parallel to PF, because both are orthogonal to AB. Consequently, $R'F'$ decreases by the same amount as $P'F'$. Since CH remains fixed, $P'F'$ is again equal to $HG' = HC + F'R'$.

The result now follows very quickly: the angle POP' is equal to QNQ' and also to ROR' (but oriented inversely). Thus, by Eucl. III.20, the inscribed angle RCR' is half the size of $QS'Q'$. This means that when Q rotates around the centre N of the nine-point circle, the Simson line, which is attached to Q

and always parallel to CR, *rotates with half the angular speed in the opposite direction*. This is precisely the property defining the deltoid.   □

Remark. Because of the succinct presentation by Steiner, many other proofs have been published in the meantime. We refer to M. de Guzmán (2001) for a particularly elementary, but not short proof and for additional references.

Steiner's ortholine of the complete quadrilateral. The above results provide an elegant proof of another surprising discovery of Steiner.

Theorem 7.27 (J. Steiner, 1827/1828). *The orthocentres of the four triangles of a complete quadrilateral are collinear (see Fig. 7.30).*

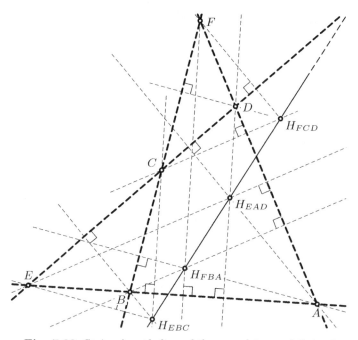

Fig. 7.30. Steiner's ortholine of the complete quadrilateral

Proof. We know from Theorem 4.15 that the circumcircles of the four triangles have a common point M (see Fig. 4.17). This point, lying on all four circumcircles, possesses a Simson line for each of these triangles. But any two of these lines have two points in common, hence all these Simson lines coincide. From the proof of Theorem 7.26 we know that the midpoints between M and the four orthocentres all lie on this line. Hence by Thales, the four orthocentres are also collinear, on a line parallel to the first line and located twice as far from M.   □

Remark. For an excellent account of this important paper by Steiner, with a translation and complete proofs, we refer to Ehrmann (2004).

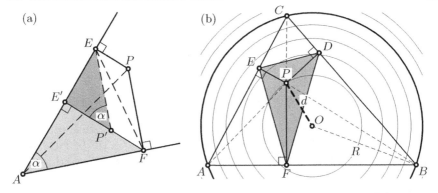

Fig. 7.31. Proof of Steiner's formula (a); Sturm's theorem (b)

Sturm's circles. Charles-François Sturm (1803–1855), before becoming famous for Sturm sequences in algebra and Sturm–Liouville theory in analysis, wrote very neat articles on elementary geometry in the Gergonne Journal. The following result of Sturm (1823/24) was inspired by Servois' paper on the Simson line.

Let ABC be a triangle and P a point arbitrarily chosen inside or outside of ABC. We draw the perpendiculars PD, PE and PF to the (possibly extended) sides of ABC and call DEF the *pedal triangle* of P. We want to find its area (see Fig. 7.31 (b)).

Theorem 7.28. *The area $\mathcal{A}'$ of the pedal triangle DEF remains constant if P moves on circles concentric to the circumcircle of ABC. More precisely, this area is given by*

$$\mathcal{A}' = \frac{\mathcal{A}}{4}\left(1 - \frac{d^2}{R^2}\right), \tag{7.52}$$

where R is the radius of the circumcircle, d the distance PO and $\mathcal{A}$ the area of ABC.

Proof. Sturm's original proof was simplified (and generalised to n-gons) by Steiner (1826b) as follows: consider only one vertex A with angle α and perpendiculars PF and PE, see Fig. 7.31 (a). Then

$$\square - 2\triangle = PA^2 \frac{\sin 2\alpha}{4}, \tag{7.53}$$

where $\square$ is the area of the quadrilateral $AFPE$ and $\triangle$ the area of the triangle FPE. We show Steiner's formula by extending the triangle FPE to the parallelogram $FPEP'$ of area $2\triangle$. Then $\square - 2\triangle$ is the area of the two grey triangles AFE' and $P'EE'$, whose areas are $AF^2 \frac{\sin\alpha\cos\alpha}{2}$ and $FP^2 \frac{\sin\alpha\cos\alpha}{2}$, respectively. Formula (7.53) now follows from Pythagoras and (5.8). Adding up (7.53) for all three vertices gives

$$\mathcal{A} - 2\mathcal{A}' = PA^2 \frac{\sin 2\alpha}{4} + PB^2 \frac{\sin 2\beta}{4} + PC^2 \frac{\sin 2\gamma}{4}. \tag{7.54}$$

We now consider the triangle ABC together with its angles and area as fixed and the point $P = (x, y)$ together with the area $\mathcal{A}'$ as variable. Inserting $PA^2 = (x - x_a)^2 + (y - y_a)^2$ and similar expressions for PB^2 and PC^2, we obtain after simplification

$$\mathcal{A}' = -T(x^2 + y^2) + 2x_o x + 2y_o y + U = -T((x - x_o)^2 + (y - y_o)^2) + V,$$

where T, x_o, y_o, U and V are constants. The level curves are thus concentric circles with centre (x_o, y_o). If P moves to one of the vertices, the pedal triangle shrinks to a segment with area 0. We conclude that the level curve through a vertex must be the circumcircle and $(x_o, y_o) = O$, the circumcentre. Thus we obtain

$$\mathcal{A}' = V - Td^2. \tag{7.55}$$

We finally set $d = 0$, where DEF becomes the triangle of the medial reduction of Fig. 4.8 (b) whose area is one quarter of $\mathcal{A}$. This, together with $\mathcal{A}' = 0$ for $d = R$, finally leads to formula (7.52).   □

Remarks. (a) This also gives a second, analytic proof for the existence of the Simson line.

(b) Soon after his discovery, Sturm sent an "Addition" to the Gergonne Journal. By placing the point P at the incentre I, the parallelogram $FPEP'$ becomes a rhomboid and $\mathcal{A}' = \frac{\rho^2}{2}(\sin\alpha + \sin\beta + \sin\gamma)$. This, together with (7.52) and (5.19), leads to $2R\rho = R^2 - d^2$, an elegant proof of the Chapple–Euler formula (7.51) for the distance between the incentre and the circumcentre.

7.9 The Erdős–Mordell Inequality and the Steiner–Lehmus Theorem

 nce upon a time, mathematicians like P. Erdős would submit problems in elementary geometry to the *American Mathematical Monthly* and mathematicians like L.J. Mordell would propose their solutions. Problem 3740 then became the following famous theorem. A sharper result was contributed as a second solution by David F. Barrow; it implies Erdős' inequality and we prove it below. Many modifications and generalisations of these theorems were published later (see e.g. Oppenheim (1961) and Satnoianu (2003)).

Theorem 7.29 (Erdős–Mordell, 1937). *Let P be any interior point of a triangle ABC, let p, q, r be the distances of P from the vertices A, B, C and x, y, z the distances of P from the three sides (see Fig. 7.32 (a)). Then*

$$2(x + y + z) \leq p + q + r. \tag{7.56}$$

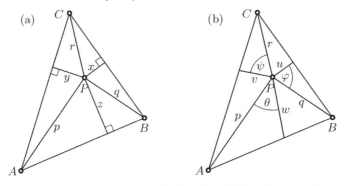

Fig. 7.32. Inequality of Erdős–Mordell (a); Barrow (b)

Theorem 7.30 (Barrow, 1937). *Let* p, q, r *be the distances of* P *from the vertices of a triangle* ABC, *and let* u, v, w *be the lengths of the bisectors of the angles* BPC, CPA *and* APB, *respectively (see Fig. 7.32 (b)). Then*

$$2(u + v + w) \leq p + q + r. \tag{7.57}$$

Proof (of Theorem 7.30). If θ denotes the angle between the segments p, w and between w, q, we have from formula (5.21) on page 121

$$w = \frac{2}{\frac{1}{p} + \frac{1}{q}} \cdot \cos\theta \leq \sqrt{pq} \cdot \cos\theta. \tag{7.58}$$

The last inequality is part of the familiar inequality $h \leq g \leq a$ between h the *harmonic*, g the *geometric* and a the *arithmetic mean* of two numbers $p, q > 0$

$$h = \frac{2}{\frac{1}{p} + \frac{1}{q}}, \quad g = \sqrt{pq}, \quad a = \frac{p + q}{2}. \tag{7.59}$$

The inequality between the means is visible in the picture to the right, which goes back to Pappus (*Collection*, Book III, Chap. XI). We draw a semicircle above the segment of length $p + q$. Since its diameter is $p + q$, the radius is a. The altitude is g by Eucl. II.14 (Fig. 2.14, right, page 39), and since $ah = g^2$ by Thales' theorem applied to the two right-angled triangles, one inside the other, we have the stated value for h.

Our aim is to prove that $p + q + r - 2u - 2v - 2w \geq 0$. If we use (7.58), and similar estimates for u and v, it suffices to show that

$$p + q + r - 2\sqrt{rq}\cos\varphi - 2\sqrt{rp}\cos\psi - 2\sqrt{pq}\cos\theta \geq 0. \tag{7.60}$$

We have to use somewhere that $\varphi + \psi + \theta = \pi$. We do this by writing $\cos\varphi = -\cos(\pi - \varphi) = -\cos(\psi + \theta) = -\cos\psi\cos\theta + \sin\psi\sin\theta$. Inserting this into the left side of (7.60) and simplifying, we get

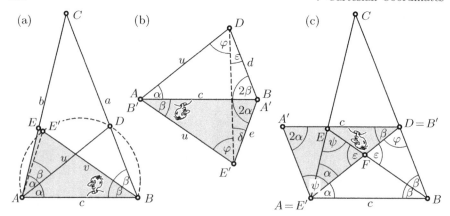

Fig. 7.33. Steiner–Lehmus theorem; Thébault's proof (a); Steiner's proof (b); Hesse's proof (c)

$$(\sqrt{p} - \sqrt{q}\cos\theta - \sqrt{r}\cos\psi)^2 + (\sqrt{q}\sin\theta - \sqrt{r}\sin\psi)^2 \qquad (7.61)$$

which, being the sum of two squares, is indeed nonnegative. □

Remarks. (a) If you find this proof tricky, then you are right; the original proofs of Mordell and Barrow are much longer. The above proof was found by simplifying Lu's proof (Lu, 2008).

(b) Another proof of the "harmonic-geometric" inequality can be obtained from the hyperbola (see Apoll. III.43 in Fig. 10.18 on page 314): the intersection point of the tangents to the hyperbola, for which the product pq is constant, with the angle bisector, has greatest distance from the origin P if $p = q = \sqrt{pq}$ are equal. For still another proof see Exercise 6 of Chap. 11 on page 343.

The Steiner–Lehmus theorem. In 1840 Lehmus asked for an elementary proof of the following result:

Theorem 7.31 (Steiner–Lehmus). *If two angle bisectors of a triangle have the same length, then the triangle is isosceles.*

The fact that such a simple geometric result, which could have been a proposition of Euclid's Book I, was apparently so difficult to prove by geometric means challenged mathematicians for many decades. Steiner, who had constructed such a proof, did not find it worth publishing. Many other proofs, which appeared in the meantime, finally convinced him to publish the paper Steiner (1844), where he declared his proof to be the simplest one.

Algebraic proof. Stewart's theorem (4.8) on page 91 gives us a quick algebraic proof of this theorem. If the bisectors of the angles α and β are of the same length, we obtain by applying (4.8) to both of these

$$a : b = \left(1 - \left(\frac{a}{b+c}\right)^2\right) : \left(1 - \left(\frac{b}{a+c}\right)^2\right).$$

This algebraic equation only allows the solution $a = b$. If $a > b$, say, the left hand side would be greater than 1 and the right hand side less than 1. □

Thébault's proof. The proof by Victor Thébault (1930), see also Coxeter and Greitzer (1967), uses Euclid's Book III. In this and the subsequent proofs we denote the angles BAC and CBA by 2α and 2β respectively. We suppose that $\alpha > \beta$. We first draw the circle through ADB (see Fig. 7.33 (a)) and define E' to be the intersection of this circle with EB. Then, the angle $E'AD$ is equal to $E'BD$, i.e. to β. Because of $\beta < \alpha$ this point lies between E and B and, by Eucl. III.20, its position on the circle ADB is such that the arc AE' is equal to the arc $E'D$ and shorter than the arc DB. Hence the arc AD is shorter than the arc $E'B$. Thus, by Eucl. III.20 and Eucl. I.18, $E'B > AD$, a fortiori $EB > AD$ or $v > u$. □

Steiner's proof. This proof assumes that $u = v$ and $\alpha > \beta$ and only uses results of Euclid's Book I. Then "zur bequemeren Übersicht" (for a more convenient visualisation) we attach the triangle AEB in a reversed position as $A'E'B'$ along the common side c of equal length to the triangle ABD (see Fig. 7.33 (b)). By Eucl. I.5 the two angles φ are equal, $e < d$ by Eucl. I.24 (which we have not discussed in detail) and $\varepsilon < \delta$ by Eucl. I.18. Now we write Eucl. I.32 for the two triangles:

$$\alpha + \varphi + \varepsilon + 2\beta = \beta + \varphi + \delta + 2\alpha \quad \Rightarrow \quad \varepsilon + \beta = \delta + \alpha,$$

which contradicts the above inequalities. □

Hesse's proof. This proof, attributed by Dörrie (1943, §43) to O. Hesse, appears to be more direct. This time we attach the triangle AEB as $A'E'B'$ along the angle bisector of equal length u to ABD (see Fig. 7.33 (c)). We denote the angle ADB by φ and the angle AEB, which is equal to $A'E'B'$, by ψ. By Eucl. I.15 the two angles ε at F are equal. From Eucl. I.32 applied to both triangles AFE and BFD, we conclude that $\alpha + \varphi = \beta + \psi$. Hence the quadrilateral $ABB'A'$, which has two opposite sides of the same length c, has also two opposite equal angles. Consequently, it is a parallelogram and we conclude that the second pair of opposite angles must be equal, i.e. $2\alpha = 2\beta$ or $\alpha = \beta$. □

7.10 The Butterfly

According to Coxeter and Greitzer (1967, p. 45), this theorem of geometry "has been around for quite a while ...". The formula for the length of a cevian, given in Sect. 5.3, provides a nice trigonometric proof.

Theorem 7.32. *If, in the configuration of Fig. 7.34, FF' is perpendicular to CO, then C bisects FF'.*

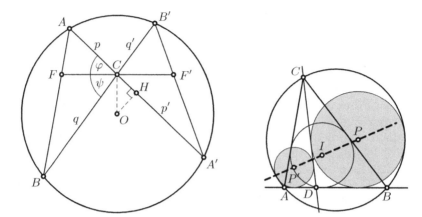

Fig. 7.34. Left: the butterfly; right: Thébault's theorem

Proof. We denote the angles ACF and FCB by φ and ψ, respectively. We use formula (5.20) on page 121 for $w = CF$ and $w' = CF'$, and subtract, to get

$$w - w' = \frac{\sin(\varphi + \psi)}{\frac{\sin\psi}{p} + \frac{\sin\varphi}{q}} - \frac{\sin(\varphi + \psi)}{\frac{\sin\psi}{p'} + \frac{\sin\varphi}{q'}}. \tag{7.62}$$

We factor out $\sin(\varphi + \psi)$ and expand the remaining difference of fractions. This leads to a fraction with numerator

$$\sin\psi\, \frac{p - p'}{pp'} + \sin\varphi\, \frac{q - q'}{qq'}.$$

This expression is zero because $pp' = qq'$ (Eucl. III.35) and $p' - p = 2\,CH = 2\,CO\sin\varphi$ (see Fig. 7.34; H bisects AA' and the angle HOC is the angle ACF) and $q - q' = 2\,CO\sin\psi$.   □

Remark. A particularly elementary proof, using only Eucl. III.21, Eucl. III.35 and Thales, is given in Coxeter and Greitzer (1967, p. 46). Another proof is possible by applying property (6.32c) of the Möbius transformation to the map $A \mapsto B \mapsto B' \mapsto A' \mapsto A$. If we choose the Cartesian coordinates (d, w), $(d, 0)$ and $(d, -w)$ for the points F, C and F' respectively, then according to (6.39) on page 182, we have to multiply the matrices

$$\begin{bmatrix} w & d-1 \\ d+1 & -w \end{bmatrix} \begin{bmatrix} 0 & d-1 \\ d+1 & 0 \end{bmatrix} \begin{bmatrix} -w & d-1 \\ d+1 & w \end{bmatrix} \begin{bmatrix} 0 & d-1 \\ d+1 & 0 \end{bmatrix} = Const \cdot \begin{bmatrix} 1 & 0 \\ 0 & 1 \end{bmatrix}$$

(start by multiplying the first two and the last two matrices) so that this map returns to A for any initial point.

7.11 Thébault's Theorem

> "This computer proof took some 44 hours of CPU on a symbolic 3600 machine (...). The theorem almost has the status of a benchmark problem in Groebner theory."   (R. Shail, 2001)

Thébault's theorem, conjectured in 1938 (see Thébault, 1938), has the remarkable property that it required three decades to obtain a first proof, which needed 24 pages of calculations. Shorter, but not easier proofs were later published, mainly in Elemente der Mathematik (see Stärk, 1989 and Turnwald, 1986) and in Dutch journals. We refer to Ayme (2003) and Kulanin and Faynshteyn (2007) for complete accounts of these and other proofs.

Theorem 7.33. *Let ABC be a triangle with incentre I and let D be an arbitrary point on the line AB. Then the centres P and P′ of the circles which touch (a) the line CD, (b) the line AB and (c) the circumcircle of ABC are collinear with I (see Fig. 7.34, right).*

The proof presented below (due to Ostermann and Wanner, 2010) was difficult to find (see the remarks above) but once found it is, we hope, a pleasure to follow. It is based on the following "machine":

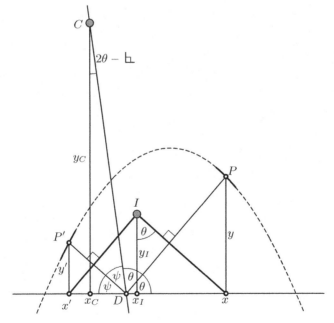

Fig. 7.35. The TTT-machine

Definition 7.34 (The TTT-machine).[18] Suppose we have, in the plane, a fixed x-axis, and two fixed points C and I such that $y_C > 2y_I$. We let a point D move on this axis, and cause the line CD to rotate around C. Our machine should produce the two (mutually orthogonal) bisectors of the angles between CD and the x-axis. The point I is projected onto the x-axis, orthogonally to these bisectors, giving the points x and x'. The points P and P' then lie above these points, on the respective angle bisectors.

For more inspiration, you can construct such a machine from rusty pieces of iron and let it rotate noisily, or, more quietly, write a Java applet (Bernard Gisin, `www.juggling.ch/gisin/geogebra/Thebault_theorem.html`).

Lemma 7.35. *When DC rotates around C, the points P and P' of the TTT-machine describe the same parabola, which has a vertical axis and is open downwards. The points $P'IP$ are always collinear and $p \cdot P'I = \frac{1}{p} \cdot IP$, where $p = \tan\theta$.*

Proof (of the lemma). We denote the coordinates of I by (x_I, y_I), those of C by (x_C, y_C) and take the slope of DP, $p = \tan\theta$, as parameter. The angle θ at D reappears three more times as an orthogonal or parallel angle to produce the four similar white triangles of Fig. 7.36 (a). If we denote by a and b the sides of the rectangle with diagonal DI and apply Thales' theorem to these four triangles, we obtain the lengths indicated in that figure, and see that the two grey triangles are similar with similarity factor p^2. Since PI and IP' are parallel, this proves the second statement.

The same angle θ also reappears as orthogonal angle at I (see Fig. 7.35). We thus have by Thales' theorem

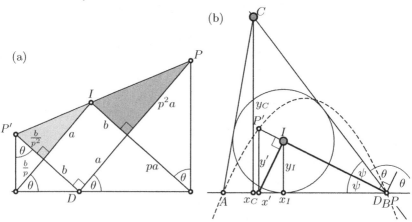

Fig. 7.36. Why $P'IP$ are collinear (a); why I is the incentre (b)

---

[18] The first T stands for Thébault, the second for Turnwald, whose corrected version of a formula of Thébault was our main motivation, the third T stands for Jean Tinguely and emphasises the dynamic thinking of our proof.

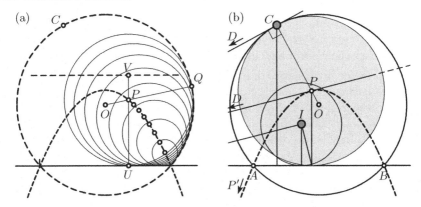

Fig. 7.37. The enveloping circle (a) and the circumcircle (b)

$$\text{(I)} \qquad x = x_I + p\,y_I\,, \qquad y = p\,(x - x_D)\,,$$

$$\text{(II)} \qquad x' = x_I - \frac{1}{p}\,y_I\,, \qquad y' = -\frac{1}{p}\,(x' - x_D)\,.$$

We compute the abscissa of D from the right-angled triangle below CD as

$$x_D = x_C + y_C \tan(2\theta - \llcorner) = x_C + y_C\,\frac{p^2 - 1}{2p} = x_C + \frac{y_C}{2}\left(p - \frac{1}{p}\right)$$

(because $\tan 2\theta = \frac{2p}{1-p^2}$; see the third formula of (5.6)). We insert the formula for x_D into (I) to obtain

$$y = px - px_D = px - px_C - \frac{y_C}{2}(p^2 - 1)\,.$$

Elimination of p with $p = \frac{x-x_I}{y_I}$ then yields an expression for y which is quadratic in x; as we know by (3.2), this represents a parabola. Exchanging p and $-\frac{1}{p}$ leaves x_D invariant and replaces equations (I) by equations (II). Therefore, we obtain the same quadratic expression in x' for y' as in x for y, and the same parabola. Since $y_C > 2y_I$, our parabola is open downwards. $\square$

Proof (of the theorem). We have to show that the points I, P and P' of the lemma are the points with the properties required in the theorem. We do this in three steps by cleverly running our machine.

First step. Denote by A and B the intersection points of the parabola with the x-axis (see Fig. 7.36 (b)). We first let the point D move towards B. We see that the point P', and hence also the point I, lie on the bisector of the angle ABC. Then moving D to A, we arrive at the conclusion that *the points A and B have the property that I is the incentre of the triangle ABC.*

Second step. We now consider the family of circles centred at points P on the parabola and tangent to the x-axis, i.e. of radius $PQ = PU = y$ (see Fig. 7.37 (a)). The enveloping curve of these circles is a circle centred at O, the focus of the parabola. This is the converse of a result that has been known since Pappus (see the explanations for Exercise 3 of Chap. 3). Each of these circles touches the enveloping circle at the point Q, which is on OP produced.

Third step. Since, by construction, the points P and P' lie on the bisectors of the angles between CD and the x-axis, the circles centred at P and P' and tangent to the x-axis are also tangent to CD. We now run our machine until the tangent CD, which rotates around C, becomes orthogonal to CO (see Fig. 7.37 (b)). In this case, the circle centred at P can only touch the line CD at the point $Q = C$ and the enveloping circle must therefore pass through C. Since it also passes through A and B (here all distances are equal to 0), we conclude that *this circle is identical with the circumcircle of triangle* ABC. $\square$

Remark. For nice particular cases of this proof see Exercises 27 and 28 below.

7.12 Billiards in an Ellipse

Consider a billiard table of elliptical form (see Fig. 7.38). We know from Sect. 3.2 that a ball passing through a focus is reflected into the other focus. Given the angle φ_1 under which the ball leaves F (or F'), we wish to find the angle φ_2 under which it arrives at F' (or F).

In order to solve this problem, we may assume that the foci of the ellipse have abscissas -1 and 1. The semi-major axis is then $a = \frac{1}{e}$, where e denotes the eccentricity. Then formulas (7.8) become

$$\ell_{1,2} = \frac{1}{e} \pm ex\,,$$

where x is the abscissa of P. The idea is now to set $c_i = \cos\varphi_i$. By definition of the cosine, we have

$$c_1 = \frac{x+1}{ex + \frac{1}{e}}\,, \qquad c_2 = \frac{x-1}{-ex + \frac{1}{e}}\,,$$

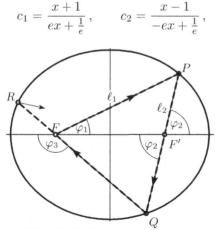

Fig. 7.38. Billiards in an ellipse

which are Möbius transformations. To express c_2 as a function of c_1, we invert the first Möbius transformation and insert it into the second one. Using matrices, we obtain

$$\begin{bmatrix} 1 & -1 \\ -e & \frac{1}{e} \end{bmatrix} \begin{bmatrix} 1 & 1 \\ e & \frac{1}{e} \end{bmatrix}^{-1} = \frac{1+e^2}{1-e^2} \cdot \begin{bmatrix} 1 & -\theta \\ -\theta & 1 \end{bmatrix}, \qquad \theta = \frac{2e}{e^2+1},$$

which results, by ignoring the constant factor, in

$$c_2 = \frac{c_1 - \theta}{-\theta\, c_1 + 1} \qquad \text{with matrix} \qquad A = \begin{bmatrix} 1 & -\theta \\ -\theta & 1 \end{bmatrix}. \tag{7.63}$$

The following angles φ_3, φ_4, etc. are determined by the *powers* of A. This matrix has eigenvectors $[1,1]^\mathsf{T}$ and $[-1,1]^\mathsf{T}$ with eigenvalues $1 \mp \theta$. In non trivial situations (i.e. when the ellipse is not a circle and $\varphi_1 \neq 0$) the vector $[c_n, 1]^\mathsf{T}$ will converge to the eigenvector with maximal eigenvalue, i.e. $c_n \to -1$ for $n \to \infty$, and the angles φ_n converge to π.

7.13 Urquhart's 'Most Elementary Theorem of Euclidean Geometry'

> "Urquhart considered this to be the 'most elementary theorem', since it involves only the concepts of straight line and distance. The proof of this theorem by purely geometrical methods is not elementary. Urquhart discovered this result when considering some of the fundamental concepts of the theory of special relativity."
>
> (D. Elliott, *J. Australian Math. Soc.* 1968, p. 129)

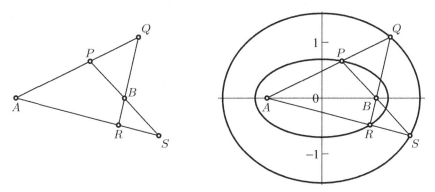

Fig. 7.39. Urquhart's theorem

M.L. Urquhart (1902–1966) was a highly appreciated lecturer in mathematics and physics at several Australian universities; he communicated his mathematical discoveries only to some of his friends. The following theorem became known by his obituary notice through Elliott (1968) and gained wider popularity through Tabachnikov's book (1995)[19] on billiards.

Theorem 7.36 (Urquhart). *Let the points A, B, P, Q, R, S be disposed as in Fig. 7.39 (left). Then,*

$$AP + PB = BR + RA \qquad implies \qquad AQ + QB = BS + SA. \qquad (7.64)$$

Proof. We give a *backwards* proof, i.e. we suppose that *both* relations in (7.64) are satisfied and conclude that S, R, A are aligned. These relations mean that the points P, R and Q, S lie on two confocal ellipses with foci A and B (Fig. 7.39, right). The "billiards" of these ellipses are determined by formula (7.63) with two different constants θ_1 and θ_2, originating from the two eccentricities of these ellipses. Nevertheless, the trajectories

$$A \mapsto P \mapsto B \mapsto S \mapsto A \qquad \text{and} \qquad A \mapsto Q \mapsto B \mapsto R \mapsto A$$

return to A under the *same angle* φ_3, because the matrices

$$\begin{bmatrix} 1 & -\theta_1 \\ -\theta_1 & 1 \end{bmatrix} \qquad \text{and} \qquad \begin{bmatrix} 1 & -\theta_2 \\ -\theta_2 & 1 \end{bmatrix}$$

commute.   □

7.14 Exercises

1. Give another analytical proof of the property of the altitudes (Theorem 4.2 of Chap. 4) and of the Euler line (Theorem 4.10 of Chap. 4) by placing the side AB on the x-axis and the vertex C on the y-axis. Denote the coordinates of A by $(-a, 0)$, those of B by $(b, 0)$, and those of C by $(0, c)$.

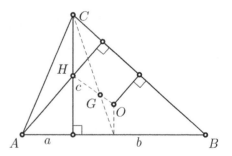

2. (*The three-circle problem of Apollonius*). Find circles which are tangent to three given circles (see Fig. 7.40).

Remark. This is one of the famous classical problems, attributed by Pappus (*Collection*, introduction to Book VII) to Apollonius' lost books *On Contacts*. It fascinated the mathematicians for centuries. At the end of his *Responsum* (1593b), after having brilliantly solved Adriaan van Roomen's

---

[19]The authors are grateful to Pierre de la Harpe for drawing their attention to Urquhart's theorem and to Tabachnikov's book.

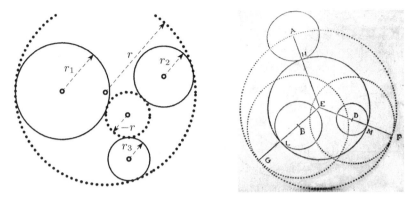

Fig. 7.40. The three-circle problem of Apollonius; right: Viète's illustration from van Schooten's edition 1646

problem (see Section 6.5), Viète challenged van Roomen with this problem of Apollonius. *Adrianus Romanus* fell into the trap and presented in 1596 a solution using the hyperbolas of Exercise 4 of Chap. 3. Viète (1600) objected that this was not a solution with ruler and compass and would have been refused by Euclid and all his school ("vero nemo pronunciabit Geometra. Reclamaret Euclides, & tota Euclideorum schola"). Viète then gave a geometrical solution which relied on Thales' theorem (Viète's picture of Fig. 1.5 is part of this solution). This led to the first work of J. Steiner (1826a), where this problem and similar ones are solved geometrically in great detail (Aufg. II, p. 175). Euler (E648, 1790) and Gauss (*Werke*, vol. 4, p. 400, remark VI) gave solutions with the help of identities for trigonometric functions. Find an easy algebraic version.

3. (An identity of Euler, 1750, §30) Prove that for any convex quadrilateral in the plane (or in space),

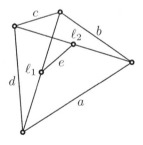

$$a^2 + b^2 + c^2 + d^2 = \ell_1^2 + \ell_2^2 + 4e^2 \,,$$

where ℓ_1 and ℓ_2 are the lengths of the diagonals and e is the distance of the midpoints of the diagonals (see the picture).

4. Give an analytic proof of Proclus' construction of the ellipse "with a stick" in Fig. 3.8 on page 68 (a) by Thales' and Pythagoras' theorems, (b) by a trigonometric calculation (see Fig. 7.41). Also verify the "trammel construction" (c), which I. Newton discovered at the very beginning of his mathematical studies, where the point P lies outside the interval GF whose endpoints glide along the axes.

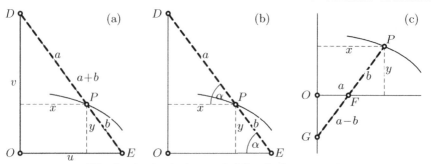

Fig. 7.41. Analytic proof of Proclus' construction

5. (*Regiomontanus' maximum problem in the version attributed to Napoleon Bonaparte.*) A statue of height h of an emperor stands on a socle of height s above eye-level. At which distance d from the socle is the emperor seen under the greatest angle? Also find a property of the bisector of the angle under which the emperor is seen.

6. Solve another of the problems from Einstein's *Maturitätsexamen* in 1896 (see Hunziker, 2001): Cut a circle of radius 1 by parallel segments (the dashed lines in Fig. 7.42 (a)), and draw the family of circles with these segments as diameters. Show that these circles possess an ellipse with semi-axes $\sqrt{2}$ and 1 as common envelope.

7. Find a proof of the *13-wine-bottles theorem*:[20] If 13 wine-bottles of diameter d are placed in a box of rectangular shape of arbitrary width w satisfying $3d \le w \le (2 + \sqrt{3})d$, then the topmost three bottles all lie on the same level (see Fig.7.43 left).

8. (Fermat, third problem of article V, *Oeuvres*, vol. I, pp. 157–158; also by Joh. Bernoulli, see Hairer and Wanner, 1997, p. 97). Determine the rectangle of maximal area in Fig. 7.44 (I), where the point (x, y) lies on the circle of radius 1.

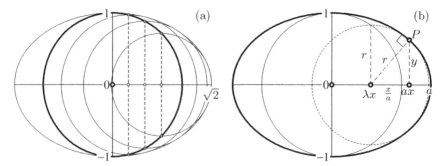

Fig. 7.42. Circles tangent to an ellipse, from Einstein's examination

---

[20]The authors are grateful to J.M. Sanz-Serna, Valladolid, for suggesting this problem.

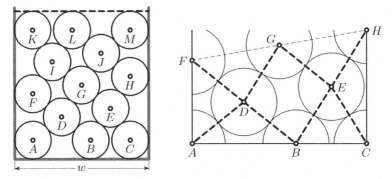

Fig. 7.43. The 13-wine-bottles theorem

9. (Fermat, auxiliary problem for solving the problem of the cylinder of maximal surface, in article VII, *Oeuvres*, vol. I, p. 167). Find the "*H*" of maximal total length inscribed in a circle (Fig. 7.44 (II)).

10. (D. Laugwitz, Aufg. 815, *Elem. Math.* 33, 1978). The golden ratio is still good for a surprise: find the "Swiss cross" of maximal area inscribed in a circle (Fig. 7.44 (III)).

11. Find the cylinder of maximal volume inscribed in a given sphere (Fig. 7.44 (IV)).

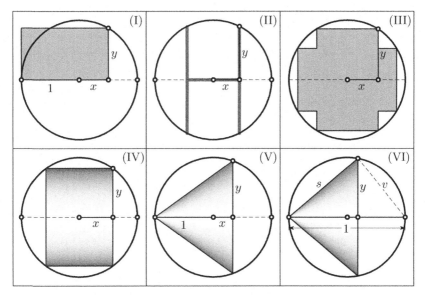

Fig. 7.44. Min-max problems; (I): rectangle of maximal area; (II): "*H*" of maximal length; (III): Swiss cross of maximal area inscribed in a fixed circle; (IV): cylinder of maximal volume; (V): cone of maximal volume; (VI): cone of maximal surface inscribed in a fixed sphere

12. Find the cone of maximal volume inscribed in a given sphere (Fig. 7.44 (V)).

13. (Fermat, second problem of article V, *Oeuvres*, vol. I, pp. 155–157). Find the cone of maximal surface area inscribed in a given sphere (Fig. 7.44 (VI)).
    Hint. This is the most difficult of the problems solved by Fermat, because the surface area of a cone depends on s, which is related to x and y by a square root. Without particular care you will run into messy computations. Fermat found an elegant manner of avoiding this trap: use the distance v of the figure as the unknown quantity.

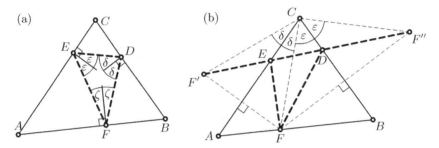

Fig. 7.45. The fourth Fagnano problem, triangle with minimal perimeter (a); and its solution by Fejér (b)

14. Solve the fourth of the problems invented by Fagnano (1779, see Fig. 7.45 (a)): Given a triangle ABC, find points D, E, F on its sides such that the perimeter of DEF is minimal, i.e. such that the angles marked δ, ε and ζ are equal.

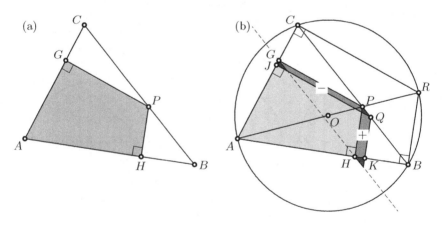

Fig. 7.46. Maximal area of a quadrilateral in a triangle

15. (A problem proposed by M. Hajja and P. Krasopoulos, *Elem. Math.* 65 (2010), p. 37.) A triangle ABC is given (see Fig. 7.46 (a)). Let P be a

point on BC, and let H and G be its orthogonal projections onto AB and AC, respectively. Find the position of P for which the quadrilateral $AHPG$ has maximal area.

16. Find for the *Cardan joint* of Fig. 5.22, with given bend angle γ, the values of a and b (which must satisfy (5.42)), for which the deviation $|a - b|$ is maximal. Express this maximal deviation $|a - b|_{\max}$ as a function of γ.

17. Although the folium of Descartes can be defined by the simple algebraic equation $x^3 + y^3 = 3xy$ (see page 200), calculating its *area* remained a challenge for a very long time. Even the calculations of Jac. Bernoulli were incorrect (see J.E. Hofmann, 1956, p. 20). Equipped with some knowledge of integral formulas, find the area inside the curve in the first quadrant ($x > 0$ and $y > 0$).

Hint. You could simplify the computations, following C. Huygens (see the autograph drawing in Fig. 7.47, left), by expressing v as a function of θ (solution (b)). The most elegant solution (c) is obtained by cutting the folium by a family of lines passing through the origin.

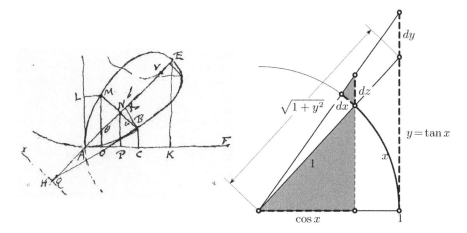

Fig. 7.47. Huygens' drawing of Descartes' folium (1691, left); differentiation of tan and arctan (right)

18. In calculus you learn the differentiation formulas

$$y = \tan x \quad \Rightarrow \quad y' = \frac{1}{\cos^2 x} = 1 + \tan^2 x$$

$$y = \arctan x \quad \Rightarrow \quad y' = \frac{1}{1 + x^2} \, . \tag{7.65}$$

Give a geometric explanation of these formulas.

19. For two given points F_1 and F_2 and with $\ell_1 = PF_1$, $\ell_2 = PF_2$, the condition $\ell_1 + \ell_2 = C$ leads to ellipses, the condition $\ell_1 - \ell_2 = C$ to

hyperbolas, the condition $\ell_1 \cdot \ell_2 = C$ to Cassini curves; to which terribly exotic curves will the condition $\ell_1/\ell_2 = C$ lead?

20. Find the area inside a loop of the lemniscate (7.34) or, better, (7.36).

21. (a) Express the points of the "equilat-
    eral" hyperbola $x^2 - y^2 = 1$ in polar
    coordinates (ρ, φ). Then replace ρ by
    $r = \frac{1}{\rho}$, i.e. map the point P of the hy-
    perbola to the point Q in harmonic
    position w.r. to the unit circle. Show
    that all these points Q lie on a lem-
    niscate. (b) Let R be the orthogonal
    projection of the origin onto the tan-
    gent at P. Show that all these points
    R lie on a lemniscate (the same one).

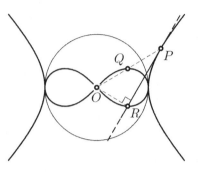

22. (*The original motivation of Jac. Bernoulli*) Parametrise the lemniscate
    (7.34) by r, the distance from the origin, satisfying $x^2 + y^2 = r^2$. Show
    that the (infinitesimal) arc length of the lemniscate then satisfies

$$ds = \frac{a^2}{\sqrt{a^4 - r^4}}\, dr\,. \tag{7.66}$$

This was the first example of what later became famous under the name
elliptic integrals.

23. The object of this exercise is to prove the nice formula

$$OH^2 = R^2(1 - 8\cos\alpha\cos\beta\cos\gamma) \tag{7.67}$$

for the distance OH on the Euler line.

Hint. An elegant solution is due to R. Müller (1905) (see Fig. 7.48, left):
(a) Show that $HC = 2R\cos\gamma$, (b) $HF = 2R\cos\alpha\cos\beta$, (c) $HF = FM$
(where M is the intersection of the extended altitude with the circumcircle). Then apply Eucl. III.35.

24. To generalise the nice result of Exercise 24 of Chap. 1, we ask whether for
    every regular n-star, rotating around a fixed point C in a circle, the sum
    of the squares of the ray lengths is always constant, independent of the
    position of C and the angle of rotation. For example, for $n = 6$ this would
    mean that

$$p^2 + p'^2 + q^2 + q'^2 + r^2 + r'^2 = 6R^2 \tag{7.68}$$

(see Fig. 7.48, right). A numerical computer simulation indicated that this
is true for $n = 4, 6, 8, 10, 12, 14, \ldots$, but wrong for $n = 2$ and for odd n.
Give an analytical proof.

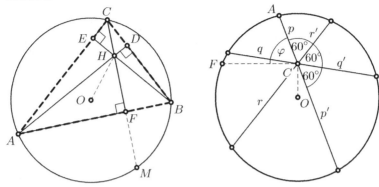

Fig. 7.48. Proof of Müller's formula (left); Nelsen's generalised result (right)

25. Prove the following identity of Giulio Fagnano (1750, vol. II, Appendice: *Nuova et generale proprietà de' Poligoni*, Corollario II): let G be the barycentre of the triangle ABC, then

$$GA^2 + GB^2 + GC^2 = \frac{1}{3}(AB^2 + BC^2 + CA^2).$$

26. (A challenging problem from the Int. Math. Olympiad, Hanoi 2007, submitted by Luxembourg.) Consider five points A, B, C, D and E such that $ABCD$ is a parallelogram and $BCED$ is a cyclic quadrilateral. Let ℓ be a line passing through A. Suppose that ℓ intersects the interior of the segment DC at F and the line BC at G so that $EF = EG = EC$. Prove that ℓ is the bisector of angle DAB.

27. (The $h/2$-circle.) Show, by cleverly running the TTT-machine, that: *If the incentre of a triangle is moved upwards to the midpoint of the altitude, and the radius of the incircle is increased accordingly, then the resulting circle is tangent to the circumcircle.*

28. Find the circle which is tangent to two sides and the circumcircle of a given triangle.

    Hint. Use the TTT-machine.

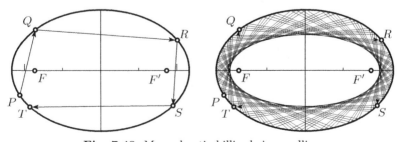

Fig. 7.49. More chaotic billiards in an ellipse

29. If we play billiards in an ellipse and the "ball", starting from P, reflected at Q, R, S etc., does *not* pass through a focus, then its behaviour is more chaotic (see Fig. 7.49, left). However, if we draw one hundred reflections (Fig. 7.49, right), then the trajectories show a mysterious curve as their envelope. Guess the shape of this curve and prove your conjecture.

To be Constructible, or not to be

...that is the question...

"Magnopere sane est mirandum, quod, quum iam Euclidis tempo-
ribus circuli divisibilitas geometrica in tres et quinque partes nota
fuerit, nihil his inventis intervallo 2000 annorum adiectum sit, ...
[It is greatly to be wondered at that although the geometric divisibility of the circle into 3 and 5 parts was known already in Euclid's
time, nothing was added for 2000 years after these discoveries ...]"
(C.F. Gauss, *Disquisitiones Arithmeticae*, 1801, Art. 365)[1]

"Die Geschichte dieser Entdeckung ist bisher nirgends von mir
öffentlich erwähnt, ich kann es aber sehr genau angeben. Es war
der 29. März 1796, ... [I have not yet communicated the story of
this discovery, but I can do it very precisely. It was on March 29,
1796 ...]" (C.F. Gauss, 1819, in a letter to his
former student C.L. Gerling, Gauss' *Werke*, vol. X, pp. 121–126)

According to Descartes, his new geometry should have replaced all the "thick
books" of the Greeks (including Euclid, Apollonius, Pappus, and the efforts
to find constructions with ruler and compass; see the quotations in Chap. 6),
with the only exception of the theorems of Thales and Pythagoras.

Quite unexpectedly, it was precisely the revival by the young Gauss of
questions concerning constructions with ruler and compass that led to remarkable results in algebra *and* geometry. On 29 March 1796, Gauss discovered
that the regular 17-gon can be constructed with ruler and compass. Deeply
moved by his discovery, the *first* new result in that field for 2000 years (see
the quotations above), he decided to become a mathematician. He started his
Notizenjournal (a sort of mathematical diary) the day after his discovery.[2]
There he made short notes of many of his discoveries and lonely studies. The

---

[1] Engl. translation by Waterhouse and Clarke, Springer-Verlag, 1986
[2] First published by F. Klein, Math. Annalen 57 (1903), pp. 1–34

A. Ostermann and G. Wanner, *Geometry by Its History*, 241
Undergraduate Texts in Mathematics, DOI: 10.1007/978-3-642-29163-0_8,
© Springer-Verlag Berlin Heidelberg 2012

"Principia quibus innititur sectio circuli, ac divisibilitas eiusdem geometrica in septemdecim partes &c." is the very first of his notes (see Fig. 8.1).

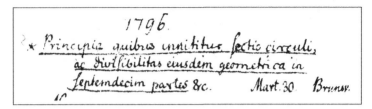

Fig. 8.1. Reproduction of the beginning of Gauss' *Notizenjournal* concerning the construction of the regular 17-gon

8.1 The Complex Plane and the Logarithmic Spiral

> "Les *imaginaires*, en Géométrie pure, présentent de graves difficultés ... [The *imaginary* numbers cause serious difficulties in pure Geometry ...]"
> (M. Chasles, *Traité de Géométrie supérieure*, 2e éd., 1880, p. xiii)

We start by extending the previous constructions with ruler and compass (see Lemma 7.1) to the complex plane.

The complex plane. The main idea is to identify the real plane

$$\mathbb{R}^2 = \big\{ (x,y) \mid x,y \in \mathbb{R} \big\} \simeq \mathbb{C} = \big\{ x + iy \mid x,y \in \mathbb{R} \big\} \tag{8.1}$$

with the complex plane (Wessel 1799, Gauss 1799, Argand 1806). The symbol i slowly came into use to denote the "number" $\sqrt{-1}$ (first by Euler in 1777), with the property

$$i = \sqrt{-1} \qquad \text{i.e.} \qquad i^2 = -1, \tag{8.2}$$

see Fig. 8.2 (left). For a given complex number $z = x + iy$, the real numbers x and y are called the *real* and *imaginary part* of z, respectively. The complex number $\overline{z} = x - iy$ is called the *complex conjugate* of z. Further, let φ denote the angle (taken with a sign) between the positive x-axis and the ray from the origin through the point (x,y). This angle, normalised by $-\pi < \varphi \leq \pi$, is denoted by $\arg z$ and called the *argument* of z. Finally, the non-negative number $|z| = r = \sqrt{x^2 + y^2}$, which is the distance from z to the origin, is called the *modulus* (or *absolute value*) of z. With this notation, we have

$$z = r\big(\cos\varphi + i\sin\varphi\big) \tag{8.3}$$

in *polar coordinates*.

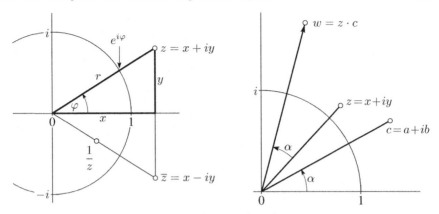

Fig. 8.2. The complex plane

Multiplication. Taking (8.2) into account, the product of two complex numbers is seen to be

$$c = a + ib \atop z = x + iy \quad \Rightarrow \quad cz = ax - by + i(bx + ay)\,. \tag{8.4}$$

Division. Division is based on the fact that by (8.4) the product

$$z \cdot \bar{z} = (x + iy)(x - iy) = x^2 + y^2 = r^2 \tag{8.5}$$

is real. So we simply multiply the numerator and the denominator of a fraction by the complex conjugate of the denominator. For example,

$$\frac{1}{6 + 2i} = \frac{6 - 2i}{(6 + 2i)(6 - 2i)} = \frac{6 - 2i}{6^2 + 2^2} = \frac{3}{20} - \frac{i}{20}\,.$$

In this manner, every complex number can be divided by any complex number $\neq 0$ and, with the above addition and multiplication, $\mathbb{C}$ becomes a *field*.

The logarithmic spiral. This beautiful curve originates from one of Jacob Bernoulli's "meditations" (*Meditatio LI, Werke*, vol. 2, p. 289; written around 1684; *Linea curva infinitarum dimensionum*). Motivated by the question of trisecting a given angle ("bi-tri-quadrisectum"), Jacob Bernoulli considered a sequence of similar triangles, having each a side in common with the preceding one ("triangula inter se similia & proportionalia", see Fig. 8.3, left).

Algebraic formulas. In order to determine recursively the coordinates of the points $D, E, F, \ldots$, we suppose triangle ABC to be given, with side $AB = 1$ along the x-axis, and denote the coordinates of C by (a, b). Then, if D with coordinates (x, y) is the lower vertex of one of these triangles, we want to compute the coordinates $(\tilde{x}, \tilde{y})$ of the next point E. The answer to this question is explained in Fig. 8.4 and leads to the formulas

$$\begin{aligned} \tilde{x} &= ax - by\,, & x &= \tfrac{1}{a^2 + b^2}(a\tilde{x} + b\tilde{y})\,, \\ \tilde{y} &= bx + ay\,, & y &= \tfrac{1}{a^2 + b^2}(-b\tilde{x} + a\tilde{y})\,, \end{aligned} \tag{8.6}$$

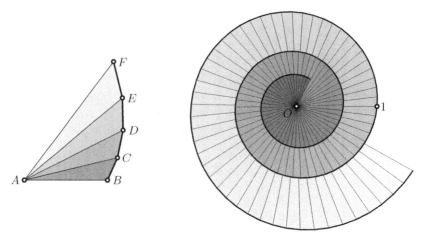

Fig. 8.3. Jac. Bernoulli's similar triangles (left); the logarithmic spiral $z = e^{ct}$ with
$c = 0.5 + 17.5i$, $-\frac{2}{3} \le t \le \frac{1}{3}$ (right)

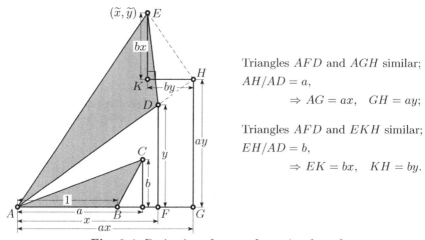

Triangles AFD and AGH similar;
$AH/AD = a$,
$$\Rightarrow AG = ax, \quad GH = ay;$$

Triangles AFD and EKH similar;
$EH/AD = b$,
$$\Rightarrow EK = bx, \quad KH = by.$$

Fig. 8.4. Derivation of a transformation formula

(the formulas in the column on the right bring us "down" the spiral, from E
to D). We observe a wonderful relation of these formulas to the formulas (8.4)
for complex multiplication and division. We thus understand the geometric
significance of complex multiplication (multiply the absolute values and add
the arguments) explained in Fig. 8.2 (right). These formulas are closely connected to the addition theorems for trigonometric functions (5.6). This is no
miracle, since Figs. 8.4 and 5.7 (left), on page 118, are almost identical.

Analytic formulas. If one decreases more and more the angles at the vertex A,
by increasing the number of triangles, the broken line $B, C, D, E, \ldots$ tends to
a curve ("... in curvam degenerabunt", see Fig. 8.3, right), which extends to
infinity in both directions and reproduces its form over and over again ("quae

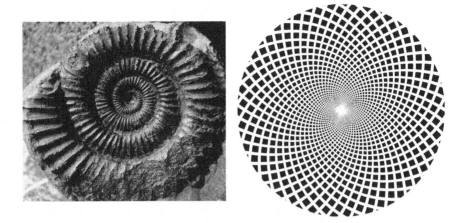

Fig. 8.5. Ammonite Dactylioceras, 165 million years old (left); 34 logarithmic spirals of slope Φ and 55 logarithmic spirals of slope $-1/\Phi$ (right)

infinitarum erit dimensionum, utpote cui describendae infinitae reperiendae mediae proportionales"). In particular this curve intersects each of the rays to the origin under the same angle. For the deeply religious Jacob Bernoulli, this curve had great mystical importance, symbolising eternal life, and he asked that it be engraved on his tombstone with the inscription "eadem mutata resurgo".

For many species of ammonites (Fig. 8.5, left), life in the form of a logarithmic spiral — if not eternal — lasted at least for many millions of years. Fig. 8.5 (right) offers the pure pleasure of looking at a picture in which this marvellous curve is combined with the beauty of the golden ratio and the Fibonacci numbers.

In order to derive analytic formulas, we replace in (8.6) $a \mapsto 1 + \frac{a}{N}$ and $b \mapsto \frac{b}{N}$, with a and b fixed and N a large number. Then by (8.6), the J-th point on the spiral becomes

$$\left(1 + \frac{a + ib}{N}\right)^J = \left(1 + \frac{c}{N}\right)^J, \qquad \text{where } c = a + ib.$$

Letting both N and J tend to infinity in such a way that $J = N \cdot t$ with t fixed, this becomes

$$\left(1 + \frac{c}{N}\right)^{N \cdot t} = \left(\left(1 + \frac{c}{N}\right)^N\right)^t \to e^{ct}. \tag{8.7}$$

Any good book on analysis tells you that

$$e^c = \lim_{N \to \infty} \left(1 + \frac{c}{N}\right)^N = 1 + c + \frac{c^2}{1 \cdot 2} + \frac{c^3}{1 \cdot 2 \cdot 3} + \dots \tag{8.8}$$

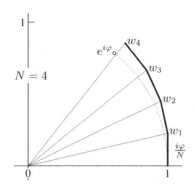 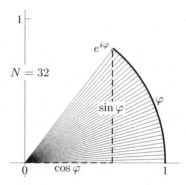

Fig. 8.6. Exponential function for purely imaginary $i\varphi$

is the famous exponential function; the last identity is obtained from the binomial expansion (Euler, 1748; see also Hairer and Wanner, 1997, pp. 25–27).

Euler's formula. The idea is to replace c in (8.7) by a purely imaginary value $i\varphi$. The triangle ABC becomes right angled at B and the logarithmic "spiral" tends to the unit circle where φ represents the arc length (see Fig. 8.6). This figure exhibits at once Euler's famous formula

$$e^{i\varphi} = \cos\varphi + i\sin\varphi. \tag{8.9}$$

With it the polar representation (8.3) can be written as

$$z = x + iy = r \cdot e^{i\varphi} \tag{8.10}$$

and the product of $z = r \cdot e^{i\varphi}$ with $c = s \cdot e^{i\alpha}$ becomes

$$cz = rs \cdot e^{i(\alpha+\varphi)}, \tag{8.11}$$

where once again, we see the geometric meaning of complex multiplication. Particular cases of Euler's formula are

$$e^{\frac{i\pi}{2}} = i, \qquad e^{i\pi} = -1, \qquad e^{2i\pi} = 1. \tag{8.12}$$

8.2 Constructions with Ruler and Compass

Complex roots. Let z be a complex number with modulus r and argument φ. From the properties of the product, it is obvious that a complex root $w = \sqrt{z}$ of z (i.e. a solution of $w^2 = z$) has modulus $|w| = \sqrt{r}$ and argument $\arg w = \frac{\varphi}{2}$ (see Fig. 8.7, left). However, we must be careful since $e^{2i\pi} = 1$. Therefore, a second square root of z exists with argument $\frac{\varphi}{2} + \pi$.

Since a complex square root is obtained from a real square root and the bisection of an angle, it can be constructed with ruler and compass (Eucl. I.9). Together with Fig. 6.1 we get the following result.

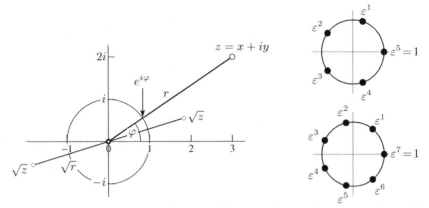

Fig. 8.7. Complex square roots (on the left); roots of unity for $n=5$ and $n=7$ (on the right)

Lemma 8.1. *Each composition of rational operations and square roots in the complex plane corresponds to a construction with ruler and compass.*

n-th roots. Let z be a complex number with modulus r and argument φ. As before, the n-th root of z has the modulus $\sqrt[n]{r}$ and any one of the arguments $\frac{\varphi}{n} + \frac{2k\pi}{n}$, $k = 0, 1, \ldots, n-1$. In particular, for $z = 1$ we have

$$\sqrt[n]{1} = \varepsilon^k = e^{\frac{2ki\pi}{n}}, \qquad k = 0, 1, \ldots, n - 1 . \tag{8.13}$$

These values are called the nth *roots of unity*, see Fig. 8.7 on the right. The points ε^k thus represent the vertices of a regular n-gon inscribed in the unit circle, and any geometric construction for solving the equation

$$z^n - 1 = 0 \tag{8.14}$$

provides us with a construction of this regular polygon.

8.3 The Method of Gauss and Vandermonde

Before attacking the famous 17-gon, we will explain the method of Gauss and Vandermonde for computing 5th roots. The same ideas will be useful for $n = 17$.

The long tradition of solving algebraic equations (Tartaglia and Cardano 1545, Lagrange 1770) motivates us to look for certain combinations of roots that satisfy an equation of lower degree. For the roots of unity, a good choice are sums of the following type (Vandermonde 1771)

$$\varepsilon + \varepsilon^k + \varepsilon^{k^2} + \ldots \tag{8.15}$$

Gauss used the same method a quarter of a century later, but made no mention of Vandermonde.

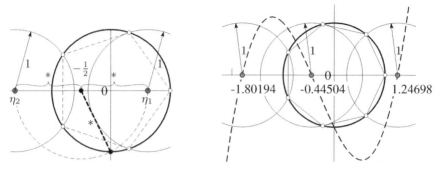

Fig. 8.8. Construction of the regular pentagon inscribed in the unit circle (left), and of the regular heptagon (right)

The regular pentagon. Here we want to solve $\varepsilon^5 = 1$. Try $k = 4$ in (8.15). Then the powers k^j are

$$1 \to 4 \to 16 \to 64 \to 256 \to \dots$$

Since $\varepsilon^5 = 1$, we can take the residues modulo 5 of these powers (i.e. their *remainders* on division by 5); we obtain

$$1 \to 4 \to 1 \to 4 \to 1 \to \dots$$

The theory of such "power residues" was established by Euler (1761) in [E262]. But 4 is not a good choice, since not all (non-zero) residues modulo 5 appear in this list. However, by choosing $k = 2$, we obtain the complete sequence of (non-zero) residues modulo 5:

$$1 \to 2 \to 4 \to 3 \to 1 \,. \tag{8.16}$$

Such values of k were called *primitive roots* by Euler (1774) in [E449]: 2 is a primitive root modulo 5. These are the good choices for our method. We group the terms by pairs and set

$$\eta_1 = \varepsilon + \varepsilon^4, \qquad \eta_2 = \varepsilon^2 + \varepsilon^3. \tag{8.17}$$

Since the sum of all the roots of the polynomial $\varepsilon^5 - 1$ is 0, we obtain

$$\begin{aligned}
\eta_1 + \eta_2 &= \varepsilon + \varepsilon^2 + \varepsilon^3 + \varepsilon^4 = -1 \,, \\
\eta_1 \cdot \eta_2 &= \varepsilon^3 + \varepsilon^4 + \varepsilon + \varepsilon^2 = -1 \,.
\end{aligned} \tag{8.18}$$

By Viète, η_1 and η_2 are the roots of the quadratic polynomial

$$\eta^2 + \eta - 1 = 0 \qquad \Rightarrow \qquad \eta_{1,2} = \frac{-1 \pm \sqrt{5}}{2} \,, \tag{8.19}$$

and we have managed to reduce the problem of finding the roots of a polynomial of degree 5 to that of finding those of a polynomial of degree 2.

In our case, ε^4 and ε^3 are complex conjugate to ε and ε^2, respectively. Because of (8.17), η_1 and η_2 are just twice the real parts of ε and ε^2, respectively. This leads to the construction of the regular pentagon indicated in Fig. 8.8 (left).[3]

The regular heptagon. Relying on the same ideas as before, we now calculate the 7th roots of unity. The powers of 2 modulo 7 are

$$1 \to 2 \to 4 \to 1\,. \tag{8.20}$$

As this sequence is too short, we try $k = 3$ and find

$$1 \to 3 \to 2 \to 6 \to 4 \to 5 \to 1\,. \tag{8.21}$$

We thus have a primitive root modulo 7. Unfortunately, it turns out that there exists no partition, like e.g. $\eta_1 = \varepsilon + \varepsilon^2 + \varepsilon^4$ and $\eta_2 = \varepsilon^3 + \varepsilon^6 + \varepsilon^5$, which would make the product of η_1 and η_2 independent of ε.

Therefore, one has to define *three* quantities

$$\eta_1 = \varepsilon + \varepsilon^6, \qquad \eta_2 = \varepsilon^3 + \varepsilon^4, \qquad \eta_3 = \varepsilon^2 + \varepsilon^5. \tag{8.22}$$

A straightforward calculation shows that

$$\eta_1 + \eta_2 + \eta_3 = -1\,, \qquad \eta_1\eta_2 + \eta_2\eta_3 + \eta_3\eta_1 = -2\,, \qquad \eta_1\eta_2\eta_3 = 1\,, \tag{8.23}$$

which means that η_1, η_2, η_3 are the roots of

$$\eta^3 + \eta^2 - 2\eta - 1 = 0\,. \tag{8.24}$$

We will see later that the roots of this equation cannot be constructed with ruler and compass. Hence we used a numerical calculation to produce the drawing on the right of Fig. 8.8.

8.4 The Regular 17-Sided Polygon

We now come to Gauss' great discovery, the construction of the regular 17-gon, i.e. the solution of the equation

$$\varepsilon^{17} - 1 = 0\,, \tag{8.25}$$

by solving a sequence of equations of the *second* degree. We start with powers of 2, and reduce them modulo 17:

$$1 \to 2 \to 4 \to 8 \to 16 \to 15 \to 13 \to 9 \to 1\,.$$

This sequence is too short. The next choice, however, works:

$$\begin{aligned}1 \to 3 \to 9 \to 10 \to 13 \to 5 \to 15 \to 11 \to 16 \to \\ \to 14 \to 8 \to 7 \to 4 \to 12 \to 2 \to 6 \to 1\,.\end{aligned} \tag{8.26}$$

We take every other exponent for η_1 (the black points in Fig. 8.9, left) and the remaining ones for η_2 (the grey points). This leads to

---

[3]This construction is similar to Ptolemy's construction in Fig. 6.11, but is not exactly the same.

$$\eta_1 = \varepsilon^1 + \varepsilon^9 + \varepsilon^{13} + \varepsilon^{15} + \varepsilon^{16} + \varepsilon^8 + \varepsilon^4 + \varepsilon^2 \,,$$
$$\eta_2 = \varepsilon^3 + \varepsilon^{10} + \varepsilon^5 + \varepsilon^{11} + \varepsilon^{14} + \varepsilon^7 + \varepsilon^{12} + \varepsilon^6 \,. \tag{8.27}$$

It is easy to see that $\eta_1 + \eta_2 = -1$, since the sum of all the roots is zero. The product $\eta_1 \cdot \eta_2$ contains 64 terms (see the right picture of Fig. 8.9), and each power "miraculously" appears exactly four times. Thus

$$\eta^2 + \eta - 4 = 0 \quad \Rightarrow \quad \eta_1 = \frac{-1 + \sqrt{17}}{2} \,, \quad \eta_2 = \frac{-1 - \sqrt{17}}{2} \tag{8.28}$$

(the grey points tend to be more to the left; this determines the sign).

One continues with

$$\mu_1 = \varepsilon^1 + \varepsilon^{13} + \varepsilon^{16} + \varepsilon^4 \,, \qquad \mu_3 = \varepsilon^3 + \varepsilon^5 + \varepsilon^{14} + \varepsilon^{12} \,,$$
$$\mu_2 = \varepsilon^9 + \varepsilon^{15} + \varepsilon^8 + \varepsilon^2 \,, \qquad \mu_4 = \varepsilon^{10} + \varepsilon^{11} + \varepsilon^7 + \varepsilon^6 \,, \tag{8.29}$$

and finds

$$\mu_1 + \mu_2 = \eta_1, \quad \mu_1\mu_2 = -1, \quad \Rightarrow \quad \mu_1 = \frac{\eta_1 + \sqrt{\eta_1^2 + 4}}{2} \,, \tag{8.30}$$

$$\mu_3 + \mu_4 = \eta_2, \quad \mu_3\mu_4 = -1, \quad \Rightarrow \quad \mu_3 = \frac{\eta_2 + \sqrt{\eta_2^2 + 4}}{2} \,. \tag{8.31}$$

In the next step, we set

$$\beta_1 = \varepsilon^1 + \varepsilon^{16} \,, \qquad \beta_2 = \varepsilon^{13} + \varepsilon^4 \tag{8.32}$$

to get

$$\beta_1 + \beta_2 = \mu_1, \quad \beta_1\beta_2 = \mu_3 \quad \Rightarrow \quad \beta_1 = \frac{\mu_1 + \sqrt{\mu_1^2 - 4\mu_3}}{2} \,, \tag{8.33}$$

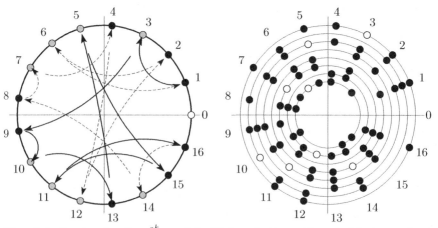

Fig. 8.9. Roots of unity ε^{3^k} modulo 17 (quadratic residues in black, quadratic non-residues in grey, left); product $\eta_1 \cdot \eta_2 = -4$ (right)

and finally we obtain

$$\varepsilon^1 + \varepsilon^{16} = \beta_1, \quad \varepsilon^1 \varepsilon^{16} = 1, \quad \Rightarrow \quad \varepsilon = \frac{\beta_1 + \sqrt{\beta_1^2 - 4}}{2}, \qquad (8.34)$$

which is the desired result.

Theorem 8.2 (Gauss 1796). *The regular 17-gon can be constructed with ruler and compass.*

Gauss also knew that the same method works for any number of the form

$$F_k = 2^{(2^k)} + 1, \qquad (8.35)$$

if it is a prime number. This is the case for the first five:

$$F_0 = 3, \quad F_1 = 5, \quad F_2 = 17, \quad F_3 = 257, \quad F_4 = 65537. \qquad (8.36)$$

Fermat's conjecture (*"tous* ces nombres sont des nombres premiers [*all* these numbers are primes]") was disproved by the genius Euler, who found the counterexample $F_5 = 2^{(2^5)} + 1 = 4294967297 = 641 \cdot 6700417$. The five numbers given in (8.36) are today still the only known prime numbers of the form (8.35).

Remark. The explicit calculations for the 257-gon were carried out by Richelot and fill 84 pages, see Richelot (1832). Hermes dedicated ten years of his life to the 65537-gon. His solution is kept in a big box in the mathematical seminar at Göttingen, see Hermes (1895).

Remark. By combining these considerations with Eucl. IV.16 (see Exercise 3 below) and Eucl. I.9, it is evident that all regular n-gons with n of the form

$$n = 2^\ell p_1 \cdots p_k, \qquad (8.37)$$

where $\ell \geq 0$, and the p_i are *pairwise different* primes of the form (8.36), can be constructed with ruler and compass.

8.5 Constructions Impossible with Ruler and Compass

> "Ce principe est annoncé par M. Gauss à la fin de son ouvrage, mais il n'en a pas donné la démonstration. [This principle is announced by Gauss at the end of his book, but he has not given a proof.]"
>
> (P.L. Wantzel, *J. de Math.*, vol. 2 (1837), pp. 366–372)
>
> "... will ich Ihnen hier einen Fall vorführen – um so lieber, als im grossen Publikum so *wenig Verständnis für Beweise dieser Art* vorhanden ist. [I shall put before you here an example of this important *proof of impossibiliy* — the more willingly because there

is such a lack of *understanding for proofs of this sort* by the great
public.]"

(F. Klein, *Elementarmathematik vom höheren Standpunkte
aus*, 1924, p. 55; Engl. transl. by Hedrick and Noble, 1932, p. 51)

Towards the end of his *Disquisitiones Arithmeticae*, Gauss states in capitals
("OMNIQUE RIGORE DEMONSTRARE POSSUMUS ...") that *no* regular *n*-
gon with n not of the form (8.37) can be constructed with ruler and compass.
However, he did not give a rigorous proof; this was done by P. L. Wantzel
(1837, see the quotation). We will prove here only one particular case.

Theorem 8.3. *The regular heptagon cannot be constructed with ruler and
compass.*

Proof. Our proof is an adaption of that given by F. Klein in 1908.
First step. One shows that equation (8.24) cannot have a *rational* root: as-
suming $\eta = \frac{n}{m}$ to be a root, with n and m relatively prime, gives

$$n^3 + m\left(n^2 - 2nm - m^2\right) = 0. \tag{8.38}$$

Thus, any prime factor of m divides n, and conversely. Since m and n are
relatively prime, this implies that $\eta = \pm 1$. This gives a contradiction since ± 1
are certainly not roots of (8.24).
Second step. Suppose that (8.24) has a root of the form

$$\eta_1 = \frac{\alpha + \beta\sqrt{R}}{\gamma + \delta\sqrt{R}}, \tag{8.39}$$

where $\alpha, \beta, \gamma, \delta$ and R are rational numbers. We must have $\gamma - \delta\sqrt{R} \neq 0$, for
otherwise $\sqrt{R}$ would be rational and with it η_1; this is excluded by the first
step. We multiply the numerator and denominator of (8.39) by $\gamma - \delta\sqrt{R}$ to
get

$$\eta_1 = \frac{(\alpha + \beta\sqrt{R})(\gamma - \delta\sqrt{R})}{\gamma^2 - \delta^2 \cdot R} = P + Q\sqrt{R}, \tag{8.40}$$

where P and Q are again rational numbers. Inserting this into (8.24) gives

$$\eta_1^3 + \eta_1^2 - 2\eta_1 - 1 = (P + Q\sqrt{R})^3 + (P + Q\sqrt{R})^2 - 2(P + Q\sqrt{R}) - 1 = 0. \tag{8.41}$$

Multiplying out, we obtain a relation of the form

$$M + N\sqrt{R} = 0, \tag{8.42}$$

where M and N are again rational numbers. If $N \neq 0$, we have $\sqrt{R} = -\frac{M}{N}$
and once more, this square root is rational. Hence $M = N = 0$ and thus
$M - N\sqrt{R} = 0$. Consequently (doing the same calculations with $\sqrt{R}$ replaced
by $-\sqrt{R}$)

$$\eta_2 = P - Q\sqrt{R} \tag{8.43}$$

is *also* a root of (8.24). However, by Viète

$$\eta_1 + \eta_2 + \eta_3 = -1 \qquad \Rightarrow \qquad \eta_3 = -1 - \eta_1 - \eta_2 = -1 - 2P \qquad (8.44)$$

is a rational root of (8.24). This is impossible by the first step of our proof.

General step. The rest is now easy. If η_1 contains *several* square roots (iterated or not), one eliminates one root after another by repeatedly applying the second step of this proof. After each step, the roles of η_1 and η_3 must be interchanged. □

Doubling the cube. We now return to the classical Greek problems of Sect. 1.8. The problem of doubling the cube consists in constructing the real root of $x^3 - 2 = 0$. This equation does not have a rational root: if m, n are relatively prime and $n^3 - 2m^3 = 0$ then n must be even; setting $n = 2\ell$ gives $m^3 = 4\ell^3$, so m must be even, a contradiction. The rest of the above proof applies almost without any changes, so we conclude that *doubling the cube with ruler and compass is impossible.*

Trisecting an angle. We have seen in Sect. 6.3 that this problem leads to the cubic equation (6.9) $x^3 - \frac{3}{4}x + \frac{d}{4} = 0$. Here, for *some* values of d there exist constructible solutions (for instance $3\alpha = 90°$ gives $d = 1$ with solution $x = \frac{1}{2}$, or $3\alpha = 180°$ gives $d = 0$ and the equation becomes quadratic). There are, however, angles for which *no* construction is possible with ruler and compass. Despite of this fact, the trisection with ruler and compass has always been a flourishing field for mathematical amateurs. An impressive collection of fruitless attempts is given by Dudley (1987).

Squaring the circle.

> "Ich kann mit einigem Grunde zweifeln, ob gegenwärtige Abhandlung von denjenigen werde gelesen, oder auch verstanden werden, die den meisten Antheil davon nehmen sollten, ich meyne von denen, die Zeit und Mühe aufwenden, die Quadratur des Circuls zu suchen. Es wird sicher genug immer solche geben ... die von der Geometrie wenig verstehen ... [I have good reasons to doubt that the present article will be read, or even understood, by those who should profit most by it, namely those who spend time and efforts in trying to square the circle. There will always be enough such persons ... who understand very little of geometry ...]"
>
> (J.H. Lambert, 1770)

The proof of the impossibility of squaring the circle with ruler and compass is more difficult. The problem consists in finding a construction for π or $\sqrt{\pi}$ and gave rise to innumerable fruitless attempts. Finally, the opinion prevailed that the problem is impossible (Lambert 1770, see the quotation). Lambert himself found that π is irrational. A rigorous proof of irrationality is due to Legendre 1794 (see also Hairer and Wanner, 1997, Sect. I.6).

Irrationality is not sufficient, since certain irrational numbers can be constructed with ruler and compass (for example $\sqrt{2}$). The proof of impossibility was finally achieved by F. Lindemann in 1882. He showed that π is *transcendental*, i.e. is not the root of any polynomial with rational coefficients (not all zero). His proof is involved.

8.6 Exercises

1. Use complex arithmetic to prove a theorem, somewhat similar to Napoleon's theorem, which is attributed to either H.H. van Aubel (1878) or E. Collignon (see Kritikos, 1961): Let A_1, A_2, A_3, A_4 be the vertices of an arbitrary quadrilateral and let B_1, B_2, B_3, B_4 be the centres of the four squares constructed on its sides. Then the segments B_1B_3 and B_2B_4 have the same length and are perpendicular.

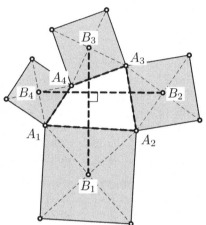

2. Show that the length of the side of the regular 17-gon inscribed in the unit circle is

$$\frac{1}{4}\sqrt{34 - 2\sqrt{17} - 2\sqrt{34 - 2\sqrt{17}} - 4\sqrt{17 + 3\sqrt{17} + \sqrt{170 - 26\sqrt{17}} - 4\sqrt{34 + 2\sqrt{17}}}}.$$

3. Show that the regular 51-gon is constructible with ruler and compass.

4. Very simple geometric questions can lead to problems whose solutions are not constructible with ruler and compass (D. Laugwitz, *Elem. Math.* 26 (1971), p. 135): Find the point $P_1 = (x_1, y_1)$ on the hyperbola $x^2 - y^2 = 1$ which has shortest distance from a given point $P_0 = (x_0, y_0)$. Tackle the question (a) by using the formula for the tangent at P_1 and (b) by using the Fermat–Leibniz method to minimise $(x - x_0)^2 + (y - y_0)^2$ under the condition $x^2 - y^2 = 1$. Show that the solution is not constructible with ruler and compass.

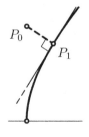

5. Verify a discovery of C. Huygens (1724, vol. 2, p. 391): Let AB be a diameter of a circle of radius 1, the angle BAD be 45°, the angle ABF be 60° and let E be the intersection of AD and BF (see Fig. 8.10). Then the distance AE solves the problem of doubling the cube with an error less than $\frac{1}{2000}$.

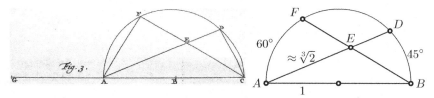

Fig. 8.10. Huygens' approximate doubling of the cube; left: facsimile from Huygens (1724)

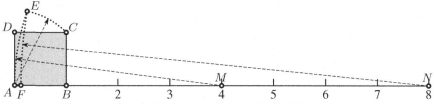

Fig. 8.11. Finsler's construction for the doubling of the cube

6. Verify the approximate construction for the doubling of the cube given in Finsler (1937/38). Let A, B, C, D in Fig. 8.11 be a unit square, a face of the cube which we want to double in volume. Let M and N be at distance 4 and 8 from A, respectively. Construct E by drawing two circles; the first centred at A with radius AC, the second centred at M with radius AM. Then construct F by drawing the circle centred at N with radius NE. Show that the distance $10 \cdot AF$ is an excellent approximation to $\sqrt[3]{2}$.

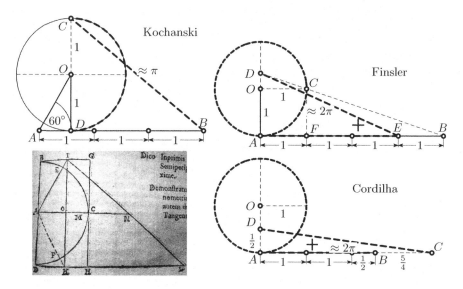

Fig. 8.12. π-approximations; left: Kochanski and his original construction, arranged slightly differently; right: Finsler and Cordilha

7. This exercise spans four centuries and three continents. It is inspired by A. Steiner and G. Arrigo (2008, "... è un bell' esercizio ... [it's a beautiful exercise]"). In Fig. 8.12 three approximate constructions using the unit circle are sketched: (a) by A.A. Kochanski (1685, a Polish priest) BC for π; (b) by P. Finsler (1937/38, Zürich) $AE+ED$ for 2π; (c) J. Cordilha (Rio de Janeiro 1932; in a letter to P. Finsler) $AB + CD$ for 2π. Determine for each of these constructions the error of the corresponding approximation to $\pi = 3.14159265358979323846264338327 9\ldots$ Add to this comparison (d) the approximation $\pi \approx \frac{355}{113}$ (Adrianus Metius 1571–1635); finally the approximations

$$\text{(e)} \quad \pi \approx \sqrt[4]{9^2 + \frac{19^2}{22}}, \qquad \text{(f)} \quad \pi \approx \frac{63}{25}\left(\frac{17 + 15\sqrt{5}}{7 + 15\sqrt{5}}\right), \qquad (8.45)$$

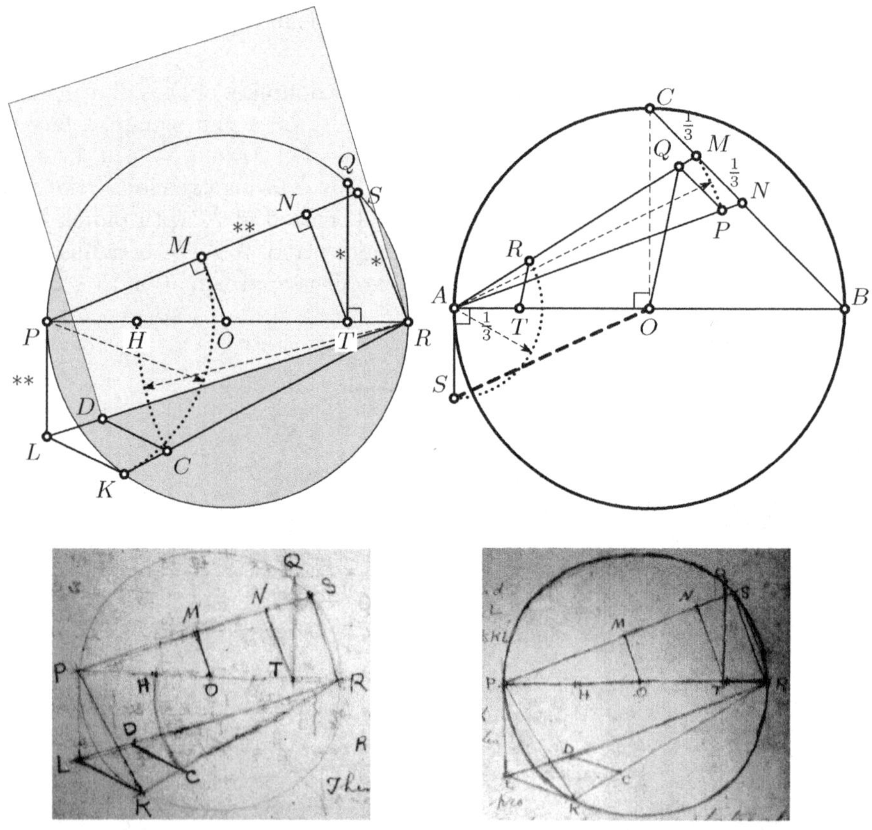

Fig. 8.13. Constructions of Ramanujan for π-approximations; below: Drawings from Ramanujan's Notebooks (vol. 1, p. 66 and vol. 2, p. 225; facsimile publication Tata Institute, Bombay)

both found by the Indian mathematician Ramanujan (1914). The approximations (d) and (e) are the basis for the constructions of Exercises 8 and 9 below.

8. Verify the following construction for the *approximate* quadrature of the circle, found by S. Ramanujan (1913): Let a circle be given, with radius 1, centre O and diameter PR (see Fig. 8.13, left). Then construct H and T on PR such that $OH = \frac{1}{2}$ and $OT = \frac{2}{3}$, Q on the circle vertically above T, S on the circle such that $RS = TQ$, N and M on PS such that TNP and OMP are right angles, L vertically below P such that $PL = MN$, K on the circle such that $PK = PM$, C on KR such that $CR = HR$, D on LR such that $DC \parallel LK$. Show that then $RD = \sqrt{\frac{355}{113}}$, which is Metius' approximation for $\sqrt{\pi}$ from Exercise 7 above. Ramanujan apparently liked his construction, as we can see from his note-books (see Fig. 8.13).

9. The tireless Ramanujan also found a relatively simple construction for the even better approximation (e) in (8.45) (see Fig. 8.13, upper right): Let AB be a diameter of the unit circle and C the "North Pole". Draw points T, M, N such that $\frac{1}{3} = AT = CM = MN$ as indicated in the figure. Then determine P on AN such that $AP = AM$ and Q on AM such that $PQ \parallel NM$. Finally determine R on AM such that $TR \parallel OQ$, and S on the perpendicular to AO such that $AR = AS$. Then $\sqrt{SO}$ is an approximation to $\pi/3$ corresponding to (8.45 (e)).

9

Spatial Geometry and Vector Algebra

> *vector*, a quantity having direction as well as magnitude (from Latin *vector*, carrier; stems from *vehere*, to carry)
>
> (Hoad, *The Concise Oxford Dictionary of English Etymology*)

Spatial geometry starts in Book XI of Euclid's *Elements* with a long list of definitions and boring propositions. It is here that analytic geometry shows its full power: one simply adds a third coordinate z. If you know how to calculate with two variables, you can also calculate with three.

Just as easily, one then adds a fourth coordinate, then a fifth one, etc. The only constraint is the limited supply of letters. It is thus judicious to write

$$x_1, x_2, \ldots, x_n \tag{9.1}$$

for the coordinates.

Vector notation. A second revolution, comparable to that of Descartes, took place towards the end of the 19th century with the introduction of *vectors*.

At that time, one began to consider n-tuples of coordinates as *new mathematical objects*[1]

$$x = (x_1, x_2, \ldots, x_n) \qquad \text{or} \qquad a = (a_1, a_2, \ldots, a_n). \tag{9.2}$$

With these objects, one can perform *algebraic operations*, such as calculating their sum and their product with a scalar, by performing them componentwise:

$$a + b = (a_1 + b_1, a_2 + b_2, \ldots, a_n + b_n), \qquad \lambda a = (\lambda a_1, \lambda a_2, \ldots, \lambda a_n). \tag{9.3}$$

The vector notation results in much shorter and clearer proofs. Moreover, the proofs are the same for *any dimension*.

---

[1] Many authors distinguish vectors from scalars by a special notation. For example, vectors are often denoted as $\mathbf{a}, \mathbf{x}$ or $\vec{a}, \vec{x}$ or $\overrightarrow{a}, \overrightarrow{x}$ or $\underline{a}, \underline{x}$ or $\mathfrak{a}, \mathfrak{x}$. Like Banach (see Fig. 9.2) we use ordinary letters in the following.

A. Ostermann and G. Wanner, *Geometry by Its History*,
Undergraduate Texts in Mathematics, DOI: 10.1007/978-3-642-29163-0_9,
© Springer-Verlag Berlin Heidelberg 2012

Historical development of vectors. The introduction of vectors can be traced back to several origins.

(a) Grassmann (extensive Größen [extensive quantities]).

> "... il est très utile d'introduire la considération des nombres complexes, ou nombres formés avec plusieurs unités, ... [it is very useful to introduce the consideration of complex numbers, or numbers composed of several quantities]"
>
> (Peano, Math. Ann. 32 (1888), p. 450)

> "... e il lavoro più profondo che abbiasi su questo soggetto è senza dubbio l'*Ausdehnungslehre* pubblicato dal Grassmann ... [and the profoundest work which we have on this subject is without doubt the *Ausdehnungslehre* published by Grassmann]"
>
> (Peano 1894, *Opere Scelte* III, p. 340)

The German theologian and linguist Hermann Grassmann (1809–1877), self-taught in mathematics, published in 1844 his work *Die lineale Ausdehnungslehre*, an unreadable book, interspersed with mystic and abstract considerations. In 1862, a revised edition appeared, which did not attract more attention. Grassmann's ideas became more widely known in mathematics only 20 to 30 years later (see the quotations from Peano).

(b) Hamilton (quaternions).

> "At the age of five Hamilton could read Latin, Greek, and Hebrew. At eight he added Italian and French; at ten he could read Arabic and Sanskrit and at fourteen, Persian."
>
> (M. Kline, 1972, p. 777)

> "Tomorrow will be the fifteenth birthday of the Quaternions. They started into life, or light, full grown, on the 16th of October, 1843, as I was walking with Lady Hamilton to Dublin, and came up to Brougham Bridge. That is to say, I then and there felt the galvanic circuit of thought closed ... I felt a *problem* to have been at that moment *solved*, an intellectual *want relieved*, which had *haunted* me for at least *fifteen years* before."
>
> (Hamilton; quoted by M. Kline, 1972, p. 779)

In 1837, William R. Hamilton (1805–1865), a celebrated Irish physicist (optics, mechanics) and mathematician, introduced the complex numbers

$$a + ib \;\leftrightarrow\; (a, b)$$

as *pairs of real numbers*. This definition is still used today. Later, he struggled mightily but unsuccessfully (see the quotation) to generalise these numbers, which can be multiplied and divided, to *three* dimensions. Finally, in 1843 he found a generalisation to *four* dimensions

$$a + ib + jc + kd \,,$$

the *quaternions*. With the noncommutative multiplication rules

$$i^2 = j^2 = k^2 = -1 \,, \qquad \begin{matrix} ij = k \,, & jk = i \,, & ki = j \,, \\ ji = -k \,, & kj = -i \,, & ik = -j \,, \end{matrix} \qquad (9.4)$$

the *product* of two quaternions $e \cdot f$ (written in matrix notation) is

$$
\begin{aligned}
(x_0 + ix_1 + jx_2 + kx_3) \\
\cdot (y_0 + iy_1 + jy_2 + ky_3)
\end{aligned}
=
\begin{bmatrix}
x_0 & -x_1 & -x_2 & -x_3 \\
x_1 & x_0 & -x_3 & x_2 \\
x_2 & x_3 & x_0 & -x_1 \\
x_3 & -x_2 & x_1 & x_0
\end{bmatrix}
\begin{bmatrix}
y_0 \\ y_1 \\ y_2 \\ y_3
\end{bmatrix}
\qquad (9.5)
$$

and is again a quaternion. The fact that the product with the "conjugate"

$$(a + ib + jc + kd) \cdot (a - ib - jc - kd) = a^2 + b^2 + c^2 + d^2$$

turns out to be real, allows the definition of a *division* of quaternions in a way similar to that in (8.5) for complex numbers.

Towards vectors. We see in (9.5) (in grey) that a skew-symmetric matrix appears in dimensions $1, 2, 3$ whose structure will soon become familiar to us. This part of the matrix changes sign if the two factors are exchanged. Hamilton called this three-dimensional part of a quaternion

$$ix_1 + jx_2 + kx_3 \;\leftrightarrow\; (x_1, x_2, x_3)$$

a *vector* (1845, an object transporting something).

Hamilton (and his successors) were very proud of the invention of quaternions. It may be a surprise for many mathematicians to see that quaternion multiplication had already been published in 1760 by Euler in a work on the representation of integers as a sum of four squares (E242, see Fig. 9.1).

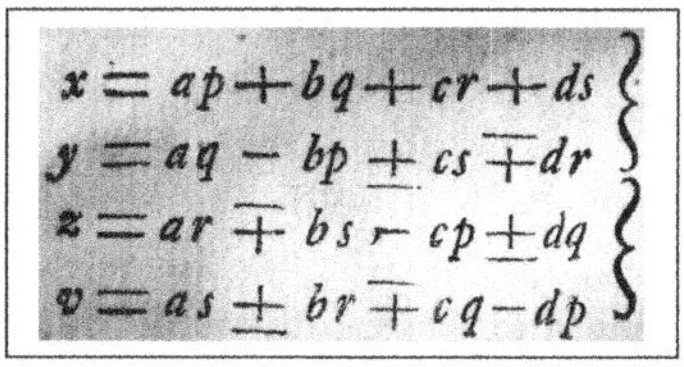

Fig. 9.1. Publication from 1760 of quaternion multiplication in Euler E242. The correspondence with formula 9.5 is obtained by setting $a = y_0$, $b = -y_1$, $c = -y_2$, $d = -y_3$, $p = x_0$, $q = x_1$, $r = x_2$, $s = x_3$ and taking the upper signs.

(c) Heaviside, Gibbs (vectors in physics).

> "In mathematics and especially in physics two very different kinds of quantity present themselves. Consider, for example, mass, time, density, temperature, force, displacement of a point, velocity, and acceleration. Of these quantities some can be represented adequately by a single number ... A *vector* is a quantity which is considered as possessing *direction* as well as *magnitude*. A *scalar* is a quantity which is considered as possessing *magnitude* but no direction."   (Gibbs and Wilson, *Vector Analysis*, 1901, p. 1)

Certain quantities in physics such as velocity, force, electric and magnetic fields
have not only a certain *value* but also a certain *direction*. Inspired by the work
of Hamilton, physicists started to apply these ideas to mechanics, electricity
and magnetism (Clifford, Heaviside, Gibbs). They discovered, however, that
is is better to remove all traces of quaternions from the theory of vectors. In
this respect, they are much closer to Grassmann's ideas than to Hamilton's.

(d) Banach, Wiener (axioms of a "general" vector space).

> "Fréchet était très excité par le fait que Banach avait donné plu-
> sieurs mois avant Wiener un système d'axiomes de l'espace vecto-
> riel ... [Fréchet was very excited by the fact that Banach had given
> a system of axioms for vector spaces several months before Wiener
> ...]"
>   (H. Steinhaus, *Oeuvres de Banach*, p. 15)

During the first half of the 20th century, *vector spaces* were finally defined in
an axiomatic way independently by Stefan Banach (in the second chapter of
his *Théorie des opérateurs linéaires*, Lwów 1932; see Fig. 9.2) and by Norbert
Wiener (see the quotation).

$$
\begin{aligned}
&1)\quad x + y = y + x, \\
&2)\quad x + (y + z) = (x + y) + z, \\
&3)\quad x + y = x + z \ \textit{entraîne} \ y = z, \\
&4)\quad a(x + y) = ax + ay, \\
&5)\quad (a + b)x = ax + bx, \\
&6)\quad a(bx) = (ab)x, \\
&7)\quad 1 \cdot x = x.
\end{aligned}
$$

Fig. 9.2. First publication of vector space axioms in Banach (1932)

Geometric meaning of vectors. By choosing the canonical basis[2] $e_1 = (1, 0)$, $e_2 = (0, 1)$, we have the equivalence (for $n = 2$)

$$
a = (a_1, a_2) \quad \Leftrightarrow \quad a = a_1 e_1 + a_2 e_2 \quad \Leftrightarrow \quad \tag{9.6}
$$

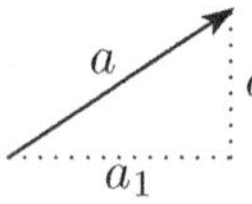

between the 'algebraic' objects of Grassmann and the 'geometric' objects of
the physicists. For $n = 3$, see Fig. 9.4 (a).

The meaning of algebraic operations. *Multiplication by a scalar* λ in
(9.3) lengthens (or shortens) the vector a (and reverses its direction if $\lambda < 0$),
see Fig. 9.3 (a) and Fig. 9.4 (b). The *sum* $a + b$ of two vectors completes a
parallelogram, see Fig. 9.3 (b) and Fig. 9.4 (c). Equivalently, this sum can be

---

[2]In this and the following chapter we tacitly assume that the reader is familiar
with some basic notions of linear algebra.

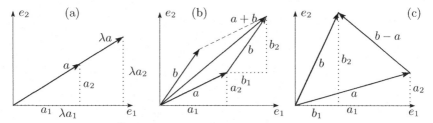

Fig. 9.3. Vectors in $\mathbb{R}^2$: multiplication by a scalar (a), addition of two vectors (b), difference (c)

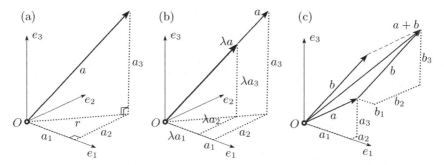

Fig. 9.4. Vectors in $\mathbb{R}^3$: coordinates (a), multiplication by a scalar (b), addition (c)

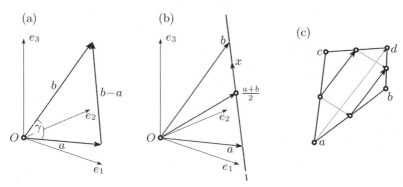

Fig. 9.5. Difference of vectors (a), representation of a straight line (b), Varignon's theorem (c)

seen as putting the two vectors a and b together in a head and tail fashion by parallel translation of one of them. In this way, it is convenient to identify two parallel vectors that have the same length and the same direction.[3] The *difference* $b - a$ is the vector that connects the point a with the point b, see Fig. 9.3 (c) and Fig. 9.5 (a).

---

[3]In the literature, one finds several definitions that take these distinctions into account ("vector", "free vector", "bound vector", "position vector", etc.). We simply call a vector "a point" if it starts at the origin O.

9.1 First Applications of Vectors

Straight line through two points; parametric form. In order to find the straight line that passes through distinct points a and b, one proceeds as follows. Starting from a, one walks in direction $b - a$ to get

$$x = a + \lambda(b - a),\tag{9.7}$$

where λ is a parameter that determines the position of the point x on the line: for $\lambda = 0$ we have $x = a$, for $\lambda = 1$ we recover $x = b$. The *midpoint* of the segment $[a, b]$ is obtained by taking $\lambda = \frac{1}{2}$, which gives

$$x = \frac{a + b}{2},\tag{9.8}$$

see Fig. 9.5 (b).

Remark. In two dimensions, by writing formula (9.7) componentwise and eliminating λ from the second equation with the help of the first, one retrieves the usual parameter-free form for the equation of a straight line (see the last formula of (7.2)).

Varignon's theorem. Pierre Varignon (1654–1722), great connoisseur of calculus in France, obtained in the course of his work in statics the following result: *the midpoints of the four sides of an arbitrary quadrilateral in $\mathbb{R}^3$ form a plane parallelogram*, see Fig. 9.5 (c).

For its proof we observe that one pair of midpoints is given by

$$\frac{a + b}{2}, \quad \frac{b + d}{2}, \qquad \text{the other by} \qquad \frac{a + c}{2}, \quad \frac{c + d}{2}.$$

The difference of these pairs is the same vector $\frac{d-a}{2}$.   □

Plane through three points; parametric form. Generalising the idea in (9.7) to three given points a, b and c forming a triangle, we start from a and walk in *two* directions, $b - a$ and $c - a$. Using two parameters λ and μ, we reach with

$$x = a + \lambda(b - a) + \mu(c - a)\tag{9.9}$$

all points in the required plane, see Fig. 9.6 (a).

Barycentric coordinates. In order to make equation (9.9) more symmetric, we write it as

$$\begin{aligned} x &= (1 - \lambda - \mu)a + \lambda b + \mu c \\ &= m_1 a + m_2 b + m_3 c, \end{aligned} \qquad m_1 + m_2 + m_3 = 1.\tag{9.10}$$

The coordinates m_1, m_2, m_3 are called *barycentric coordinates*; they were introduced by Möbius (*Der barycentrische Calcul*, 1827; see Fig. 9.7). Imagine three masses m_1, m_2, m_3 located at the vertices of the triangle abc. If $m_1 + m_2 + m_3 = 1$, then the point x is their barycentre.

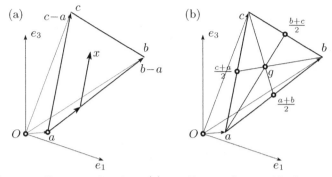

Fig. 9.6. Creation of a plane (a), medians and centroid of a triangle (b)

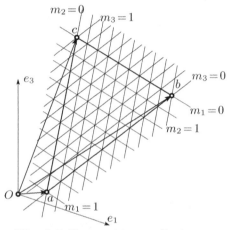

Fig. 9.7. Barycentric coordinates

Nonnegative barycentric coordinates (i.e. $m_1, m_2, m_3 \geq 0$) have the following interpretation: let x be a point as in (9.10); if

(a) $m_1 = 1$ or $m_2 = 1$ or $m_3 = 1$, then x is a vertex of the triangle;
(b) $m_1 = 0$ or $m_2 = 0$ or $m_3 = 0$, then x is on a side;
(c) $m_1 > 0$ and $m_2 > 0$ and $m_3 > 0$, then x is inside the triangle;
(d) $m_1 = m_2$ or $m_2 = m_3$ or $m_3 = m_1$, then x is on one of the medians;
(e) $m_1 = m_2 = m_3 = \frac{1}{3}$, then x is the centroid (see Fig. 9.6 (b)).

It is difficult to prove the theorem of Archimedes (Theorem 4.1 on page 85) in a more elegant way.

9.2 Gaussian Elimination, Volume and Determinant

Given three vectors a, b, c in $\mathbb{R}^3$ (or n vectors in $\mathbb{R}^n$), we want to determine the volume of the parallelepiped which they span.

Gaussian elimination. We solve this problem by considering these vectors as the *row vectors* of a matrix, and carrying out Gaussian elimination. This "algorithmus expeditissimus", used for many centuries to solve systems of linear equations, was described systematically by Gauss (1809) when considering the method of least squares. The matrix is simplified in several steps "per eliminationem vulgarem". The first step consists in subtracting multiples of the first row from the lower rows so that their first coefficients become zero:

$$\begin{bmatrix} a_1 & a_2 & a_3 \\ b_1 & b_2 & b_3 \\ c_1 & c_2 & c_3 \end{bmatrix} \Rightarrow \begin{bmatrix} a_1 & a_2 & a_3 \\ 0 & b_2 - \frac{b_1}{a_1}a_2 & b_3 - \frac{b_1}{a_1}a_3 \\ c_1 & c_2 & c_3 \end{bmatrix} \Rightarrow \begin{bmatrix} a_1 & a_2 & a_3 \\ 0 & b_2 - \frac{b_1}{a_1}a_2 & b_3 - \frac{b_1}{a_1}a_3 \\ 0 & c_2 - \frac{c_1}{a_1}a_2 & c_3 - \frac{c_1}{a_1}a_3 \end{bmatrix}$$

The coefficient a_1, which allowed this elimination and which must be different[4] from 0, is called the first *pivot*. In the second step we use the second pivot $b_2 - \frac{b_1}{a_1}a_2$ in the second row and eliminate the second coefficients of the lower row(s):

$$\Rightarrow \begin{bmatrix} a_1 & a_2 & a_3 \\ 0 & b_2 - \frac{b_1}{a_1}a_2 & b_3 - \frac{b_1}{a_1}a_3 \\ 0 & 0 & c_3 - \frac{c_1}{a_1}a_3 - \frac{c_2 - \frac{c_1}{a_1}a_2}{b_2 - \frac{b_1}{a_1}a_2}\left(b_3 - \frac{b_1}{a_1}a_3\right) \end{bmatrix} \tag{9.11}$$

If $n > 3$ the algorithm continues with the complicated expression in position $(3,3)$ of (9.11) as third pivot, and the algebraic expressions of the results soon become unmanageable. But the numerical versions of the algorithm are the basis for countless scientific calculations with thousands of variables.

Volume of a parallelepiped. The geometric meaning of Gaussian elimination is illustrated in Fig. 9.8 for $n = 2$ and in Fig. 9.9 for $n = 3$. In each of these operations, one of the vectors is moved in the direction of one of the others, and by Eucl. I.35 (or Eucl. XI.29), *the volume does not change*. After having eliminated the coefficients below the diagonal as in (9.11), we continue to eliminate the coefficients *above* the diagonal by eliminating *from below* (last picture in Fig. 9.8; second row in Fig. 9.9). For $n = 3$, we arrive at a rectangular cuboid, *whose sides are the three pivots*.

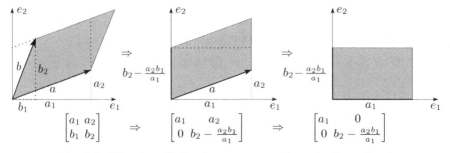

Fig. 9.8. Gaussian elimination preserves volume in dimension 2

---

[4]Otherwise, we exchange rows.

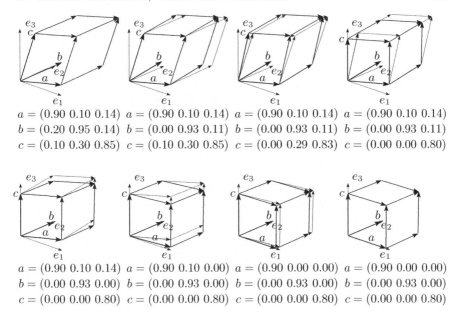

$a = (0.90\ 0.10\ 0.14)$ $a = (0.90\ 0.10\ 0.14)$ $a = (0.90\ 0.10\ 0.14)$ $a = (0.90\ 0.10\ 0.14)$
$b = (0.20\ 0.95\ 0.14)$ $b = (0.00\ 0.93\ 0.11)$ $b = (0.00\ 0.93\ 0.11)$ $b = (0.00\ 0.93\ 0.11)$
$c = (0.10\ 0.30\ 0.85)$ $c = (0.10\ 0.30\ 0.85)$ $c = (0.00\ 0.29\ 0.83)$ $c = (0.00\ 0.00\ 0.80)$

$a = (0.90\ 0.10\ 0.14)$ $a = (0.90\ 0.10\ 0.00)$ $a = (0.90\ 0.00\ 0.00)$ $a = (0.90\ 0.00\ 0.00)$
$b = (0.00\ 0.93\ 0.00)$ $b = (0.00\ 0.93\ 0.00)$ $b = (0.00\ 0.93\ 0.00)$ $b = (0.00\ 0.93\ 0.00)$
$c = (0.00\ 0.00\ 0.80)$ $c = (0.00\ 0.00\ 0.80)$ $c = (0.00\ 0.00\ 0.80)$ $c = (0.00\ 0.00\ 0.80)$

Fig. 9.9. Gaussian elimination preserves volume in dimension 3

Theorem 9.1. *The volume of a parallelepiped is equal (apart from the sign) to the product of the pivots of Gaussian elimination applied to the coefficient matrix of the generating vectors. This product is called the determinant of the matrix.*[5]

> "Je crois avoir trouvé pour cela une Règle assez commode & générale, ... On la trouvera dans l'*Appendice*, N°. 1. [I believe that I have discovered a quite convenient and general rule for this ... It is given in the Appendix, No. 1.]" (G. Cramer, 1750, p. 60)

Examples. For $n = 2$, the product of the first two pivots in (9.11) gives

$$\mathcal{A} = a_1 b_2 - a_2 b_1 = \det \begin{bmatrix} a_1 & a_2 \\ b_1 & b_2 \end{bmatrix} \tag{9.12}$$

for the area of a parallelogram, and for $n = 3$ the product simplifies to

---

[5]The theory of determinants is older than that of matrices. The first motivations came from algebra (Maclaurin, 1748, G. Cramer, 1750 (see Fig. 9.10) and Bézout, 1764); the principal architects of determinants were Vandermonde (1772) and Laplace (1772); the geometric significance as a volume was discovered by Euler, Lagrange and Jacobi (1769, 1773 and 1841, for transforming multiple integrals). A classical introduction to the theory of determinants is the book by A.C. Aitken (1964).

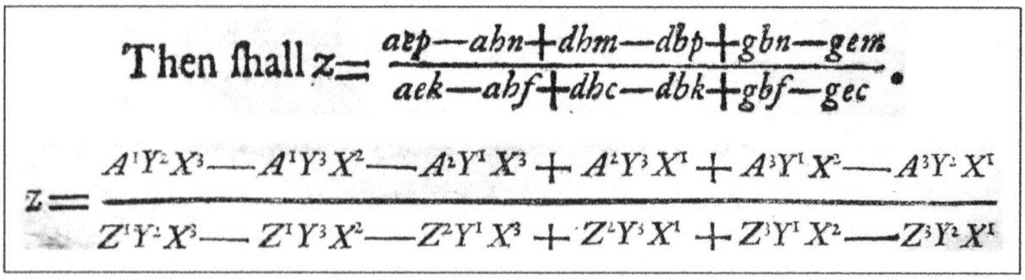

$$\text{Then fhall } z = \frac{azp - abn + dbm - dbp + gbn - gem}{aek - abf + dbc - dbk + gbf - gec}.$$

$$z = \frac{A^1Y^2X^3 - A^1Y^3X^2 - A^2Y^1X^3 + A^2Y^3X^1 + A^3Y^1X^2 - A^3Y^2X^1}{Z^1Y^2X^3 - Z^1Y^3X^2 - Z^2Y^1X^3 + Z^2Y^3X^1 + Z^3Y^1X^2 - Z^3Y^2X^1}$$

Fig. 9.10. *Cramer's rule and first determinants as denominators of solutions of linear equations*; above: from Maclaurin (1748); below: from Cramer (1750)

$$\mathcal{V} = a_1b_2c_3 + a_2b_3c_1 + a_3b_1c_2 - a_3b_2c_1 - a_2b_1c_3 - a_1b_3c_2$$

$$= \det \begin{bmatrix} a_1 & a_2 & a_3 \\ b_1 & b_2 & b_3 \\ c_1 & c_2 & c_3 \end{bmatrix}. \tag{9.13}$$

See Fig. 9.10 for reproductions of the first appearance of determinants in print. For $n = 4$ the corresponding formula has 24 terms, for $n = 5$ there are 120 terms.

9.3 Norm and Scalar Product

It can be seen from Fig. 9.4 (a) that the length of a vector a, denoted by $|a|$ and called its *norm*,[6] is given by

$$|a|^2 = r^2 + a_3^2 = a_1^2 + a_2^2 + a_3^2 = \sum_i a_i^2 \tag{9.14}$$

(apply Pythagoras' theorem twice). With this norm, the space $\mathbb{R}^3$ becomes a *normed space* and consequently a *metric space* with distance

$$d(a, b) = |b - a| = \sqrt{\sum_i (b_i - a_i)^2}.$$

Scalar product. We set $c = b - a$, so that a, b, c form a triangle as in Fig. 9.5 (a), and expand:

$$|c|^2 = |b - a|^2 = \sum_i (b_i - a_i)^2 = \sum_i b_i^2 - 2\sum_i a_i b_i + \sum_i a_i^2. \tag{9.15}$$

Definition 9.2. *The quantity* $\sum_i a_i b_i$ *in the preceding equation is called the* scalar product *of the vectors a and b:*

$$\langle a, b \rangle = \langle a \,|\, b \rangle = a \cdot b = \sum_i a_i b_i \qquad with \quad \langle a, a \rangle = a \cdot a = |a|^2. \tag{9.16}$$

The notation $a \cdot b$ is due to Gibbs and explains the name dot product.

---

[6] also very commonly denoted by $\|a\|$.

We see that the scalar product is symmetric, and *linear* in both arguments:

$$\langle a, b \rangle = \langle b, a \rangle \qquad \text{and} \qquad \langle \lambda_1 a_1 + \lambda_2 a_2, b \rangle = \lambda_1 \langle a_1, b \rangle + \lambda_2 \langle a_2, b \rangle . \qquad (9.17)$$

With this notation, equation (9.15) becomes

$$|c|^2 = |a|^2 + |b|^2 - 2\langle a, b \rangle \qquad \text{or} \qquad \langle a, b \rangle = \frac{|a|^2 + |b|^2 - |c|^2}{2} . \qquad (9.18)$$

Comparing these equations with Pythagoras' theorem, its converse and the law of cosines (formula (5.10) on page 119), we read off three important consequences:

Theorem 9.3. *Two non-zero vectors are orthogonal if and only if their scalar product vanishes.*

Theorem 9.4. *Let γ be the angle between two non-zero vectors a and b. Then*

$$\langle a, b \rangle = |a|\,|b|\cos\gamma \qquad \text{or} \qquad \cos\gamma = \frac{\langle a, b \rangle}{|a|\,|b|} . \qquad (9.19)$$

Theorem 9.5. *Let e be a unit vector ($|e| = 1$). Then the scalar product $a \cdot e = |a|\cos\gamma$ gives the length of the orthogonal projection of a onto e (see the figure), where γ is the angle between a and e.*

This is in agreement with the fact that $a_i = a \cdot e_i$ for the canonical basis vector e_i. We further find the representation

$$a = \sum_i \langle a, e_i \rangle\, e_i$$

with respect to any orthonormal basis.[7]

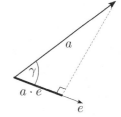

Cartesian equation of a plane. Let a be a given point and n be a vector of norm 1. Then the point x *lies in the plane passing through a and orthogonal to n* if and only if (see Fig. 9.11 (a))

$$(x - a) \cdot n = 0 \qquad \text{i.e.} \qquad n_1 x_1 + n_2 x_2 + n_3 x_3 = q , \qquad (9.20)$$

where $q = n_1 a_1 + n_2 a_2 + n_3 a_3$. On the other hand, given an equation

$$\alpha_1 x_1 + \alpha_2 x_2 + \alpha_3 x_3 = r ,$$

one knows that $(\alpha_1, \alpha_2, \alpha_3)/\sqrt{\alpha_1^2 + \alpha_2^2 + \alpha_3^2}$ is a unit vector orthogonal to the plane defined by this equation.

---

[7]A set of vectors is called orthonormal if its elements are mutually orthogonal and of unit length.

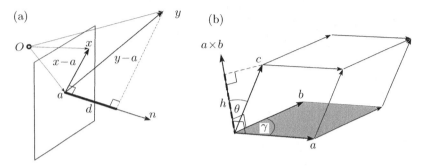

Fig. 9.11. Equation of a plane and distance of a point (a); mixed triple product (b)

Distance of a point to a plane. Let y denote an arbitrary point. Its distance to a plane, determined by (9.20), is given by

$$
\begin{aligned}
d &= (y - a) \cdot n \\
&= (y_1 - a_1) \cos \varepsilon_1 + (y_2 - a_2) \cos \varepsilon_2 + (y_3 - a_3) \cos \varepsilon_3 ,
\end{aligned}
\tag{9.21}
$$

see Fig. 9.11 (a) and Theorem 9.5. The second expression is obtained from Theorem 9.4 by inserting $n_i = \langle n, e_i \rangle = \cos \varepsilon_i$, where ε_i are the angles between n and the coordinate axes. In this form the equation was stated by Hesse (1861, pp. 15–18); the analogue for lines was given in Hesse (1865, pp. 14–17).

Angle between two planes. One calculates a normal vector to each plane and determines the angle *between these two vectors*.

Angle between a straight line and a plane. Knowing the parametric form (9.7) of the line, one calculates the desired angle from the angle between the direction vector $b - a$ of the line and a normal vector of the plane.

It is remarkable for how many geometric questions the scalar product turns out to be useful.

9.4 The Outer Product

Given two vectors a and b, we wish to find a vector x perpendicular to the plane spanned by a and b. In three dimensions, the orthogonal complement of two non collinear vectors is a line through O and hence is determined by a unique direction vector. This vector was studied by Grassmann as a certain product of a and b. The product exists, as a vector of the same dimension, only in three dimensions. It is of great importance for various applications in spatial geometry and in physics.

The two orthogonality conditions (see Theorem 9.3) give us a system of two linear equations:

$$a_1 x_1 + a_2 x_2 + a_3 x_3 = 0\,,$$
$$b_1 x_1 + b_2 x_2 + b_3 x_3 = 0\,,$$

which can be transformed by Gaussian elimination into

$$a_1 x_1 + a_2 x_2 \qquad\qquad + a_3 x_3 \qquad\qquad = 0\,,$$
$$\left(b_2 - \frac{b_1 a_2}{a_1}\right) x_2 + \left(b_3 - \frac{b_1 a_3}{a_1}\right) x_3 = 0\,.$$

In this last equation x_3 can be freely chosen. The choice $x_3 = a_1 b_2 - a_2 b_1$ results in particularly nice formulas for x_2 and x_1, namely

$$x_1 = a_2 b_3 - a_3 b_2\,, \qquad x_2 = a_3 b_1 - a_1 b_3\,, \qquad x_3 = a_1 b_2 - a_2 b_1\,. \qquad (9.22)$$

The product

$$a \times b = (a_2 b_3 - a_3 b_2,\ a_3 b_1 - a_1 b_3,\ a_1 b_2 - a_2 b_1)$$
$$= \left(\det\begin{bmatrix} a_2 & a_3 \\ b_2 & b_3 \end{bmatrix},\ \det\begin{bmatrix} a_3 & a_1 \\ b_3 & b_1 \end{bmatrix},\ \det\begin{bmatrix} a_1 & a_2 \\ b_1 & b_2 \end{bmatrix}\right) \qquad (9.23)$$

is called the *outer product* (or *cross product* or *vector product*) of a and b, see Fig. 9.11 (b). The symbol $\times$ is due to Gibbs and has been in use for about one century. Certain similar structures in modern algebra later motivated its replacement by the symbol $\wedge$.

Mixed triple product. Take the *outer* product $a \times b$, and compute its *scalar* product with a third vector c. To our great surprise, the result is the determinant (9.13):

$$(a \times b) \cdot c = \det\begin{bmatrix} a_1 & a_2 & a_3 \\ b_1 & b_2 & b_3 \\ c_1 & c_2 & c_3 \end{bmatrix} = \mathcal{V}\,. \qquad (9.24)$$

The mixed triple product, also called scalar triple product or box product, is thus the volume of the parallelepiped spanned by a, b and c. The mixed product is invariant under cyclic permutations

$$(a \times b) \cdot c = (b \times c) \cdot a = (c \times a) \cdot b\,. \qquad (9.25)$$

Norm of the outer product. Employing the relation (9.19) for the scalar product in (9.24), we obtain

$$\mathcal{V} = (a \times b) \cdot c = |a \times b| \cdot |c| \cdot \cos\theta = |a \times b| \cdot h\,, \qquad (9.26)$$

where h is the distance of the point c from the plane spanned by a and b (see Fig. 9.11 (b)). Comparing this with Eucl. XI.27 ff., i.e. the formula $\mathcal{V} = \mathcal{A} \cdot h$ in (2.7), we conclude that the *norm of the outer product is equal to the area of the parallelogram spanned by a and b*, namely

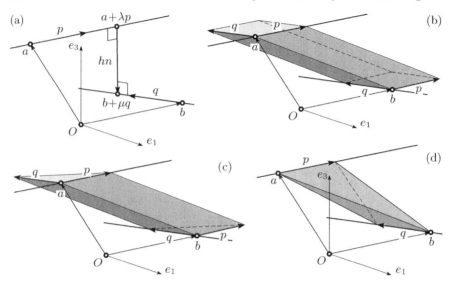

Fig. 9.12. Distance of two skew lines in space (a); volume of a parallelepiped (b); triangular prism (c); tetrahedron between two vectors (d)

$$|a \times b| = \mathcal{A} = |a| \cdot |b| \cdot \sin\gamma. \tag{9.27}$$

Plane passing through three points. The outer product allows one to transform the parametric form of a plane into the parameter-free (Cartesian) form. One obtains the interesting result that the plane passing through three points a, b and c can be written in the form (see (9.20), (9.24) and (9.25))

$$0 = (x - a) \cdot \Big((b - a) \times (c - a) \Big)$$

$$= \Big((x - a) \times (b - a) \Big) \cdot (c - a) = \det \begin{bmatrix} x_1 - a_1 & x_2 - a_2 & x_3 - a_3 \\ b_1 - a_1 & b_2 - a_2 & b_3 - a_3 \\ c_1 - a_1 & c_2 - a_2 & c_3 - a_3 \end{bmatrix}$$

$$= \det \begin{bmatrix} x_1 - a_1 & x_2 - a_2 & x_3 - a_3 & 0 \\ b_1 - a_1 & b_2 - a_2 & b_3 - a_3 & 0 \\ c_1 - a_1 & c_2 - a_2 & c_3 - a_3 & 0 \\ a_1 & a_2 & a_3 & 1 \end{bmatrix} = \det \begin{bmatrix} x_1 & x_2 & x_3 & 1 \\ a_1 & a_2 & a_3 & 1 \\ b_1 & b_2 & b_3 & 1 \\ c_1 & c_2 & c_3 & 1 \end{bmatrix}. \tag{9.28}$$

Distance of two skew lines in space. Let a and b be two points in $\mathbb{R}^3$ and let p and q be two given direction vectors. We wish to determine the shortest distance h between the lines $a + \lambda p$ and $b + \mu q$ (see Fig. 9.12 (a)).

Solution. By parallel translations of the vectors p and q we obtain a parallelepiped spanned by the vectors $b - a$, p and q (see Fig. 9.12 (b)). By (9.13), the volume $\mathcal{V}$ of this parallelepiped is

$$\mathcal{V} = \det \begin{bmatrix} b_1 - a_1 & b_2 - a_2 & b_3 - a_3 \\ p_1 & p_2 & p_3 \\ q_1 & q_2 & q_3 \end{bmatrix}. \tag{9.29}$$

By (9.27), the parallelepiped has a base of area $\mathcal{A} = |p \times q|$, while its altitude is precisely the distance h which we are looking for. Hence, by comparing (9.29) with $\mathcal{V} = \mathcal{A} \cdot h$ (Eucl. XI.27 ff.), we obtain

$$h = \frac{1}{|p \times q|} \det \begin{bmatrix} b_1 - a_1 & b_2 - a_2 & b_3 - a_3 \\ p_1 & p_2 & p_3 \\ q_1 & q_2 & q_3 \end{bmatrix}. \tag{9.30}$$

Volume of the tetrahedron between two vectors in space. If we replace the base parallelogram by a triangle to obtain a triangular prism, the volume will be divided by 2 (Fig. 9.12 (c)). Finally, as in Euclid's figure (upper picture of Fig. 2.35 on page 50), we decompose this prism into three tetrahedra of equal volume. One of these will be the tetrahedron whose opposite edges are the vectors p and q (see Fig. 9.12 (d)). The volume $\mathcal{V}_{\mathrm{Tet}}$ of this tetrahedron is thus

$$\mathcal{V}_{\mathrm{Tet}} = \frac{1}{6} \det \begin{bmatrix} b_1 - a_1 & b_2 - a_2 & b_3 - a_3 \\ p_1 & p_2 & p_3 \\ q_1 & q_2 & q_3 \end{bmatrix} = \frac{h \cdot |p \times q|}{6}. \tag{9.31}$$

This last formula, which is attributed to J. Steiner (see Dörrie, 1943, §198), allows the interesting conclusion that *the volume of a tetrahedron between two vectors in space is independent of the particular choice of a and b on the lines.*

Position of the shortest connection between two skew lines. For the shortest connection between the two lines in Fig. 9.12 (a), the vector connecting the points $a + \lambda p$ and $b + \mu q$ must be perpendicular to both p and q, hence the scalar product of $(a + \lambda p) - (b + \mu q)$ with p and q must be zero. This gives

$$\begin{aligned} (p \cdot p)\lambda - (q \cdot p)\mu &= (b - a) \cdot p, \\ (p \cdot q)\lambda - (q \cdot q)\mu &= (b - a) \cdot q, \end{aligned} \tag{9.32}$$

a linear system for the unknowns λ and μ to solve.

Orientation. By definition, a triple of vectors a, b and c has *positive orientation*, if the sign of $\det(a, b, c)$ is positive. Consequently, these vectors have the same orientation as the three basis vectors e_1, e_2 and e_3. We deduce from (9.26) that the sign of the outer product is chosen so that the triple a, b and $a \times b$ has positive orientation (since $e_1 \times e_2 = e_3$).

In dependence of the orientation of the vectors, the above formulas for volumes and distances can give negative values. Many authors therefore take absolute values. However, it is often preferable not to destroy the additional information about the orientation of the involved vectors.

9.5 Spherical Trigonometry Revisited

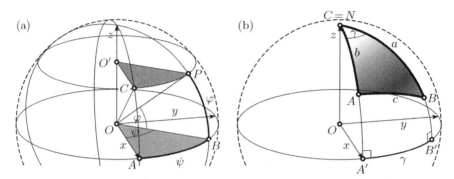

Fig. 9.13. Spherical coordinates (a); the spherical law of cosines (b)

Spherical coordinates. Spherical coordinates, used in many branches of science (geography, astronomy, analysis on the sphere) measure the position of a point P on a sphere of radius 1 by measuring along the *equator* the *longitude* ψ from a point A to a point B, and from there along the *meridian* the *latitude* φ to reach P. If we place the origin of our Cartesian coordinates at the centre of the sphere, the x-axis in direction of A and the z-axis in the direction of the north pole, we ask for the Cartesian coordinates of P (see Fig. 9.13 (a)).

Solution. If we let C be the point of latitude φ on the meridian of A, we have the coordinates

$$B = (\cos\psi, \sin\psi, 0) \quad \text{and} \quad C = (\cos\varphi, 0, \sin\varphi). \tag{9.33}$$

Since the triangles OAB and $O'CP$ are similar with similarity factor $O'C = \cos\varphi$, the x, y-coordinates of P are by Thales' theorem those of B multiplied by this factor, while its z-coordinate is that of C. Hence we have found the spherical coordinates

$$P = (\cos\varphi\cos\psi, \cos\varphi\sin\psi, \sin\varphi). \tag{9.34}$$

The spherical law of cosines. We place a spherical triangle with given arc lengths a and b and a given angle γ as indicated in Fig. 9.13 (b) with C at the North Pole and the side b on the Greenwich meridian. Then we have by (9.33) and (9.34) the coordinates $A = (\sin b, 0, \cos b)$ and $B = (\sin a \cos\gamma, \sin a \sin\gamma, \cos a)$ (the angles a and b are measured from the North Pole, which exchanges sines and cosines). By (9.19) the scalar product of these unit vectors once again gives the formula $\cos c = \sin b \sin a \cos\gamma + \cos b \cos a$.

Remark. Proofs of the spherical law of cosines and of sines for triangles in arbitrary position using more difficult vector algebra are indicated in Exercises 5 and 6 on page 288.

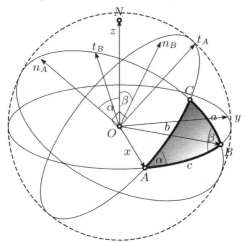

Fig. 9.14. Spherical triangle with vector algebra

Spherical trigonometry using Euler angles. Another possibility is to use *skew* spherical coordinates. Because of their similarity with Euler's method of describing the movement of rigid bodies, we call them *Euler angles*. We move from the point A along the equator on an arc of length c and arrive at $B = (\cos c, \sin c, 0)$. From there we describe an arc of length a on a great circle which makes an angle β with the equator, and arrive at the point C (see Fig. 9.14). To compute the coordinates of C, we first look for a unit vector n_B perpendicular to the plane of this great circle, i.e. n_B must be perpendicular to B and make an angle β with N. Thus the scalar product of n_B with $N = (0, 0, 1)$ must be $\cos \beta$ and its scalar product with $B = (\cos c, \sin c, 0)$ must be zero. This leads to $n_B = (-\sin c \sin \beta, \cos c \sin \beta, \cos \beta)$, a vector already of length 1. We then construct a vector t_B lying in the desired plane and orthogonal to B. This means that $t_B = B \times n_B = (\sin c \cos \beta, -\cos c \cos \beta, \sin \beta)$. Finally, the point C is a linear combination of B and t_B with $\cos a$ and $\sin a$ as coefficients:

$$C = B \cos a + t_B \sin a \tag{9.35}$$
$$= (\cos a \cos c + \sin a \sin c \cos \beta, \ \cos a \sin c - \sin a \cos c \cos \beta, \ \sin a \sin \beta).$$

We now repeat the same procedure starting directly from the point A and moving on an arc of length b under an angle α on the great circle through A and C. This gives similarly $n_A = (0, -\sin \alpha, \cos \alpha)$ and $t_A = n_A \times A = (0, \cos \alpha, \sin \alpha)$, so that

$$C = A \cos b + t_A \sin b = (\cos b, \ \cos \alpha \sin b, \ \sin \alpha \sin b). \tag{9.36}$$

If we compare, coordinate by coordinate, the results for C in (9.36) and (9.35), we obtain at one stroke the law of cosines (5.35) on page 133, formula (5.37) which leads to the law of cotangents, and the law of sines (5.36).

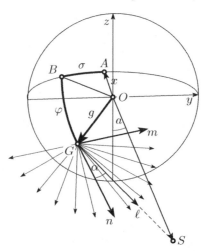

Fig. 9.15. Vector computation of the sundial (the x-coordinate OA is perpendicular to the wall and seen from below, the coordinates y, z span the wall, OB points horizontally to south, OG is the gnomon pointing south parallel to the Earth's axis, the shadow ℓ rotates around G, S is the shadow of G on the wall on March 21 at 1.30 p.m.)

The sundial again. We return to the sundial described in Example 3 on page 137 and find the required angles with vector algebra.

The computations are explained in Fig. 9.15. We choose the coordinate system as indicated there. We retain the notations φ for the latitude and σ for the declination of the wall from the south direction (measured from south to east). The coordinates of G are obtained from (9.34) and become in our situation

$$g = (\cos\varphi\cos\sigma, -\cos\varphi\sin\sigma, -\sin\varphi). \qquad (9.37)$$

We choose the orthonormal basis m, n in the plane perpendicular to the gnomon, where we find m from the conditions $m_3 = 0$ (i.e. m is horizontal) and $m \cdot g = 0$, and n as $n = m \times g$. This gives

$$m = (\sin\sigma, \cos\sigma, 0) \quad \text{and} \quad n = (-\sin\varphi\cos\sigma, \sin\varphi\sin\sigma, -\cos\varphi). \qquad (9.38)$$

Then the direction of the shadow $\ell = (\ell_1, \ell_2, \ell_3)$ which rotates around G is given by

$$\ell = n\cos\alpha + m\sin\alpha, \qquad (9.39)$$

where α has the same meaning as before. The shadow S on the wall is of the form $S = g + \lambda\ell$, where λ is determined by $s_1 = 0$, hence $\lambda = -\frac{g_1}{\ell_1}$. Thus

$$S = \left(0, \; g_2 - \frac{g_1}{\ell_1}\cdot\ell_2, \; g_3 - \frac{g_1}{\ell_1}\cdot\ell_3\right). \qquad (9.40)$$

At the equinox (March 21 or Sept. 23) this point moves on a straight line. If we want to compute the beautiful hyperbolas on which S moves during the rest of the year, we replace the ℓ above by a linear combination of ℓ and g. If we are interested only in the angle a, we compute from (9.40)

$$\cot a = -\frac{s_3}{s_2} = -\frac{g_3\ell_1 - g_1\ell_3}{g_2\ell_1 - g_1\ell_2} = \frac{(g \times \ell)_2}{(g \times \ell)_3}. \tag{9.41}$$

We simplify this by using (9.39) and obtain $g \times \ell = (g \times n)\cos\alpha + (g \times m)\sin\alpha = m\cos\alpha - n\sin\alpha$ (the relation $g \times n = m$ has been used above, $g \times m$ must be perpendicular to g and m, hence is either n or $-n$; you can also use Exercise 4). So we finally obtain

$$\cot a = \frac{m_2\cos\alpha - n_2\sin\alpha}{m_3\cos\alpha - n_3\sin\alpha}. \tag{9.42}$$

Inserting the coordinates for n and m from (9.38), we obtain precisely formula (5.43) on page 138 and, indirectly, a new proof of Euler's cotangent formula (5.40).

9.6 Pick's Theorem

We now apply formula (9.12) for the area of parallelograms and if divided by 2, of triangles to the special case of *lattice polygons*, i.e. *polygons all of whose vertices have integer coordinates*. This condition will lead to a surprisingly beautiful result, discovered by G. Pick (1899). His paper remained unnoticed for more than half a century, until it was revived by Steinhaus' growing interest in mathematical education. An excellent list of references is given in Grünbaum and Shephard (1993). In order to discover this result, we have drawn in Fig. 9.16 several lattice triangles and a polygon.

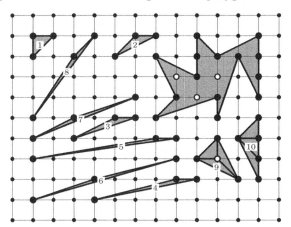

Fig. 9.16. Discovering Pick's theorem

The crucial observation is that the triangles numbered 1 through 8 all have the same area, $\frac{1}{2}$. This is evident from Eucl. I.41 for triangles $1, 2, 3, 4$ and 5. For triangle 6 we have with (9.12)

$$2\mathcal{A} = \det\begin{bmatrix} 7 & 2 \\ 3 & 1 \end{bmatrix} = 7 - 6 = 1 \tag{9.43}$$

and similarly for triangles 7 and 8. All these triangles are characterised by the fact that the only lattice points on or inside them are the three vertices. Therefore they cannot be divided into smaller lattice triangles and we call them, what else, ἄτομος.

In a second step we join atoms to form molecules. Triangles 9 and 10 are both composed of three atoms and both have area $\frac{3}{2}$. This results from *one* additional interior lattice point for triangle 9, and *two* additional boundary lattice points for triangle 10. We see that *an interior lattice point is worth two boundary lattice points* and we conjecture the following theorem.

Theorem 9.6 (Pick's theorem). *Let a simple closed lattice polygon (i.e. one without self-intersections) have i interior lattice points and b boundary lattice points. Then its area is*

$$\mathcal{A} = i + \frac{b-2}{2} \, . \tag{9.44}$$

For example the spider in Fig. 9.16 has area $\mathcal{A} = 3 + \frac{14-2}{2} = 9$.

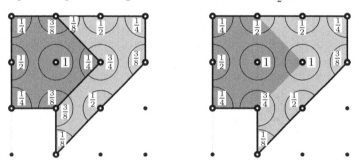

Fig. 9.17. Additivity of $\mathcal{A}$ and $\mathcal{W}$

Proof. Probably the most elegant proof was found by Varberg (1985). It does not require number theory nor chemistry but relies on the following idea: *consider for each interior or boundary lattice point the "visibility angle" w measuring the view from the lattice point into the polygon.* This angle is normalised as follows: a 360° panoramic view counts for 1, a view of 180° counts for $\frac{1}{2}$, a right angle for $\frac{1}{4}$ and so on (see Fig. 9.17). Then add up all these angles to obtain the number $\mathcal{W} = w_1 + w_2 + \dots$. For example, the polygon on the right of Fig. 9.17 has $\mathcal{W} = 2 \cdot 1 + \frac{3}{4} + 4 \cdot \frac{1}{2} + \frac{3}{8} + 3 \cdot \frac{1}{4} + \frac{1}{8} = 6$. The quantity $\mathcal{W}$ is *additive*, that is, if West and East Germany are united, then (see Fig. 9.17 again)

$$W = W_1 + W_2 . \qquad (9.45)$$

In order to see this, we need only add up the visibility angles along the Iron Curtain $\frac{3}{8} + \frac{3}{8} = \frac{3}{4}$, $\frac{1}{4} + \frac{3}{4} = 1$, $\frac{3}{8} + \frac{1}{8} = \frac{1}{2}$. The area $\mathcal{A}$ shares the same property and we will see that

$$\mathcal{A} = W . \qquad (9.46)$$

After dividing the polygon into triangles (in precisely the same way as Proclus' result (2.16) on page 54 was found), it suffices to prove formula (9.46) for lattice triangles. This last verification is carried out in three steps (see Fig. 9.18). Firstly it is immediately seen to hold for rectangles with horizontal and vertical sides (a). Each lattice point sits in a dotted region of exactly the same area as the indicated w. Secondly, upon division by 2, it holds for right-angled triangles with legs parallel to the lattice (b). The only calculation needed here is that $\frac{1}{4} - w + w = \frac{1}{4}$. Finally, since W is additive, it is also *subtractive*, so we can represent an arbitrary triangle in (c) as the difference of a parallel triangle with two parallel triangles and one parallel rectangle. This concludes the proof of Pick's theorem, because the sum in W over the interior lattice points equals i, and the sum over the boundary points, by using Proclus' formula (2.16), equals $\frac{b-2}{2}$. □

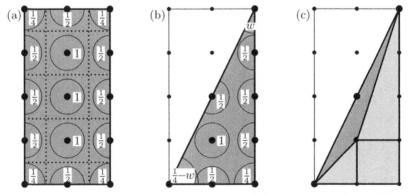

Fig. 9.18. Validity of $\mathcal{A} = W$ for lattice rectangles (a); right-angled lattice triangles with legs parallel to the lattice (b); arbitrary lattice triangles (c)

Remark. The above theorem can be extended to a *lattice annulus*. Consider two lattice polygons, one inside the other, as the borders of an annulus, with b_1 lattice points on the outer border, b_2 on the inner border, i_1 interior lattice points in the annulus, and i_2 interior points in the inner polygon. We consider the annulus as the difference between the outer and the inner polygons. By subtracting the corresponding formulas (9.44) we get for its area

$$\mathcal{A} = i_1 + b_2 + i_2 + \frac{b_1 - 2}{2} - \left(i_2 + \frac{b_2 - 2}{2} \right) = i_1 + \frac{b_1 + b_2}{2} . \qquad (9.47)$$

This last result can also be obtained from (9.46) and the fact that the sum of the *outer* angles of a b-gon is $\frac{b+2}{2}$, a result that according to Heath (1926), p. 322, was already known to Aristotle.

9.7 A Theorem on Pentagons in Space

> "Pólya disclaimed any previous knowledge of the theorem and added 'if van der Waerden didn't know about it then it wasn't known to mathematics'."
>
> (footnote in Dunitz and Waser, 1972)

Van der Waerden (1970) published the following surprising theorem:

Theorem 9.7. *A pentagon in $\mathbb{R}^3$ with all side lengths and all angles equal (see Fig. 9.19) must be planar.*

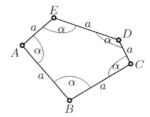

Fig. 9.19. Van der Waerden's pentagon theorem

Remark. The theorem is not true for isogonal equilateral n-gons with $n = 4$ or $n \geq 6$ (see Fig. 9.20). Van der Waerden mentioned in his paper that he had been led to this theorem by discussions with a chemist at the University of Zürich, J.D. Dunitz. Van der Waerden's proof was elegant, but not elementary. Soon after van der Waerden's publication, many more elementary proofs were found (H. Irminger (1970), S. Šmakal (1972), J.D. Dunitz and J. Waser (1972)). Dunitz and Waser explain that the theorem had been discovered in chemistry 25 years earlier "in the course of an electron-diffraction study of gaseous arsenomethane $(AsCH_3)_n$" and present several of the classical proofs

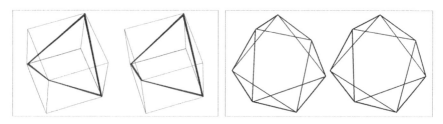

Fig. 9.20. Equilateral and isogonal, but not planar, polygons for $n = 4$ (left) and $n = 7$ (right)

by J. Waser. S. Šmakal gives a proof using vector calculus (see page 299) and remarks that the theorem, which was thought to be unknown to mathematicians (see the quotation) had been published in 1961 in the Russian journal *Prosveshchenie*, inspired by a problem proposed in 1957 by V.I. Arnol'd.

Proof. (a) In order to help our geometric intuition, we start with some calculations: we take all distances AB, BC, CD, DE and EA equal to 1. If all angles α are fixed, then by Eucl. I.4 all diagonals $AC = BD = CD = CE = DA = d$ are determined. They satisfy $d = 2 \sin \frac{\alpha}{2}$ or, by (5.9) on page 118, $d^2 = 2 - 2c$ where $c = \cos \alpha$. Thus the shape of the triangle ABC with sides $1, d, 1$ is fixed; we choose the coordinate system so that

$$A = (-\tfrac{d}{2}, 0, 0), \quad B = (0, -r, 0), \quad C = (\tfrac{d}{2}, 0, 0) \quad \text{with} \quad r = \sqrt{1 - \tfrac{d^2}{4}}.$$

The point D is determined by the conditions $CD = 1$, $AD = d$ and $BD = d$. Using (9.14), this gives three equations for the coordinates x, y, z of this point:

$$(x - \tfrac{d}{2})^2 + \quad y^2 \quad + z^2 = 1,$$
$$(x + \tfrac{d}{2})^2 + \quad y^2 \quad + z^2 = d^2,$$
$$x^2 \quad + (y + r)^2 + z^2 = d^2.$$

Subtracting the second equation from the first, then the third from the second, and finally inserting x and y into the first equation, we obtain, after simplification,

$$x = \frac{d}{2} - \frac{1}{2d}, \quad y = \frac{d^2 - 3/2}{2r}, \quad z^2 = -\frac{(d^2 + d - 1)(d^2 - d - 1)(d^2 + 1)}{d^2(4 - d^2)}.$$
$$(9.48)$$

The results for the fifth point E are precisely the same, except that x is replaced by $-x$. We see that x and y are uniquely determined for all d with $0 < d < 2$. The variable z appears as z^2 with a formula containing the expressions (1.3) from the golden ratio. Thus for $d = \Phi$ and $d = \frac{1}{\Phi}$ we obtain $z = 0$. In this case our pentagon is planar, once the ordinary regular pentagon, once the "Soviet style" pentagon of page 9. For d between these two values, we have $z^2 > 0$ and there are two solutions for D, one above the xy-plane and one symmetrically below this plane. The same two possibilities hold for the z-coordinates of the point E. We conclude, in particular, that

> *if four points of our pentagon lie in one plane (i.e. z is zero),*
> *then the fifth point also lies in this plane.* $\qquad (9.49)$

(b) We now calculate analytically the distance DE: if the sign of z is the same for D and E, we have $DE = d - \frac{1}{d}$, twice the value of x in (9.48). Clearly $DE \neq 1$ if d is not the golden ratio or its inverse. If the sign of z is different for D and E, we compute $DE^2 - 1 = (d - \frac{1}{d})^2 + 4z^2 - 1$ which simplifies to $\frac{5(-d^4+3d^2-1)}{4-d^2}$, so that again $DE \neq 1$ away from the golden ratio. This concludes the proof in all cases.

(c) But all the calculations of step (b), and all the details of the calculations of step (a), become superfluous if we follow the elegant reasoning of Irminger (1970): suppose that the z-coordinates of D and E are the same: then the points $ACDE$ lie in one plane. By (9.49), the fifth vertex must also lie in the same plane. Suppose, secondly, that D lies below and E lies above the plane x, y spanned by ABC. Then both points A and E lie *above* the plane spanned by BCD. So if we start our calculations with *this* triangle, we are back at the first situation.   □

9.8 Archimedean Solids

The description of the five regular, or Platonic, solids was the culmination of Euclid's last Book XIII (see Chap. 2). What can we do better than conclude this chapter with the solids whose faces are regular polygons of *more* than one kind. More precisely, we look for convex solids such that each vertex is surrounded by regular polygons arranged in the same way. Pappus, in his *Collection*, Book V,[8] briefly described 13 such solids which he attributed to Archimedes. However, the work of Archimedes to which Pappus refers has not been found.

The slow rediscovery of these semiregular, or Archimedean solids required considerable effort by scientists and artists (for example Piero della Francesca, Luca Pacioli, Leonardo da Vinci, and Albrecht Dürer). Only Kepler finally published the complete collection of these "Archimedêa Corpora" in his *Harmonices mundi* (1619, Liber II, Propositio XXVIII). He also gave them names, whose English translations have become standard (see Fig. 9.21 and Table 9.1).

A detailed account of this rediscovery is given in Field (1996). A recent discovery of printing blocks for woodcuts by an unknown artist dating from about 1550 is described in Schreiber, Fischer and Sternath (2008).

To construct these solids, we can start from one of the Platonic solids and cut off vertices and/or truncate edges in five different ways, as follows:

1. The easiest method is to cut off the vertices in a symmetric way. Take for example a cube of side length 1 (see Fig. 9.22, left) and shorten each edge around a vertex by an undetermined length u; then the vertices are replaced by equilateral triangles of side length $\sqrt{2}\,u$, while the square faces become octagons with sides of length $\sqrt{2}\,u$ and $1-2u$. If $1-2u = \sqrt{2}\,u$, i.e. if $u = \frac{1}{2+\sqrt{2}}$, these octagons are regular and we obtain our (and Kepler's) first Archimedean solid, the *truncated cube* ("Cubus truncus"), whose beauty can be admired in three dimensions in Fig. 9.22, right. For the truncated dodecahedron we have $u = \frac{1}{2+\Phi}$ (number 3 in Fig. 9.21), and for the truncated tetrahedron,

---

[8] in the 1660 edition on page 129 without title; in Hultsch's edition on page 351 under the title *Libri quinti pars secunda, In Archimedis solidorum doctrinam*

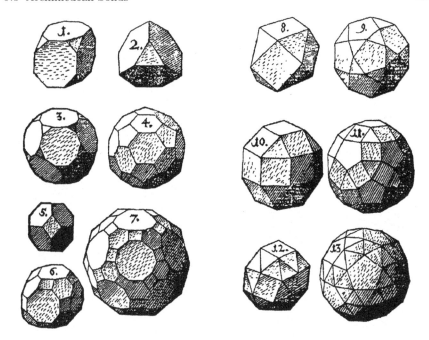

Fig. 9.21. Archimedean solids (drawings by Kepler, *Harmonices mundi* 1619)

Table 9.1. Kepler's list of the Archimedean solids

n	name	composed of	faces	edg.	vert.
1	trunc. cube	8 triang., 6 octog.	14	36	24
2	trunc. tetrahedron	4 triang., 4 hexag.	8	18	12
3	trunc. dodecahedron	20 triang., 12 decag.	32	90	60
4	trunc. icosahedron	12 pent., 20 hexag.	32	90	60
5	trunc. octahedron	6 squares, 8 hexag.	14	36	24
6	trunc. cuboctahedron	12 squares, 8 hexag., 6 octog.	26	72	48
7	trunc. icosidodecahedron	30 squares, 20 hexag., 12 decag.	62	180	120
8	cuboctahedron	8 triang., 6 squares	14	24	12
9	icosidodecahedron	20 triang., 12 pent.	32	60	30
10	rhombicuboctahedron	8 triang., 18 squares	26	48	24
11	rhombicosidodecahedron	20 triang., 30 squares, 12 pent.	62	120	60
12	snub cube	32 triang., 6 squares	38	60	24
13	snub dodecahedron	80 triang., 12 pent.	92	150	60

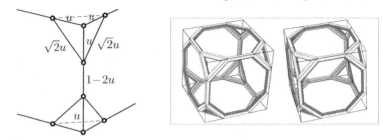

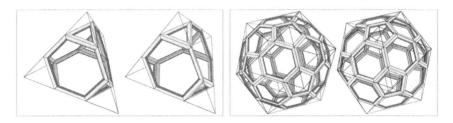

Fig. 9.22. First method of truncation (left); truncated cube (right)

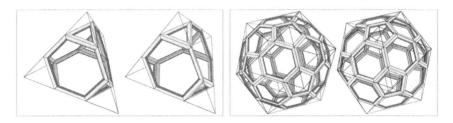

Fig. 9.23. truncated tetrahedron (left), truncated icosahedron (right)

icosahedron and octahedron (numbers 2, 4 and 5 respectively) we have $u = \frac{1}{3}$ (two of them in Fig. 9.23).

2. If we cut off the vertices up to the midpoints of the edges (i.e. if we set $u = \frac{1}{2}$), we again obtain regular faces. If we start from the cube or the octahedron, this leads to the same solid shown in Fig. 9.24 (left), called a *cuboctahedron*. Starting from the dodecahedron or the icosahedron this procedure gives the *icosidodecahedron* (Fig. 9.24, right). In Kepler's catalogue (Fig. 9.21) the two solids are numbered 8 and 9. The cuboctahedron was probably the very first of the Archimedean solids to be discovered ("according to Heron, [...] Archimedes [...] said that Plato also knew one of them"; Heath (1921), vol. I, p. 295). In a famous engraving of Dürer from 1514, a lady, after having cut off *two* of the eight vertices from a cube, gave up, threw saw and plane away and fell into *Melencolia* (Fig. 9.25).

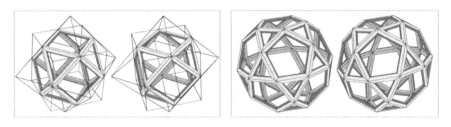

Fig. 9.24. Truncating up to the midpoint of the edges; cuboctahedron from the cube and the octahedron (left); icosidodecahedron (drawing from A. Abdulle and G. Wanner, *Elem. Math.* 57 (2002), right)

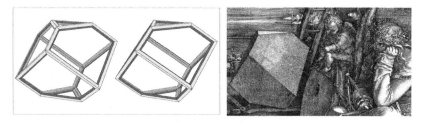

Fig. 9.25. Dürer's Melencolia and its interpretation based on serious geometry

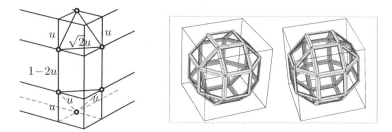

Fig. 9.26. Truncating the edges; the rhombicuboctahedron (right)

3. The third possibility is to truncate the *edges*, as shown in Fig. 9.26 for the cube. The edges are then replaced by rectangles of side lengths $\sqrt{2}\,u$ and $1 - 2u$. These are squares if $u = \frac{1}{2+\sqrt{2}}$. The solid corresponding to this choice of u can be admired in Fig. 9.26 (right) and has the complicated name *rhombicuboctahedron*. The solid obtained similarly from the dodecahedron bears the name *rhombicosidodecahedron* (number 11 in Fig. 9.21).

4. We now truncate the edges *and* the vertices of the cube by two undetermined lengths u and v (see Fig. 9.27, left). The vertices, which earlier became triangles, now become hexagons with side lengths $\sqrt{2}\,u$ and $\sqrt{2}\,v$. The rectangles which replace the edges have sides of length $\sqrt{2}\,u$ and $1 - 2u - 2v$. The faces become octagons with sides of length $\sqrt{2}\,v$ and $1 - 2u - 2v$. All these polygons are regular if $u = v = \frac{1}{4+\sqrt{2}}$. The solid thus obtained, shown in

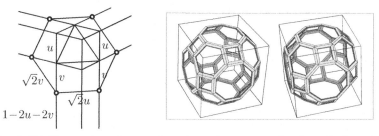

Fig. 9.27. Truncating edges and vertices; the truncated cuboctahedron (right)

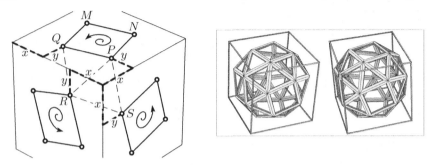

Fig. 9.28. Skew truncations (left); the snub cube (right)

Fig. 9.27 (right), is the *truncated cuboctahedron*[9] (number 6 in Kepler's list). The solid obtained analogously from the dodecahedron is called the *truncated icosidodecahedron* (number 7 in Fig. 9.21).

5. The last two Archimedean solids were the most difficult to find. We shrink the faces by an undetermined factor *and rotate them* by an undetermined angle. We demonstrate the calculations for the cube. Suppose that its sides have length 1 and choose an orientation on the cube. For two unknown lengths x and y we then place points $P, Q, R, S, \ldots$ on each face at distance (x, y) from the vertices by respecting the chosen orientation (see Fig. 9.28, left). For any x and y the polygon $QPNM$ (and all others around the cube constructed similarly) are squares and the triangles PRS etc. are equilateral. We then calculate the distances RQ, PQ and PR using Pythagoras or (9.14) and obtain

$$RQ^2 = y^2 + (1 - 2x)^2 + y^2$$
$$PQ^2 = (1 - x - y)^2 + (x - y)^2$$
$$PR^2 = x^2 + (x - y)^2 + y^2$$

which must all be equal. We subtract the last equation from the other two and obtain

$$PQ^2 - PR^2 = 1 - 2x - 2y(1 - x) = 0 \quad \Rightarrow \quad 2y = \frac{1 - 2x}{1 - x}$$
$$\Downarrow$$
$$RQ^2 - PR^2 = 1 - 4x + 2x^2 + x \cdot 2y = 0 \quad \Rightarrow \quad 1 - 4x + 4x^2 - 2x^3 = 0.$$

This last equation has one real solution, not constructible with ruler and compass, for which we obtain numerically $x = 0.352201128739$, whence $y = 0.228155493654$. With these values, PQR and all other triangles around the cube constructed similarly will be equilateral and the construction of the *snub cube* ("Cubus simus"; see Fig. 9.28, right; number 12 in Fig. 9.21) is complete. The shrinking factor of the squares is 0.43759 and the rotation angle is $16°28'$.

---

[9]The name is slightly misleading; if you simply truncate the corners of a cuboctahedron without further care, you obtain rectangles and not squares.

The same procedure, just with much more complicated formulas, applied to the dodecahedron leads to the shrinking factor 0.56212 and the rotation angle 13°6′ for Kepler's *snub dodecahedron* (see Fig. 9.29, left; number 13 in Fig. 9.21, also the last one in Pappus' list).

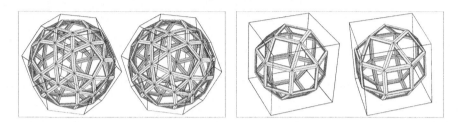

Fig. 9.29. The snub dodecahedron (left); the pseudorhombicuboctahedron (right)

A last surprise. This was the state of the art concerning Archimedean solids after Kepler until the 20th century, when several mathematicians discovered independently that *another* body also satisfies the condition concerning the arrangement of the faces around each vertex: if we look at the rhombicuboctahedron in Fig. 9.26, we see that its central part consists of an octagonal wall of squares, which allows us to twist the "roof" by 45°, without rotating the "basement". The solid obtained in this manner is called — the longest name in this book — a *pseudorhombicuboctahedron* and is drawn in Fig. 9.29. It has however lost the beautiful global symmetry of the previous solids under rotations. An account of the very slow process by which this discovery eventually became better known is given in Grünbaum (2009), with many references to the literature.

9.9 Exercises

1. On the occasion of the publication of an excellent book on geometry, a reception is organised at the *Hofburg* in Vienna, in the presence of high-ranking representatives of politics, academia and the media. The illustrious guests are served *Dom Pérignon* in champagne glasses that have the shape of a paraboloid of revolution. They are allowed to drink until the vertex of the paraboloid becomes visible (see Fig. 9.30 (a)). Then they are asked by Albert Stadler (see *Elem. Math.* 64 (2009), p. 129): "What percentage of the champagne is left in the glass?"

2. (An exercise from *Elem. Math.* 12 (1957), p. 47) Let ABC be a fixed triangle. For a point P inside ABC let A', B' and C' be the centroids of the triangles PBC, APC and ABP respectively. Show that the shape of the triangle $A'B'C'$ is independent of the position of P.

(a)   (b)

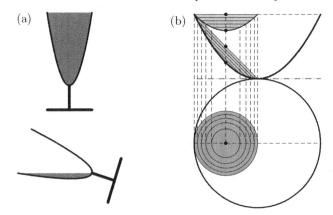

Fig. 9.30. The question in *Wiener Hofburg* (a); its solution (b)

3. The outer product is distributive over addition of vectors:

$$a \times (b + c) = a \times b + a \times c.\tag{9.50}$$

The algebraic proof of (9.50) is a straightforward application of the defining relations (9.23). Give a *geometric* proof of (9.50).

4. Verify algebraically the formulas

$$(a \times b) \times c = b\,(a \cdot c) - a\,(b \cdot c),\tag{9.51}$$

$$(a \times b) \cdot (c \times d) = (a \cdot c)(b \cdot d) - (b \cdot c)(a \cdot d) = \det \begin{bmatrix} a \cdot c & b \cdot c \\ a \cdot d & b \cdot d \end{bmatrix}.\tag{9.52}$$

5. Consider the spherical triangle ABC shown in Fig. 9.31. By Theorem 9.4 we have

$$(u \times v) \cdot (u \times w) = |u \times v|\,|u \times w|\,\cos \alpha.$$

Deduce the spherical law of cosines from this identity.
Hint. Use (9.52).

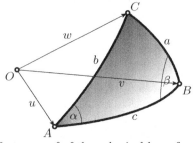

Fig. 9.31. Vector proof of the spherical law of cosines and sines

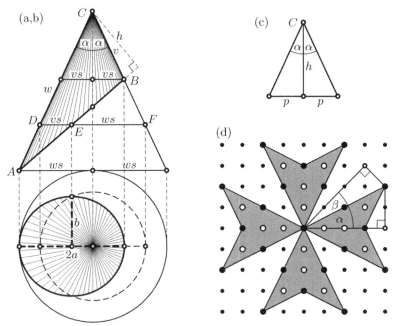

Fig. 9.32. Surface and volume of a truncated cone (a,b); Bernoulli's *Novum Theorema* (c); the Maltese cross and Pick's theorem for multiple boundary points (d)

6. In the situation of the previous exercise apply (9.27) to get

$$|(u \times v) \times (u \times w)| = |u \times v| \, |u \times w| \sin \alpha \,.$$

Deduce the spherical law of sines from this identity.
Hint. Start from the identity

$$(u \times v) \times (u \times w) = -\langle v, u \times w \rangle \, u \,,$$

which is a consequence of (9.51).

7. Suppose that a cone with apex C and given angle α between the surface and the axis of rotation is truncated by an oblique plane from A to B. Denote by v and w the lengths of the shortest and longest generatrix (Fig. 9.32 (a,b)).

(a) Show that the remaining lateral surface of the cone is given by

$$S = \pi \frac{v + w}{2} \sqrt{vw} \sin \alpha \tag{9.53}$$

(G. Pólya, *Elem. Math.* 26 (1971), p. 115).

(b) Show that the volume of the truncated cone above the plane AB is given by

$$V = \frac{\pi}{3} (vw)^{\frac{3}{2}} \sin^2 \alpha \cos \alpha. \tag{9.54}$$

(c) Prove the *Novum Theorema per Jacobum Bernoulli*, announced in the *Acta Eruditorum* 1689, p. 586: if the cone is truncated horizontally at height h, where h is the distance of the cone's apex from the plane AB, then the intersection is a circle whose diameter is $2p$, the *latus rectum Coni-Sectionis* (see Fig. 9.32 (a,b) and (c)), i.e.

$$p = h \cdot \tan \alpha. \tag{9.55}$$

8. Extend Pick's theorem to a lattice domain where the boundary polygon returns *several times* (say m times) to the same lattice point (see Fig. 9.32 (d)).

   Remark. It has apparently been overlooked for the past 250 years that the Maltese cross represents a nice geometric proof of Euler's famous discovery

   $$\arctan \tfrac{1}{2} + \arctan \tfrac{1}{3} = \arctan 1 = \tfrac{\pi}{4};$$

   indeed, for the two angles denoted by α and β in Fig. 9.32 (d) we see that $\tan \alpha = \tfrac{1}{2}$, $\tan \beta = \tfrac{1}{3}$ and $\alpha + \beta = \tfrac{\pi}{4}$.

9. Show that the vertices of an Archimedean solid lie on a sphere; or in the words of A. Dürer's *Underweysung* from 1525: *rüren in einer holen kugel mit all iren ecken an* [touch a hollow sphere with all their vertices; quoted from Schreiber, Fischer and Sternath, 2008].

10. Compute the angles under which the edges of the Archimedean solids are seen from the centre. (The related results for the Platonic bodies are given in (5.33) on page 131.) The problem is again related to the computation of the radius of the circumscribed sphere.

11. For the Archimedean solids, compute the radius ρ of the *inscribed* sphere and find out which of the solids is the "roundest", i.e. for which the ratio ρ/R is maximal.

10

Matrices and Linear Mappings

"I certainly did not get the notion of a matrix in any way through quaternions; it was either directly from that of a determinant or as a convenient way of expressing the equations

$$x' = ax + by$$
$$y' = cx + dy \quad "$$

(Cayley 1855; quotation from M. Kline, 1972, p. 805)

"Über der hartnäckigen Verfolgung des vorgesetzten Weges haben aber die Quaternionisten tiefer liegende Probleme von wahrhaftem Interesse übersehen; ... Diese tiefere Einsicht in die Verhältnisse verdanken wir Cayley. In *A Memoir on the Theory of Matrices* (Phil. Trans. 1858) entwickelt er einen Matrixkalkül ... [In persistently pursuing the preset way the quaternionists have overlooked deeper problems of real interest; ... We owe this deeper insight into the relations to Cayley. In *A Memoir on the Theory of Matrices* (Phil. Trans. 1858) he developed a matrix calculus ...]"

(F. Klein, 1926, p. 189)

The controversy between "Grassmannians" and "Quaternionists" finally ended in the victory of a third "competitor", A. Cayley and his *theory of matrices* (see the quotations). This theory turned mathematics upside down to such an extent that some books on "geometry" from the beginning of the 20th century (for example that of Schreier and Sperner) were written entirely in the language of vectors and matrices.

10.1 Changes of Coordinates

"Cette recherche peut être presque toujours rendue plus facile par des transformations analytiques qui simplifient les équations, en faisant évanouir quelques-uns de leurs termes ... [This investigation can almost always be made easier through analytic transformations, which simplify the equations by eliminating some of their terms ...]"

(J.-B. Biot, *Essai de Géométrie analytique*, Paris 1823, p. 145)

A. Ostermann and G. Wanner, *Geometry by Its History*,
Undergraduate Texts in Mathematics, DOI: 10.1007/978-3-642-29163-0_10,
© Springer-Verlag Berlin Heidelberg 2012

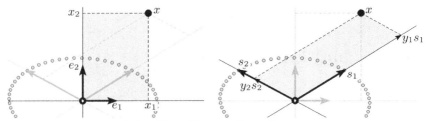

Fig. 10.1. Change of coordinates

The systematic use of changes of coordinates (G. Cramer, 1750, Chap. II, Lacroix, Biot; see the quotation) dates back to the 18th and the beginning of the 19th century.

Let[1]

$$x = \begin{bmatrix} x_1 \\ x_2 \end{bmatrix} = x_1 e_1 + x_2 e_2 \qquad \text{with} \qquad e_1 = \begin{bmatrix} 1 \\ 0 \end{bmatrix}, \qquad e_2 = \begin{bmatrix} 0 \\ 1 \end{bmatrix},$$

and let

$$s_1 = \begin{bmatrix} b_{11} \\ b_{21} \end{bmatrix}, \qquad s_2 = \begin{bmatrix} b_{12} \\ b_{22} \end{bmatrix}$$

be two *linearly independent*[2] vectors (see Fig. 10.1). Expressing the vector x in the new basis s_1, s_2 as $x = y_1 s_1 + y_2 s_2$ (see Fig. 10.1, right) gives the relations

$$\begin{aligned} x_1 &= b_{11} y_1 + b_{12} y_2 \\ x_2 &= b_{21} y_1 + b_{22} y_2 \end{aligned} \qquad \Leftrightarrow \qquad \underbrace{\begin{bmatrix} x_1 \\ x_2 \end{bmatrix}}_{x} = \underbrace{\begin{bmatrix} b_{11} & b_{12} \\ b_{21} & b_{22} \end{bmatrix}}_{B} \underbrace{\begin{bmatrix} y_1 \\ y_2 \end{bmatrix}}_{y}. \qquad (10.1)$$

These formulas mark the beginning of matrix notation (see the quotation).

Theorem 10.1. *The coordinates y of a point x with respect to the basis s_1, s_2 satisfy $x = By$, where the elements of the ith column of the matrix B are the coordinates of the vector s_i with respect to the basis e_1, e_2.*   □

Example. In the new coordinate system based on two conjugate diameters, the ellipse in Fig. 10.1 is described by $y_1^2 + y_2^2 = 1$, the equation of a circle.

10.2 Linear Mappings

There is another interpretation of matrix-vector multiplication. Let A be a given matrix

$$A = \begin{bmatrix} a_{11} & a_{12} \\ a_{21} & a_{22} \end{bmatrix}, \qquad \text{so that} \qquad Ae_1 = \begin{bmatrix} a_{11} \\ a_{21} \end{bmatrix}, \qquad Ae_2 = \begin{bmatrix} a_{12} \\ a_{22} \end{bmatrix}. \qquad (10.2)$$

---

[1]In linear algebra, it is preferable to write a vector as a *column vector*.

[2]Being *linearly independent* means that the vectors span a parallelepiped of nonzero area (volume).

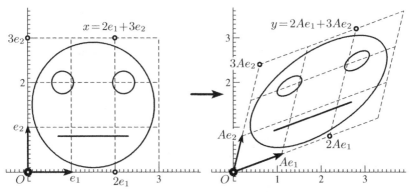

Fig. 10.2. A linear mapping

Define a map $\alpha : x \mapsto y = Ax$ by

$$
\begin{aligned}
y_1 &= a_{11}x_1 + a_{12}x_2 \\
y_2 &= a_{21}x_1 + a_{22}x_2
\end{aligned}
\qquad \text{or} \qquad
\underbrace{\begin{bmatrix} y_1 \\ y_2 \end{bmatrix}}_{y} = \underbrace{\begin{bmatrix} a_{11} & a_{12} \\ a_{21} & a_{22} \end{bmatrix}}_{A} \underbrace{\begin{bmatrix} x_1 \\ x_2 \end{bmatrix}}_{x}.
\tag{10.3}
$$

Instead of considering y as in (10.1) as the coordinates of the *same* point x with respect to *another* basis, one now considers $y = Ax$ as the coordinates of *another* point with respect to the *original* basis e_1, e_2.

Example. The action of the linear mapping $y = Ax$ for the matrix

$$
A = \begin{bmatrix} 1.1 & 0.2 \\ 0.4 & 0.8 \end{bmatrix}
\tag{10.4}
$$

is illustrated in Fig. 10.2. One observes that $\alpha(2e_1 + 3e_2) = 2\alpha(e_1) + 3\alpha(e_2)$. This property, when valid for arbitrary coefficients, characterises a *linear mapping*.[3] Once the images of the basis vectors are known, all other values of y are obtained by linear combinations. The circles of the face of Mona Lisa in Fig. 10.2 are transformed to ellipses, all with the same eccentricity.

Theorem 10.2. *The formula $y = Ax$ defines a linear mapping $y = \alpha(x)$, where the columns of the matrix A are the coordinates of the images $Ae_i = \alpha(e_i)$ of the basis vectors e_i (with respect to this basis).* $\qquad\square$

Rotations in two dimensions. Suppose that the coordinates x_1, x_2 are replaced by new coordinates y_1, y_2, by rotating the axes through an angle α (see Fig. 10.3, left). The similarity of this drawing with Fig. 5.7 (left) on page 118 is *not accidental* but intended. Indeed, in the same way as we proved there the formulas (5.6), we now obtain

---

[3]The notion of a linear mapping is due to S. Banach 1922, *Oeuvres* II, p. 321; Banach's "opérateurs linéaires" did certainly not transform two dimensional faces or cats; they transformed Lebesgue integrable functions in higher analysis.

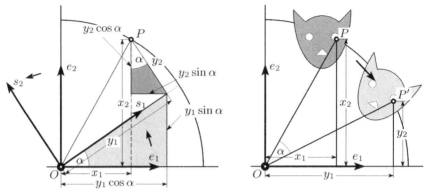

Fig. 10.3. Rotation of a coordinate system; (left) the coordinate axes turn left, (right) the cat turns right

$$
\begin{aligned}
x_1 &= y_1 \cos \alpha - y_2 \sin \alpha \\
x_2 &= y_1 \sin \alpha + y_2 \cos \alpha
\end{aligned}
\quad \text{or} \quad
\begin{bmatrix} x_1 \\ x_2 \end{bmatrix} =
\begin{bmatrix} c & -s \\ s & c \end{bmatrix}
\begin{bmatrix} y_1 \\ y_2 \end{bmatrix}
\tag{10.5a}
$$

with $c = \cos \alpha$ and $s = \sin \alpha$.

Another interpretation of the same situation would be to *rotate the cat through an angle α in the opposite direction* by keeping the basis (Fig. 10.3, right). This yields

$$
\begin{aligned}
y_1 &= x_1 \cos \alpha + x_2 \sin \alpha \\
y_2 &= -x_1 \sin \alpha + x_2 \cos \alpha
\end{aligned}
\quad \text{or} \quad
\begin{bmatrix} y_1 \\ y_2 \end{bmatrix} =
\begin{bmatrix} c & s \\ -s & c \end{bmatrix}
\begin{bmatrix} x_1 \\ x_2 \end{bmatrix}.
\tag{10.5b}
$$

These formulas are obtained either from (10.5a) (multiply the equations alternatively by $\sin \alpha$ and $\cos \alpha$ and add or subtract them) or from Theorem 10.2 (the columns of the matrix in (10.5b) are the images of e_1 and e_2).

Affine maps and translations. One also considers coordinate changes in which *the origin moves*. A typical example for this situation is a translation: for a chosen point $C = (c_1, c_2)$, consider

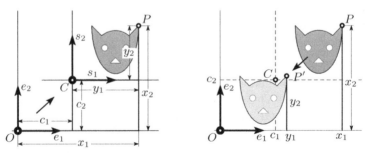

Fig. 10.4. Translation of a coordinate system; the coordinate system moves up (left), the cat moves down (right)

$$x_1 = c_1 + y_1 \qquad \text{or} \qquad \begin{bmatrix} x_1 \\ x_2 \end{bmatrix} = \begin{bmatrix} c_1 \\ c_2 \end{bmatrix} + \begin{bmatrix} y_1 \\ y_2 \end{bmatrix}. \tag{10.6}$$
$$x_2 = c_2 + y_2$$

These formulas can be interpreted in two different ways: either as a translation of the coordinate system (see Fig. 10.4, left) or as a translation of the points in the plane *in the opposite direction* (Fig. 10.4, right). Compositions of linear mappings with translations are called *affine mappings*.

Composition of linear mappings. Consider two linear mappings

$$y_k = \sum_i a_{ki} x_i \quad \text{and} \quad z_\ell = \sum_k b_{\ell k} y_k$$

with matrices $y = Ax$ and $z = By$. Their composition

$$z_\ell = \sum_k b_{\ell k} \sum_i a_{ki} x_i = \sum_i \left(\sum_k b_{\ell k} a_{ki} \right) x_i = \sum_i c_{\ell i} x_i \tag{10.7}$$

is again a linear mapping, with matrix C given by

$$c_{\ell i} = \sum_k b_{\ell k} a_{ki}. \tag{10.8}$$

This famous formula for the *product* of two matrices $C = BA$ is symbolised by

An example for the product of two matrices is

$$A = \begin{bmatrix} 1.1 & 0.2 \\ 0.4 & 0.8 \end{bmatrix}, \quad B = \begin{bmatrix} 0.8 & -0.4 \\ -0.1 & 1.3 \end{bmatrix} \quad \text{which gives} \quad BA = \begin{bmatrix} 0.72 & -0.16 \\ 0.41 & 1.02 \end{bmatrix}.$$

Inverse mapping. We next try to determine x from its image $y = Ax$ for a given y. For this we have to solve a linear system:

$$
\begin{array}{lll}
1.1\,x_1 + 0.2\,x_2 = y_1 & \text{Gaussian} & 1.1\,x_1 + 0.200\,x_2 = y_1 \\
& \Rightarrow & \\
0.4\,x_1 + 0.8\,x_2 = y_2 & \text{elimination} & \qquad + 0.727\,x_2 = -0.364\,y_1 + y_2
\end{array}
$$

$$
\begin{array}{lll}
\text{Gaussian} & 1.1\,x_1 & = 1.100\,y_1 - 0.275\,y_2 \\
\Rightarrow & & \\
\text{elimination} & & + 0.727\,x_2 = -0.364\,y_1 + y_2.
\end{array} \tag{10.9}
$$

After performing the divisions, we arrive at

$$
\begin{array}{ll}
x_1 = 1.000\,y_1 - 0.25\,y_2 & \\
& \Leftrightarrow \qquad x = A^{-1}y. \tag{10.10} \\
x_2 = -0.5\,y_1 + 1.376\,y_2 &
\end{array}
$$

We see that Gaussian elimination can be used to compute the inverse matrix.

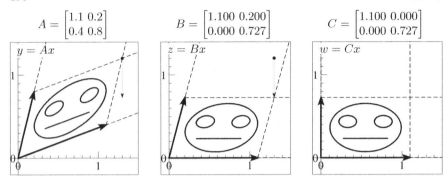

$$A = \begin{bmatrix} 1.1 & 0.2 \\ 0.4 & 0.8 \end{bmatrix} \qquad B = \begin{bmatrix} 1.100 & 0.200 \\ 0.000 & 0.727 \end{bmatrix} \qquad C = \begin{bmatrix} 1.100 & 0.000 \\ 0.000 & 0.727 \end{bmatrix}$$

Fig. 10.5. A linear mapping under Gaussian elimination

Areas and volumes. The *geometric significance* of the above procedure is illustrated in Fig. 10.5: a parallelepiped generated by the *columns* of A is reshaped by *shear transformations* parallel to the axes. *This changes neither area nor volume.* In contrast to the preceding reasoning, this fact is not derived here from Eucl. XI.27 ff., but follows directly from the ideas of Archimedes, see Fig. 1.11. At the end, the area (volume) of the image of the unit square (cube) equals the product of the pivots, i.e. the determinant of A.

Theorem 10.3. *The determinant of the matrix A, whose columns are the images of the basis vectors e_i, represents the factor by which the area or volume of a figure is increased by the linear transformation $y = Ax$. Its sign indicates whether the columns of A have the same orientation as the basic vectors e_i (positive sign) or not (negative sign).*

Comparing this with Theorem 9.1, we obtain a geometric proof of the interesting formula

$$\det(A^{\mathsf{T}}) = \det A, \tag{10.11}$$

which is known from algebra. By the *composition* of two linear mappings BA, the volume is enlarged first by the factor $\det A$ and then by the factor $\det B$. We thus get another important formula

$$\det(BA) = \det B \cdot \det A. \tag{10.12}$$

10.3 Gram's Determinant

Let a and b denote two vectors in $\mathbb{R}^3$. We want to find a formula for the area of the parallelogram generated by the two vectors.

Inspired once again by the determinant (9.24) we choose the vector c orthogonal to both a and b, with length 1. Thus, the area $\mathcal{A}$ of the parallelogram generated by a and b is

$$\mathcal{A} = \det \begin{bmatrix} a_1 & a_2 & a_3 \\ b_1 & b_2 & b_3 \\ c_1 & c_2 & c_3 \end{bmatrix}.$$

After multiplying by the transposed matrix and using (10.11) and (10.12), we get

$$\mathcal{A}^2 = \det \begin{bmatrix} a_1 & a_2 & a_3 \\ b_1 & b_2 & b_3 \\ c_1 & c_2 & c_3 \end{bmatrix} \cdot \det \begin{bmatrix} a_1 & b_1 & c_1 \\ a_2 & b_2 & c_2 \\ a_3 & b_3 & c_3 \end{bmatrix} = \det \begin{bmatrix} a \cdot a & a \cdot b & 0 \\ b \cdot a & b \cdot b & 0 \\ 0 & 0 & 1 \end{bmatrix}. \qquad (10.13)$$

The area $\mathcal{A}$ of the parallelogram generated by two vectors a and b is thus given by

$$\mathcal{A}^2 = \det \begin{bmatrix} a \cdot a & a \cdot b \\ b \cdot a & b \cdot b \end{bmatrix} =: G(a, b). \qquad (10.14)$$

This determinant is called *Gram's determinant* of a and b. The importance of (10.14), which has its origin in the method of least squares,[4] lies in the fact that it is *independent of the dimension*. It is also easily generalised to an arbitrary number of vectors. One only needs to know that there exist orthogonal vectors c, d, …

Area of a triangle. The area of the triangle spanned by a and b is half of the area given by (10.14). If we insert the formula (9.18) for the scalar products and multiply each line by 2, we obtain for the area of this triangle

$$16\mathcal{A}^2 = \det \begin{bmatrix} 2|a|^2 & -|c|^2 + |a|^2 + |b|^2 \\ -|c|^2 + |a|^2 + |b|^2 & 2|b|^2 \end{bmatrix}. \qquad (10.15)$$

By expanding this determinant we obtain the third formula of (6.25), and thus another proof of Heron's formula.

Volume of a tetrahedron. This fundamental problem asks for the volume $\mathcal{V}$ of a tetrahedron if the lengths of the six edges are given. For the solution, we place one of the vertices at the origin and denote the three edges emanating from it by the vectors a, b and c (see Fig. 10.6). The three other edges of the tetrahedron are the differences, which we denote by $d = b - a$, $e = a - c$, $f = c - b$. We know from Theorem 9.1 that the volume of the parallelepiped generated by the vectors a, b, c is the determinant (9.24). As in (9.31), this determinant is thus equal to $6\mathcal{V}$. In order to get rid of the particular coefficients of these vectors, we compute the corresponding Gram determinant:

$$36\mathcal{V}^2 = \det \begin{bmatrix} a_1 & a_2 & a_3 \\ b_1 & b_2 & b_3 \\ c_1 & c_2 & c_3 \end{bmatrix} \cdot \det \begin{bmatrix} a_1 & b_1 & c_1 \\ a_2 & b_2 & c_2 \\ a_3 & b_3 & c_3 \end{bmatrix} = \det \begin{bmatrix} a \cdot a & a \cdot b & a \cdot c \\ b \cdot a & b \cdot b & b \cdot c \\ c \cdot a & c \cdot b & c \cdot c \end{bmatrix}. \qquad (10.16)$$

To obtain our final result, we again insert the formula (9.18) for the scalar

---

[4]J.P. Gram, *Om Räkkeudviklinger, bestemte ved Hjälp af de mindste Kvadraters Methode*, København 1879; see *Ueber die Entwickelung reeller Functionen in Reihen mittelst der Methode der kleinsten Quadrate*, J. Reine Angew. Math. (Crelle) 94 (1883) 41–73.

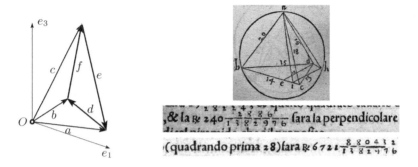

Fig. 10.6. The six edges of a tetrahedron (left); Tartaglia's example with his value for "la perpendicolare" h and the "aria corporale" $\mathcal{V}$ (right)

products and multiply each line by 2, so that

$$288\mathcal{V}^2 = \det \begin{bmatrix} 2|a|^2 & |a|^2 + |b|^2 - |d|^2 & |a|^2 + |c|^2 - |e|^2 \\ |b|^2 + |a|^2 - |d|^2 & 2|b|^2 & |b|^2 + |c|^2 - |f|^2 \\ |c|^2 + |a|^2 - |e|^2 & |c|^2 + |b|^2 - |f|^2 & 2|c|^2 \end{bmatrix}. \quad (10.17)$$

We will see in Exercise 2 below that this same formula can be written in an elegant way as the so-called "Cayley–Menger determinant"

$$288\mathcal{V}^2 = \det \begin{bmatrix} 0 & |a|^2 & |b|^2 & |c|^2 & 1 \\ |a|^2 & 0 & |d|^2 & |e|^2 & 1 \\ |b|^2 & |d|^2 & 0 & |f|^2 & 1 \\ |c|^2 & |e|^2 & |f|^2 & 0 & 1 \\ 1 & 1 & 1 & 1 & 0 \end{bmatrix}. \quad (10.18)$$

A clear derivation of these expressions, written in trigonometric and algebraic form, not as determinants, was given by Euler (E601, 1786, §8 and 9). Euler only considers the case $\mathcal{V} = 0$, as a condition for four points to lie in a plane.

Example. Niccolò Tartaglia (1560, *General trattato*, secondo libro della quarta parte, p. 34) explained an algorithm for the volume of a tetrahedron, based on the repeated use of Pythagoras' and Thales' theorems, in an entire page of Italian text. He demonstrated it on the example with edge lengths

$$|a| = 20, \quad |b| = 18, \quad |c| = 16, \quad |d| = 14, \quad |e| = 15, \quad |f| = 13, \quad (10.19)$$

whose base triangle with sides $13, 14, 15$ has the area $\mathcal{A} = 84$ already known to him. He then mistakenly[5] arrived at the altitude $h = \sqrt{240\frac{2886}{1382976}}$ (see Fig. 10.6, right). The volume is given by $\mathcal{V} = h\,\mathcal{A}/3 = h \cdot 28$. Tartaglia apparently made another mistake[6] and stated that $\mathcal{V} = \sqrt{6721\frac{880432}{1382976}}$. Many texts

---

[5]The correct value is $h = \sqrt{240\frac{615}{3136}}$.

[6]The correct value, obtained from formula (10.17), is $\mathcal{V} = \sqrt{188313\frac{3}{4}}$.

or internet sites call (10.16) or (10.17), or even (10.18) "Tartaglia's formula", which seems exaggerated.

Šmakal's proof of the pentagon theorem. As another application of Gram's determinant we present Šmakal's proof of Theorem 9.7: we denote by a_1, a_2, a_3, a_4 and a_5 the unit vectors joining A to B, B to C, etc. Since the pentagon is closed, since all side lengths are 1 and all angles are α, they satisfy

$$a_1 + a_2 + a_3 + a_4 + a_5 = 0, \quad a_i \cdot a_i = 1 \quad \text{and} \quad a_i \cdot a_{i+1} = c, \qquad (10.20)$$

where $c = -\cos\alpha$ and the indices are taken modulo 5. We multiply the sum in (10.20) in turn by a_1, a_2, a_3, a_4 and a_5 (scalar product) and obtain

$$a_1 \cdot a_3 = a_2 \cdot a_4 = a_3 \cdot a_5 = a_4 \cdot a_1 = a_5 \cdot a_2 = -c - \frac{1}{2}. \qquad (10.21)$$

With these vectors, we compute the Gram determinants $G_4 = G(a_1, a_2, a_3, a_4)$ and $G_3 = G(a_i, a_{i+1}, a_{i+2})$ and obtain, as Šmakal writes, "after an easy calculation"

$$G_4 = \det \begin{bmatrix} 1 & c & -c - \frac{1}{2} & -c - \frac{1}{2} \\ c & 1 & c & -c - \frac{1}{2} \\ -c - \frac{1}{2} & c & 1 & c \\ -c - \frac{1}{2} & -c - \frac{1}{2} & c & 1 \end{bmatrix} = \frac{5}{16} \left(4c^2 + 2c - 1 \right)^2$$

and

$$G_3 = \det \begin{bmatrix} 1 & c & -c - \frac{1}{2} \\ c & 1 & c \\ -c - \frac{1}{2} & c & 1 \end{bmatrix} = -\frac{1}{4} \left(2c + 3 \right) \left(4c^2 + 2c - 1 \right).$$

We know that $G_4 = 0$, because four vectors in $\mathbb{R}^3$ have (four dimensional) volume zero. This implies that c must be one of the solutions of $4c^2 + 2c - 1 = 0$, which means that all the Gram determinants $G_3 = 0$. All triples of vectors a_i, a_{i+1}, a_{i+2} are planar, hence the entire pentagon is planar. The two roots of $4c^2 + 2c - 1$ correspond to the two types of planar regular pentagons, as in the earlier proof. $\qquad \square$

Remark. The proof presented in Dunitz and Waser (1972), and attributed to L. Oosterhoff, is similar, but is based on the 5×5 Gram determinant of all five a's. The corresponding matrix is what is called a *circulant* matrix. This proof requires one to think in an even higher dimension, but is computationally more elegant, because one knows a closed formula for the determinant of a circulant matrix.

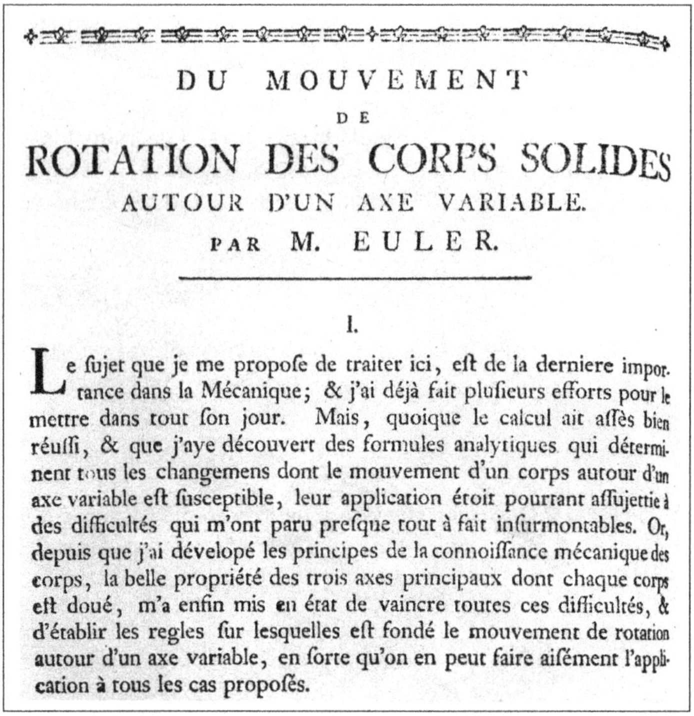

DU MOUVEMENT

DE

ROTATION DES CORPS SOLIDES

AUTOUR D'UN AXE VARIABLE.

PAR M. EULER.

I.

L e fujet que je me propofe de traiter ici, eft de la derniere importance dans la Mécanique; & j'ai déjà fait plufieurs efforts pour le
mettre dans tout fon jour. Mais, quoique le calcul ait affès bien
réuffi, & que j'aye découvert des formules analytiques qui détermi-
nent tous les changemens dont le mouvement d'un corps autour d'un
axe variable eft fufceptible, leur application étoit pourtant affujettie à
des difficultés qui m'ont paru prefque tout à fait infurmontables. Or,
depuis que j'ai dévelopé les principes de la connoiffance mécanique des
corps, la belle propriété des trois axes principaux dont chaque corps
eft doué, m'a enfin mis en état de vaincre toutes ces difficultés, &
d'établir les regles fur lesquelles eft fondé le mouvement de rotation
autour d'un axe variable, en forte qu'on en peut faire aifément l'application à tous les cas propofés.

Fig. 10.7. The beginning of Euler's classical article E292 on the movement of rigid
bodies, presented in 1758, published in 1765; Euler tells us in elegant French that
he had struggled for a long time to discover the laws of movement of a rigid body,
and that the discovery of the principal axes of the quadratic form of inertia finally
allowed him to overcome all these difficulties

10.4 Orthogonal Mappings and Isometries

> "... the problem of the linear transformation of a quadratic func-
> tion into itself has an elegant solution ..."
>
> (Cayley 1880, *Papers*, vol. 11, p. 140)

Orthogonal mappings have their origin in Euler's work E292 on rigid bodies,
1765, (see Fig. 10.7) and Cayley's paper from 1858 (see Theorem 10.9 be-
low). The aim is to characterise linear mappings $\alpha : \mathbb{R}^n \to \mathbb{R}^n$ that *preserve
distances*. We first note the following property.

Lemma 10.4. *If a linear mapping preserves distances, then it also preserves
angles.*

Proof. This result is in fact Eucl. I.8. An algebraic proof uses the identity
(9.18), that allows one to express scalar products, and consequently angles
(see (9.19)) by distances.   □

Let $y = \alpha(x)$ and $w = \alpha(v)$. We denote by Q the matrix of α, so that $y = Qx$ and $w = Qv$. If α preserves distances we get from the lemma

$$y^{\mathsf{T}}w = x^{\mathsf{T}}Q^{\mathsf{T}}Qv = x^{\mathsf{T}}v.$$

Choosing $x = e_i$ and $v = e_j$ leads to the condition

$$Q^{\mathsf{T}}Q = I. \tag{10.22}$$

A matrix Q with this property is called *orthogonal*.

Theorem 10.5. *For a square matrix Q, the following properties are equivalent:*

(a) *The columns of Q form an orthonormal basis;*
(b) *The rows of Q form an orthonormal basis;*
(c) $Q^{\mathsf{T}}Q = I$, *i.e. Q is orthogonal;*
(d) $QQ^{\mathsf{T}} = I$;
(e) Q *is invertible and $Q^{-1} = Q^{\mathsf{T}}$.*

Proof. As in Gram's matrix (10.13), the elements of the matrix $Q^{\mathsf{T}}Q = I$ are the scalar products of pairs of columns of Q. Thus, conditions (a) and (c) are equivalent. In the same way, (b) and (d) are seen to be equivalent. Since the volume of an orthonormal parallelepiped is 1, each of the conditions (a) and (b) implies that $\det Q = \det Q^{\mathsf{T}} = \pm 1$, and hence that Q is invertible. The other equivalences now follow easily by multiplying formulas (e) and (c) by Q and formula (d) by Q^{-1}. $\square$

Orientation. We distinguish two types of orthogonal matrices, those with $\det Q = 1$ (here the row and column vectors of Q are oriented positively; see page 273), and those with $\det Q = -1$ (here the row and column vectors of Q are oriented negatively).

Example 10.6. The rotations in $\mathbb{R}^2$ (see Fig. 10.3) are orthogonal with $\det Q = 1$. For $n = 3$, the identities $2^2 + 2^2 + 1^2 = 3^2$ and $2 \cdot 2 - 2 - 2 = 0$ allow one to create orthogonal matrices. One member of this family is

$$Q = \frac{1}{3}\begin{bmatrix} 2 & -2 & 1 \\ 2 & 1 & -2 \\ 1 & 2 & 2 \end{bmatrix}, \qquad \text{for which} \quad \det Q = 1. \tag{10.23}$$

Example 10.7. Reflections. Let $n = (n_1, n_2, n_3)$ be a unit vector and consider the plane that is orthogonal to n and passes through the origin. For arbitrary x, the scalar product $\langle n \mid x \rangle = n^{\mathsf{T}}x$ gives the distance to this plane. Therefore, $x - nn^{\mathsf{T}}x$ is the *orthogonal projection* of x onto the plane. Taking *twice* the distance (see Fig. 10.8, left) gives the *reflection of the point x in the plane*

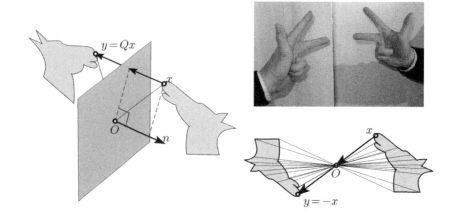

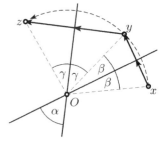

Fig. 10.8. Reflection in a plane in $\mathbb{R}^3$ (above), and in a point (below)

$$y = x - 2n\,n^\mathsf{T}x, \quad \text{i.e.} \quad y = Qx \quad \text{with} \quad Q = I - 2nn^\mathsf{T} \qquad (10.24\text{a})$$

or in detail

$$\begin{bmatrix} y_1 \\ y_2 \\ y_3 \end{bmatrix} = \begin{bmatrix} 1 - 2n_1n_1 & -2n_1n_2 & -2n_1n_3 \\ -2n_2n_1 & 1 - 2n_2n_2 & -2n_2n_3 \\ -2n_3n_1 & -2n_3n_2 & 1 - 2n_3n_3 \end{bmatrix} \begin{bmatrix} x_1 \\ x_2 \\ x_3 \end{bmatrix}. \qquad (10.24\text{b})$$

We can verify that $Q^\mathsf{T}Q = (I - 2nn^\mathsf{T})(I - 2nn^\mathsf{T}) = I - 4nn^\mathsf{T} + 4n(n^\mathsf{T}n)n^\mathsf{T} = I$, and from (9.13) that $\det Q = 1 - 2(n_1^2 + n_2^2 + n_3^2) = -1$, hence this transformation is orthogonal and changes the orientation. The corresponding mapping in two dimensions is a *reflection in a line* and also changes the orientation. On the other hand, the reflection in the origin

$$y = -x \qquad (10.25)$$

changes the orientation in three dimensions (see Fig. 10.8), but preserves it in two dimensions, since it is then just a rotation through the angle π.

Composition. The set of orthogonal transformations form a group, because if $Q_1^\mathsf{T}Q_1 = I$ and $Q_2^\mathsf{T}Q_2 = I$, then also $(Q_2Q_1)^\mathsf{T}(Q_2Q_1) = Q_1^\mathsf{T}Q_2^\mathsf{T}Q_2Q_1 = I$ and similarly for Q^{-1}. This group is called the *orthogonal group* and is denoted by $\mathrm{O}(3)$; the orthogonal transformations with $\det Q = 1$ form the subgroup denoted by $\mathrm{SO}(3)$. In contrast, the orthogonal transformations with $\det Q = -1$ do not form a group: the composition of two reflections with angle α between the two mirrors is a *rotation* through the angle 2α (since $\alpha = \beta + \gamma$, see the figure).

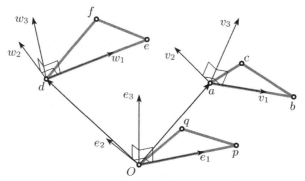

Fig. 10.9. Existence of orthogonal transformations

Existence of orthogonal transformations. If we are given two sets of points, such that all pairs of corresponding points have the same distance, we want to know if there exists a (unique) orthogonal transformation which maps one set of points into the other.

Lemma 10.8. *Let a, b, c be the coordinates of a non-degenerate triangle in $\mathbb{R}^3$ and let d, e, f be those of another with the same side lengths. Then one can construct an orthogonal matrix S, which is unique up to reflection, and a vector s such that the transformation*

$$\alpha(x) = y = s + Sx \qquad \text{satisfies} \quad \alpha(a) = d, \ \alpha(b) = e \ \text{ and } \ \alpha(c) = f. \ (10.26)$$

Proof. The proof is displayed in Fig. 10.9. We use an auxiliary triangle with coordinates $0, p, q$, again of the same side lengths, which we place in the $e_1 e_2$ plane in "Euler position" (i.e. as in the proof of Theorem 7.20, with 0 at the origin and p on the e_1-axis).

Let $u_1 = b - a$, $u_2 = c - a$, and set

$$v_1 = \frac{u_1}{|u_1|},$$

$$v_2 = \frac{\tilde{v}_2}{|\tilde{v}_2|}, \quad \text{where} \quad \tilde{v}_2 = u_2 - \alpha v_1 \quad \text{with} \quad \alpha = u_2 \cdot v_1, \qquad (10.27)$$

$$v_3 = v_1 \times v_2,$$

which are a set of orthogonal unit vectors by construction.[7] We similarly define w_1, w_2 and w_3 for the triangle d, e, f. We then take these two sets as columns of the matrices

$$Q = [\, v_1 \mid v_2 \mid v_3 \,], \qquad R = [\, w_1 \mid w_2 \mid w_3 \,], \qquad (10.28)$$

which are both orthogonal by Theorem 10.5. By Theorem 10.2 and the hypothesis on the side lengths of the triangles, the mappings

---

[7]The first two lines of equation (10.27) are the beginning of what is called the *Gram–Schmidt orthogonalisation algorithm.*

$$x = a + Qz \qquad \text{and} \qquad y = d + Rz \tag{10.29}$$

map $0, p, q$ to a, b, c and d, e, f respectively. We solve the first equation for $z = Q^\mathsf{T}(x - a)$ and insert it into the second to obtain

$$y = d + RQ^\mathsf{T}(x - a) = (d - RQ^\mathsf{T}a) + RQ^\mathsf{T}x, \tag{10.30}$$

which, with $s = d - RQ^\mathsf{T}a$ and $S = RQ^\mathsf{T}$, is the desired transformation. We could replace w_3 by $-w_3$, which would change the orientation of the image vectors. $\qquad\qquad\square$

The Bol–Coxeter proof of the pentagon theorem. A few days after sending out off-prints of his article (1970), van der Waerden received letters from G. Bol and H.S.M. Coxeter, who had independently discovered a really elegant proof (see the "Nachtrag" from 1972) of Theorem 9.7. Their proof is as follows:

Let a, b, c, d, e be the coordinates of the vertices of the pentagon $ABCDE$. By hypothesis, all distances between these points are fixed. (They are either 1 or $\sqrt{2 - 2\cos\alpha}$.) We now consider the pentagon b, c, d, e, a, which has the same property. By Lemma 10.8 there exists a map $y = s + Sx$ such that $b = s + Sa$, $c = s + Sb$ and $d = s + Sc$. We next look at the image of d. Since S is orthogonal, the image of d has the same distances from b, c and d as the point e. There are in general two points in $\mathbb{R}^3$ which have this property (as the intersection of three spheres). Hence for one of the two possibilities in the above proof we will have $e = s + Sd$. For the other point, under the hypothesis that a, b, c, d are not planar, there is only one possibility left and we have automatically $a = s + Se$. If we add up all five equations, we obtain for the centre of gravity of the pentagon $\frac{a+b+c+d+e}{5} = s + S\frac{a+b+c+d+e}{5}$. If we move the centre of gravity of the pentagon to the origin, we will have $s = 0$ and

$$b = Sa, \quad c = Sb, \quad d = Sc, \quad e = Sd, \quad a = Se. \tag{10.31}$$

Thus five applications of S move the pentagon into itself. If the pentagon were not planar, we would have $S^5 = I$. Hence it is impossible that $\det S = -1$. Therefore S preserves orientation and by Euler's theorem 10.11, which we will soon see, S must be a planar rotation of a planar pentagon. $\qquad\qquad\square$

10.5 Skew-Symmetric Matrices, the Cayley Transform

A matrix A is called *skew-symmetric*, if $A^\mathsf{T} = -A$. Such matrices have zeros in the main diagonal, and the entries not on this diagonal, say the a_{ij} with $i \neq j$, are such that $a_{ij} = -a_{ji}$.

For $n = 2$, a skew-symmetric matrix has the form

$$A = \begin{bmatrix} 0 & a \\ -a & 0 \end{bmatrix}. \tag{10.32}$$

For $n = 3$, we have

$$Ax = \begin{bmatrix} 0 & -a_3 & a_2 \\ a_3 & 0 & -a_1 \\ -a_2 & a_1 & 0 \end{bmatrix} \begin{bmatrix} x_1 \\ x_2 \\ x_3 \end{bmatrix} = \begin{bmatrix} a_2 x_3 - a_3 x_2 \\ a_3 x_1 - a_1 x_3 \\ a_1 x_2 - a_2 x_1 \end{bmatrix} = a \times x, \qquad (10.33)$$

by (9.23). Thus, a linear mapping $x \mapsto Ax$ with skew-symmetric matrix A corresponds to the *outer product* $a \times x$ (where the first factor a is fixed), and vice versa. Consequently,[8] Ax is perpendicular to x and

$$(I \pm A)x = x \pm Ax = 0 \quad \text{implies} \quad x = 0.$$

This means that for a skew-symmetric matrix A, the matrices $I \pm A$ are always invertible.

The study of orthogonal matrices was greatly simplified by the following discovery.

Theorem 10.9 (Cayley, 1846). *If a matrix A is skew-symmetric, then*

$$Q = (I + A)(I - A)^{-1} = (I - A)^{-1}(I + A) \qquad (10.34)$$

is orthogonal. Conversely, if Q is orthogonal and $\det(Q + I) \neq 0$, then Q can be written as in (10.34) with

$$A = (Q - I)(Q + I)^{-1} = (Q + I)^{-1}(Q - I), \qquad (10.35)$$

where A is skew-symmetric.

Proof. One must first understand why $I + A$ and $(I - A)^{-1}$ commute, as in (10.34). This follows from a result in linear algebra which shows that any two rational expressions formed with the same matrix (here A) commute. For a direct proof, multiply $(I + A)(I - A)^{-1}$ from the left by $I = (I - A)^{-1}(I - A)$, use $(I - A)(I + A) = I - A^2 = (I + A)(I - A)$, and simplify. The identity involving Q in (10.35), as well as $AQ = QA$, can be established similarly.

The main key to the proof is the equation

$$AQ + A - Q + I = 0, \qquad (10.36)$$

which is linear in A as well as in Q, and which can easily be solved for each of them. If we solve for Q, and if $I - A$ is invertible, we obtain (10.34); if we solve for A, and if $Q + I$ is invertible, we obtain (10.35). Thus, whenever the inverse matrices exist, either by the above discussion concerning $I \pm A$, or by hypothesis, the equations (10.34), (10.35) and (10.36) are equivalent.

We transpose the matrix in (10.36) (note that A and Q commute) and multiply by Q. This gives $A^{\mathsf{T}} Q^{\mathsf{T}} Q + A^{\mathsf{T}} Q - Q^{\mathsf{T}} Q + Q = 0$. If this is added to (10.36) we obtain an equation which can be brought to the form

---

[8]This is true for all dimensions and just as simple to prove.

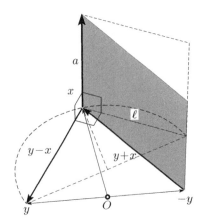
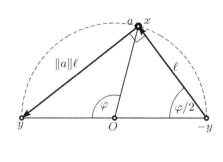

Fig. 10.10. Orthogonal transformation of $\mathbb{R}^3$ (right: seen in direction of a)

$$(A + A^{\mathsf{T}})(Q + I) = (A^{\mathsf{T}} - I)(I - Q^{\mathsf{T}}Q).\qquad(10.37)$$

Here $A + A^{\mathsf{T}} = 0$ means that A is skew-symmetric, and $I - Q^{\mathsf{T}}Q = 0$ means that Q is orthogonal. Since the matrices multiplying these expressions are invertible, we see that the two statements are equivalent. □

Example 10.10. For $n = 2$, formula (10.34) applied to the matrix in (10.32) gives

$$Q = \begin{bmatrix} 1 & a \\ -a & 1 \end{bmatrix} \begin{bmatrix} 1 & -a \\ a & 1 \end{bmatrix}^{-1} = \frac{1}{1 + a^2} \begin{bmatrix} 1 & a \\ -a & 1 \end{bmatrix} \begin{bmatrix} 1 & a \\ -a & 1 \end{bmatrix}$$

$$= \frac{1}{1 + a^2} \begin{bmatrix} 1 - a^2 & 2a \\ -2a & 1 - a^2 \end{bmatrix} = \begin{bmatrix} \cos\alpha & \sin\alpha \\ -\sin\alpha & \cos\alpha \end{bmatrix}\qquad(10.38)$$

with $\tan(\alpha/2) = a$, see Fig. 12.1 on page 347. The resulting mapping is the rotation encountered in (10.5).

Theorem 10.11 (Euler, when studying the movement of a solid). *An orthogonal mapping in three dimensions that preserves orientation* ($\det Q = 1$) *corresponds to a rotation around a vector a, through an angle φ. The components of a are the entries of A, the Cayley transform of Q, as given in (10.33). The angle of rotation is determined by* $\tan(\varphi/2) = |a|$.

Proof. In order to study the mapping $x \mapsto y = Qx$ in three dimensions, we use (10.34) and (10.33):

$$(I - A)y = (I + A)x \quad\Rightarrow\quad y - x = A(y + x) \quad\Rightarrow\quad y - x = a \times (y + x).$$

The vector $y - x$ connecting x with its image y is thus orthogonal to a and to $y + x = x - (-y)$, the vector which connects x with $-y$ (see Fig. 10.10, left). As the area of the grey parallelogram is $|a|\ell$, we obtain from Fig. 10.10 (right) the stated formula for the angle of rotation. □

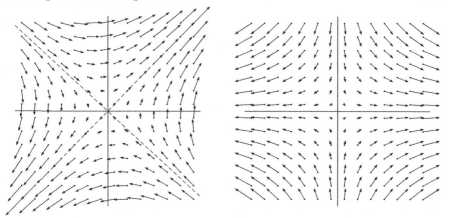

Fig. 10.11. A linear mapping as vector field (left); the same mapping in a basis of eigenvectors (right)

Remark. The above argument does not work if -1 is a (double) eigenvalue of Q. In that case we obtain a "switch", a rotation through π.

10.6 Eigenvalues and Eigenvectors

In the left part of Fig. 10.11, we recognise the linear mapping

$$y = Ax \qquad \text{with} \qquad A = \begin{bmatrix} 0.3 & 2 \\ 1.8 & 0.5 \end{bmatrix}, \tag{10.39}$$

represented as a *vector field*, i.e. the vector $y = Ax$ is attached to each point x as a small arrow.[9] Two directions attract our attention: those such that the vector Av has the same direction as the vector v, i.e. such that

$$Av = \lambda v \qquad \text{or} \qquad (\lambda I - A)v = 0, \qquad v \neq 0 \tag{10.40}$$

(see the dashed lines). Such a vector is called an *eigenvector* of the matrix A and λ is the corresponding *eigenvalue*. Obviously λ must be a root of the *characteristic polynomial*

$$\det(\lambda I - A) = \lambda^n - (a_{11} + \ldots + a_{nn})\lambda^{n-1} \pm \ldots + (-1)^n \det A. \tag{10.41}$$

If this polynomial has n distinct roots, we obtain n linearly independent eigenvectors that may be chosen as a basis.[10] In this basis, the mapping is simply

---

[9]The first occurrence of eigenvalues and eigenvectors was in the context of differential equations (Lagrange 1759, *Théorie du son*; Lagrange 1781; 6×6 matrices with the aim of calculating secular perturbations of the orbits of the six planets then known, *Oeuvres* V, pp. 125–490).

[10]If the roots are not distinct, some complications might arise (Jordan blocs).

the multiplication of the coordinates by λ_i, i.e. the matrix is transformed to a *diagonal matrix*, see the right picture of Fig. 10.11.

To derive formulas, we observe on the one hand that, by (10.1)

$$x = T\tilde{x}, \qquad y = T\tilde{y} \quad \Rightarrow \quad T\tilde{y} = AT\tilde{x} \quad \Rightarrow \quad \tilde{y} = T^{-1}AT\tilde{x},$$

where $\tilde{x}$ are the new coordinates and the columns of the matrix T are the eigenvectors v_i. On the other hand, we have

$$Av_i = \lambda_i v_i \quad \Rightarrow \quad A(v_1, v_2) = (v_1\lambda_1, v_2\lambda_2). \qquad (10.42)$$

In matrix notation, we thus have

$$AT = T\Lambda \qquad \text{or} \qquad T^{-1}AT = \Lambda, \qquad (10.43)$$

where Λ is the diagonal matrix with the λ_i on its main diagonal.

Example 10.12. For the matrix (10.39), the characteristic polynomial has the form $\lambda^2 - 0.8\lambda - 3.45$. We get

$$\lambda_1 = 2.3, \quad \lambda_2 = -1.5, \quad v_1 = \begin{bmatrix} 1 \\ 1 \end{bmatrix}, \quad v_2 = \begin{bmatrix} -2 \\ 1.8 \end{bmatrix}.$$

10.7 Quadratic Forms

At the end of this chapter, we return to the beginning of our treatment of Cartesian coordinates, the problem of Pappus in Sect. 7.2. We are interested in discovering the nature of the curves defined by quadratic equations

$$\alpha p' = A p' - H q' - G r';$$

$\flat$, la quatrième et la si

$'$, on aura pareillement

$$\alpha q' = B q' - F r' - H p';$$

me, la cinquième et la

$'$, q', on aura

$$\alpha r' = C r' - G p' - F q';$$

Fig. 10.12. Eigenvalue problem transforming a quadratic form (moments of inertia of a rigid body) in Lagrange 1788 (left); Joseph-Louis Lagrange 1736–1813 (right)

$$ax_1^2 + 2bx_1x_2 + cx_2^2 + 2dx_1 + 2ex_2 + g = 0 \qquad (10.44)$$

and in the generalisation of this problem to higher dimensions. The first clear treatment of this question was given by Euler (1748; vol. II for $n = 2$, in an appendix for $n = 3$) by brute rotations. The elegant relation with eigenvalue problems was discovered by Lagrange (see Fig. 10.12).

Example 10.13. The solutions of the equation

$$x_1^2 + 2bx_1x_2 + x_2^2 - 5x_1 - 4x_2 + g = 0, \qquad (10.45)$$

drawn in Fig. 10.13 for various values of b and g, seem to lie on conics of different kinds. How can this be verified using the tools of this chapter?

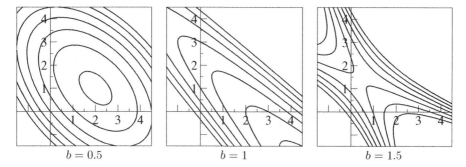

$$b = 0.5 \qquad\qquad b = 1 \qquad\qquad b = 1.5$$

Fig. 10.13. The conics corresponding to equation (10.45) for the stated values of b and various values of g

To analyse (10.44), we write the equation in matrix form[11]

$$\begin{bmatrix} x_1 & x_2 & 1 \end{bmatrix} \begin{bmatrix} a & b & d \\ b & c & e \\ d & e & g \end{bmatrix} \begin{bmatrix} x_1 \\ x_2 \\ 1 \end{bmatrix} = 0 \qquad (10.46)$$

and start by eliminating the linear terms $2dx_1 + 2ex_2$ by using a translation. For this purpose, we use equation (10.6) in the form

$$\begin{bmatrix} x_1 \\ x_2 \\ 1 \end{bmatrix} = \begin{bmatrix} y_1 \\ y_2 \\ 0 \end{bmatrix} + \begin{bmatrix} c_1 \\ c_2 \\ 1 \end{bmatrix}. \qquad (10.47)$$

Substituting (10.47) into (10.46), the symmetry of the involved matrix gives us the identity

$$\begin{bmatrix} y_1 & y_2 & 0 \end{bmatrix} \begin{bmatrix} a & b & d \\ b & c & e \\ d & e & g \end{bmatrix} \begin{bmatrix} y_1 \\ y_2 \\ 0 \end{bmatrix} + 2 \begin{bmatrix} y_1 & y_2 & 0 \end{bmatrix} \begin{bmatrix} a & b & d \\ b & c & e \\ d & e & g \end{bmatrix} \begin{bmatrix} c_1 \\ c_2 \\ 1 \end{bmatrix}$$

$$+ \begin{bmatrix} c_1 & c_2 & 1 \end{bmatrix} \begin{bmatrix} a & b & d \\ b & c & e \\ d & e & g \end{bmatrix} \begin{bmatrix} c_1 \\ c_2 \\ 1 \end{bmatrix} = 0. \qquad (10.48)$$

---

[11]The beauty becomes perfect with *homogeneous coordinates*, see Chap. 11

In order to eliminate the second term, we require that

$$\begin{bmatrix} a & b \\ b & c \end{bmatrix} \begin{bmatrix} c_1 \\ c_2 \end{bmatrix} = \begin{bmatrix} -d \\ -e \end{bmatrix}. \tag{10.49}$$

This is a linear equation that determines c_1, c_2, if $ac - b^2 \neq 0$. Denoting the third term in (10.48), which is a constant, by $-\gamma$, we arrive at

$$y^{\mathsf{T}} A y = \begin{bmatrix} y_1 & y_2 \end{bmatrix} \begin{bmatrix} a & b \\ b & c \end{bmatrix} \begin{bmatrix} y_1 \\ y_2 \end{bmatrix} = \gamma. \tag{10.50}$$

We now calculate the eigenvectors of the matrix in (10.50). The characteristic equation has the form

$$\lambda^2 - (a+c)\lambda + (ac - b^2) = 0 \qquad \Rightarrow \qquad \lambda_{1,2} = \frac{a+c}{2} \pm \sqrt{\frac{(a-c)^2}{4} + b^2}.$$

Obviously, the eigenvalues are always real. Setting

$$d = \frac{c-a}{2} \qquad \text{and} \qquad R = \sqrt{d^2 + b^2}$$

we get

$$\lambda_{1,2} I - A = \begin{bmatrix} d \pm R & -b \\ -b & -d \pm R \end{bmatrix} \qquad \Rightarrow \qquad v_1 = \begin{bmatrix} b \\ d+R \end{bmatrix}, \quad v_2 = \begin{bmatrix} d+R \\ -b \end{bmatrix}.$$

The two vectors are orthogonal to each other and, after a normalisation, the matrix T of (10.43) is orthogonal. This yields

$$T^{-1} A T = T^{\mathsf{T}} A T = \Lambda.$$

In the new variables $y = Tz$ and $y^{\mathsf{T}} = z^{\mathsf{T}} T^{\mathsf{T}}$, equation (10.50) finally becomes

$$y^{\mathsf{T}} A y = z^{\mathsf{T}} T^{\mathsf{T}} A T z = z^{\mathsf{T}} \Lambda z = \lambda_1 z_1^2 + \lambda_2 z_2^2 = \gamma. \tag{10.51}$$

In conclusion, we obtain the equation of a *conic* with centre (c_1, c_2), see (10.47). Its type[12] depends on the signs of λ_1, λ_2 and γ. If one of the two eigenvalues is zero, which is the case if $ac - b^2 = 0$, we have a parabola and (10.49) does not have a unique solution. We then proceed in a different way, as explained in (10.57) below.

Example 10.14. The conic defined by the equation

$$36x_1^2 - 24x_1x_2 + 29x_2^2 + 120x_1 - 290x_2 + 545 = 0 \tag{10.52}$$

---

[12]This conic is either an ellipse, a hyperbola or the *empty* set, e.g. $z_1^2 + z_2^2 = -1$. In the latter case, the quadratic equation (10.44) does not have (real) solutions.

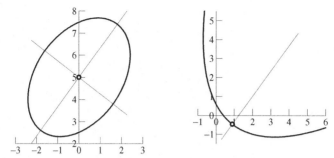

Fig. 10.14. Two examples of conics transformed to diagonal form

has the centre $c_1 = 0$, $c_2 = 5$. After a translation, one obtains

$$\begin{bmatrix} y_1 & y_2 \end{bmatrix} \begin{bmatrix} 36 & -12 \\ -12 & 29 \end{bmatrix} \begin{bmatrix} y_1 \\ y_2 \end{bmatrix} = 180 \,. \tag{10.53}$$

The eigenvalues satisfy

$$\lambda^2 - 65\lambda + 900 = 0 \quad \Rightarrow \quad \lambda_1 = 20 \,, \quad \lambda_2 = 45 \,. \tag{10.54}$$

Thus, the final result is

$$20z_1^2 + 45z_2^2 = 180 \quad \Rightarrow \quad \frac{z_1^2}{9} + \frac{z_2^2}{4} = 1 \,. \tag{10.55}$$

This is the equation of an ellipse with semi-axes $a = 3$, $b = 2$. The direction of the semi-major axis is that of the eigenvector corresponding to the *smallest* eigenvalue $\lambda_1 = 20$:

$$\begin{bmatrix} 16 & -12 \\ -12 & 9 \end{bmatrix} \begin{bmatrix} c \\ s \end{bmatrix} = 0 \quad \Rightarrow \quad c = \frac{3}{5} \,, \quad s = \frac{4}{5} \,. \tag{10.56}$$

Our ellipse is thus inclined by the angle $\arctan \frac{4}{3}$, see Fig. 10.14, left.

The case of a parabola. In the equation

$$16x_1^2 - 24x_1x_2 + 9x_2^2 - 130x_1 - 90x_2 + 50 = 0 \tag{10.57}$$

the sum of the first three terms can be written as a square:

$$(4x_1 - 3x_2)^2 - 130x_1 - 90x_2 + 50 = 0 \,. \tag{10.58}$$

We substitute $4x_1 - 3x_2 = 5y_1$, complete to an orthonormal basis $3x_1 + 4x_2 = 5y_2$, and perform a rotation. This gives

$$25y_1^2 - 50y_1 - 150y_2 + 50 = 0 \,. \tag{10.59}$$

Dividing by 25 and again completing the square, we find after a translation

$$z_1^2 = 6z_2\,.$$

The result is a parabola (see Fig. 10.14 right).

Several variables. Since every real *symmetric* matrix has real eigenvalues and a basis of orthogonal eigenvectors, we can proceed in the same way. As an example, we consider $n = 3$. Disregarding all degenerate cases, we obtain the surfaces displayed in Fig. 10.15.

(A) $\dfrac{x^2}{a^2} + \dfrac{y^2}{b^2} + \dfrac{z^2}{c^2} - 1 = 0$

(ellipsoid)

(B) $\dfrac{x^2}{a^2} + \dfrac{y^2}{b^2} - \dfrac{z^2}{c^2} + 1 = 0$

(two-sheeted hyperboloid)

(C) $\dfrac{x^2}{a^2} + \dfrac{y^2}{b^2} - \dfrac{z^2}{c^2} - 1 = 0$

(one-sheeted hyperboloid)

(D) $\dfrac{x^2}{a^2} + \dfrac{y^2}{b^2} - \dfrac{z^2}{c^2} = 0$

(cone)

(E) $\dfrac{x^2}{a^2} + \dfrac{y^2}{b^2} - 2pz = 0$

(elliptic paraboloid)

(F) $\dfrac{x^2}{a^2} - \dfrac{y^2}{b^2} - 2pz = 0$

(hyperbolic paraboloid)

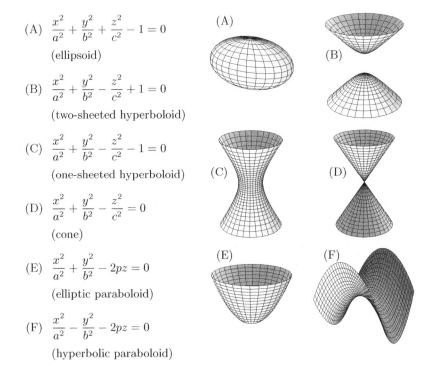

Fig. 10.15. Classification of quadratic equations in three dimensions (degenerate cases are disregarded)

Example 10.15. We encourage the reader to calculate the eigenvalues and eigenvectors (principal axes) of the corresponding matrix for the ellipsoid

$$25x_1^2 - 20x_1x_2 + 4x_1x_3 + 22x_2^2 - 16x_2x_3 + 16x_3^2 = 9 \qquad (10.60)$$

and to verify their mutual orthogonality, see Fig. 10.16.

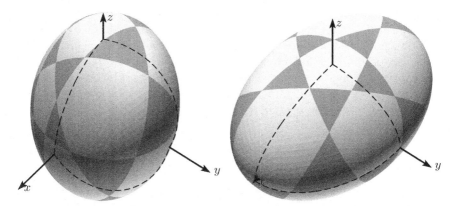

Fig. 10.16. The ellipsoid of (10.60) with semi-axes 1, $\frac{\sqrt{2}}{2}$ and $\frac{1}{2}$. The figure on the left shows the ellipsoid in arbitrary position, that on the right displays the same ellipsoid transformed to principal axes

10.8 Exercises

1. Deduce from (10.16) the formula

$$36\mathcal{V}^2 = |a|^2|b|^2|c|^2\Big(1 + 2\cos\alpha\cos\beta\cos\gamma - \cos^2\alpha - \cos^2\beta - \cos^2\gamma\Big) \quad (10.61)$$

for the volume of a tetrahedron, where α, β, γ are the angles of the three faces at the vertex O of Fig. 10.6. This expression was the starting point of the developments in Euler's E601 (1786).

2. Prove the equivalence of formulas (10.17) and (10.18) by showing that each is equivalent to

$$\det\begin{bmatrix} 0 & 0 & 0 & 0 & 1 \\ 0 & -2|a|^2 & -|a|^2-|b|^2+|d|^2 & -|a|^2-|c|^2+|e|^2 & 1 \\ 0 & -|b|^2-|a|^2+|d|^2 & -2|b|^2 & -|b|^2-|c|^2+|f|^2 & 1 \\ 0 & -|c|^2-|a|^2+|e|^2 & -|c|^2-|b|^2+|f|^2 & -2|c|^2 & 1 \\ 1 & 1 & 1 & 1 & 0 \end{bmatrix}.$$
$$(10.62)$$

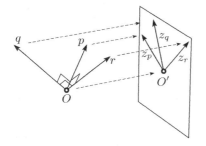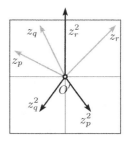

Fig. 10.17. Axonometric projection of an orthonormal basis

3. After Gauss' death hundreds of results and discoveries which he had never published or shown to anybody were found in his desk. One of these results, published posthumously in four lines without proof in Gauss' *Werke*, vol. 2, p. 309, is as follows: *if three orthonormal vectors p, q, r in $\mathbb{R}^3$ are projected by an "axonometric projection" (i.e. a parallel perspective) onto a plane (see Fig. 10.17), the image vectors z_p, z_q, z_r cannot be arbitrary, but must satisfy the relation*

$$z_p^2 + z_q^2 + z_r^2 = 0 \,, \tag{10.63}$$

where z_p, z_q, z_r are considered as numbers in $\mathbb{C}$. Prove this.

4. Fig. 10.18 presents six propositions of Apollonius concerning hyperbolas, their asymptotes, their tangents and segments parallel to these asymptotes — denoted by the original letters. Prove these propositions by a suitable affine transformation of the coordinates.

5. A hyperbola is called *equilateral* if its asymptotes are mutually perpendicular, i.e. $a = b$ in (3.13). Prove a nice discovery of Dörrie (1943, §134): *If the three vertices of a triangle lie on an equilateral hyperbola, then its orthocentre lies on this hyperbola as well.*

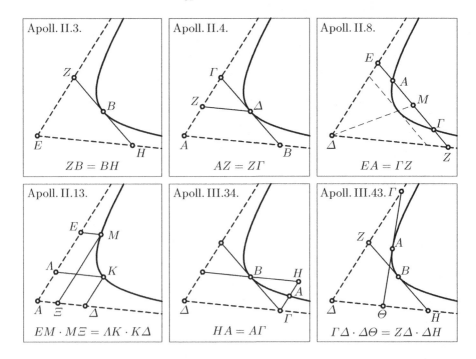

Fig. 10.18. Six propositions from Books II and III of Apollonius

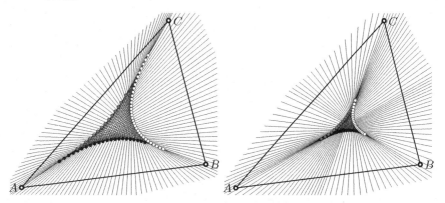

Fig. 10.19. Steinhaus' lines

6. An exercise proposed by Steinhaus (1958): In Fig. 10.19 a fixed triangle creates a family of lines which (a) divide the *perimeter* of the triangle into two equal parts (left), (b) divide the *area* of the triangle into two equal parts (right). Determine, in each case, the nature of the curves which are the envelopes of this family.

7. Isaac Barrow, the teacher and mentor of Isaac Newton, gave *Geometrical Lectures* at Cambridge University which were finally published in 1735 "Translated from the *Latin* Edition, revised, corrected and amended by the late Sir *Isaac Newton*" by Edmund Stone. In "Lecture VI, §2" Barrow solves the following problem (see Fig. 10.20, left): Let ABC be a *right-lin'd Angle*, D a given point. The points N and M move on AB and BC respectively and determine a point O such that $DOMN$ are aligned and $DO = MN$. Which curve does the point O describe? Solve this problem in two lines by a suitable affine transformation.

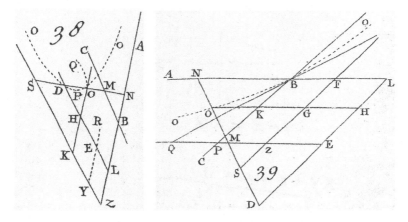

Fig. 10.20. Barrow's first (No. 38) and second problem (No. 39); reproduced by kind permission of the Syndics of Cambridge Univ. Library (classmark 7350.d.56)

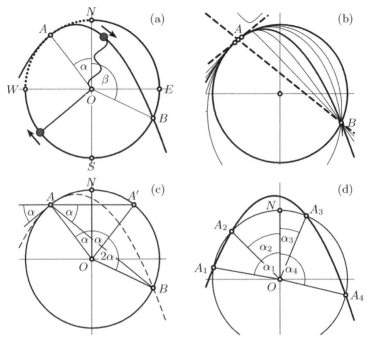

Fig. 10.21. "Wanner's first theorem" (a)–(c); generalisation (d)

8. Solve Barrow's second problem (Lecture VI, §4; see Fig. 10.20, right): in the same setting as for the previous exercise, determine the curve described by O satisfying the condition $MO = q \cdot MN$, where q is a given constant.

9. Show that a plane and an ellipsoid, determined by the equations

$$n_1 x_1 + n_2 x_2 + n_3 x_3 = d \qquad \text{and} \qquad \frac{x_1^2}{a_1^2} + \frac{x_2^2}{a_2^2} + \frac{x_3^2}{a_3^2} = 1, \qquad (10.64)$$

are tangent if and only if

$$n_1^2 a_1^2 + n_2^2 a_2^2 + n_3^2 a_3^2 = d^2. \qquad (10.65)$$

This is an extension of condition (7.11). An elegant version of this condition using matrix notation will be given in (11.24).

10. Solve the problem stated by Gergonne and Bret in the *Gergonne Journal* vol. 5, p. 172, which generalises the result of Monge in Fig. 7.5 on page 192: *If a vertex formed by three mutually perpendicular planes moves in such a way that all three planes remain tangent to the ellipsoid given in (10.64), then the vertex moves on a sphere of radius $\sqrt{a_1^2 + a_2^2 + a_3^2}$ centred at the origin.*

11. Prove "Wanner's first theorem". A boy at an Austrian secondary school plays with a stone attached to a string rotating in a vertical circle (see

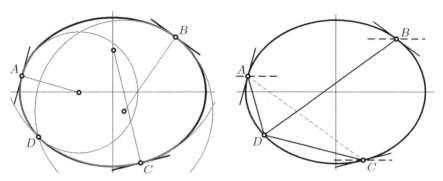

Fig. 10.22. Steiner's challenge concerning osculating circles of an ellipse

Fig. 10.21, (a)). If the speed of the stone is too small, the stone will leave the circle at a certain point A, located somewhere between W and N, and move along a parabola (Galilei *dixit*). At a second point B, located on the arc $WSEN$, the stone will return to circular movement. Determine how the position of this point B depends on that of A.

Hint. Since the arc $WSEN$ is three times as long as the arc WN, the simplest and most beautiful relation between the angles BON and NOA would be

$$\beta = 3\alpha. \tag{10.66}$$

12. Generalise Exercise 11 as follows: let A_1, A_2, A_3, A_4 be the intersections of a conic with a circle centred at O, and let α_i be the angle, counted with sign, between OA_i and ON ($i = 1, \ldots, 4$), where ON is the direction of an axis of the conic (see Fig. 10.21, (d)). Then formula (10.66) becomes

$$\alpha_1 + \alpha_2 + \alpha_3 + \alpha_4 = 0 \qquad \text{or} \qquad \pm 2\pi. \tag{10.67}$$

13. In the first three lines of Steiner (1846b) the following theorem is stated without further justification: *If D is a point on an ellipse (other than a vertex), there exist three points A, B and C on the same ellipse such that the three osculating circles at these points pass through D (see Fig. 10.22, left). Furthermore, the four points A, B, C and D are concyclic.* Give an elegant justification of Steiner's result by using ideas from the solution of Exercise 11.

11

Projective Geometry

"Unter den Leistungen der letzten fünfzig Jahre auf dem Gebiete der Geometrie nimmt die Ausbildung der *projectivischen Geometrie* die erste Stelle ein. [Among the advances of the last fifty years in the field of geometry, the development of projective geometry occupies the first place; transl. by M.W. Haskell]"

(F. Klein, first sentence of *Erlanger Programm*, 1872)

Fig. 11.1. Study of perspective; engraving by A. Dürer (two engravings of this kind by Dürer are known, one with a lute, the other with a naked woman; for obvious reasons, the authors have chosen the lute)

Scientific life in France changed notably in 1794 with the foundation of the *École Normale* and the *École Polytechnique*, "pour tirer la Nation Française de la dépendance où elle a été jusqu'à présent de l'industrie étrangère ...", where teachers like Lagrange, Laplace and Monge formed an entire generation of first-class mathematicians. Among others, we mention Fourier, Poisson, Cauchy, Liouville, Poncelet and Gergonne. The success of analytic methods

A. Ostermann and G. Wanner, *Geometry by Its History*,
Undergraduate Texts in Mathematics, DOI: 10.1007/978-3-642-29163-0_11,
© Springer-Verlag Berlin Heidelberg 2012

was then at its zenith, and for Lagrange and Laplace mathematics comprised only algebra and analysis. Only Monge lectured on *descriptive geometry*, lectures which were completely, and even proudly, ignored by the other two. But as is often the case, the future developed otherwise than had been expected, and many students of Monge, in particular Poncelet, Brianchon and Chasles started to develop *projective geometry* as a new field of geometric research, which was enthusiastically adopted by the German mathematicians Steiner and von Staudt, to such an extent that the subject, in the words of Felix Klein half a century later, "occupies the first place among the advances of the last fifty years in the field of geometry" (see the quotation).

11.1 Perspective and Central Projection

perspective   (from the Latin *perspicere*, to see through)

Perspective is concerned with the problem of representing a three dimensional object, for example the box in Fig. 11.2 (left), on a two dimensional canvas. This problem was one of the major challenges for the artists of the Italian Renaissance, beginning in the 15th Century (Brunelleschi, Piero della Francesca, Luca Pacioli, Leonardo da Vinci). The first artist from a northern country seriously interested in perspective was Albrecht Dürer (see Fig. 11.1).

Central projection. The idea indicated in Dürer's picture is the following one (see Fig. 11.2, left): we place a canvas (the *projection plane*) between the box and the artist, and draw the images of the points A', B', and so on, at

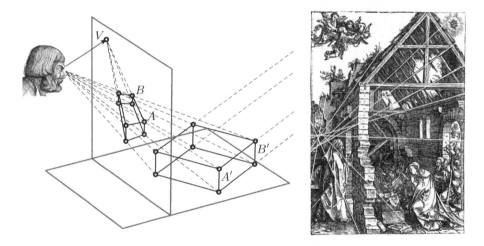

Fig. 11.2. Principle of perspective (left); (right) a woodcut of A. Dürer from 1502 (courtesy of Stift Stams, Tyrol). The added white lines should meet at the vanishing point

the intersections of the canvas with the rays of light joining these points (the *projection lines*) to the artist's eye (the *projection centre*). Since the projection lines all meet at the projection centre, this type of projection is also called a *central projection*. If the object behind the canvas is then removed, one has the impression that the box is still there.

Vanishing point. We observe in Fig. 11.2 an important result of perspective: imagine that the point A' moves towards B' and continues on a straight line until infinity; the corresponding projection line then tends more and more to the direction parallel to the segment $A'B'$. The intersection point of this parallel with the canvas is called the *vanishing point* V of this direction. We see that *the images of all lines parallel to a given segment pass through the same vanishing point.* A modern computer check in Fig. 11.2 (right) shows that 30-year-old Dürer slightly violated this principle.

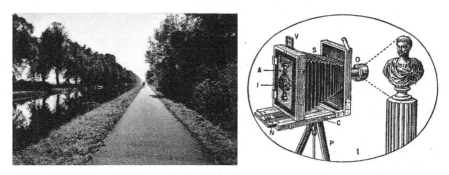

Fig. 11.3. Perspectives; on the left: a photo by C. Gressly; on the right: a modern camera seen on the outside and inside (Larousse 1929)

Remark. Since the invention of photography, modern cameras operate on the same principle, the difference is just that the screen now lies *behind* the "eye". As a consequence, the image is rotated by 180°, as demonstrated by Fig. 11.3 (right). This is why we observe the same phenomenon in photographs (see Fig. 11.3, left).

Analytic formulas. Given a point $\widetilde{x}$ (representing the eye of an artist or the focal point of a camera), a vector a (representing the projection line and the distance from $\widetilde{x}$ to the projection plane), and a spatial object x, we wish to determine the coordinates u_1, u_2 of the central projection of x onto the plane, see Fig. 11.4.

In order to fix the plane, we choose two vectors h and g that are mutually orthogonal and orthogonal to a: we first take $h = a \times (0,0,1)^{\mathsf{T}}$, which is horizontal, and then $g = h \times a$, both normalised. The vector w, joining the focus $\widetilde{x}$ to the projection of x, must be a multiple of $x - \widetilde{x}$:

$$w = \lambda(x - \widetilde{x})\,. \tag{11.1}$$

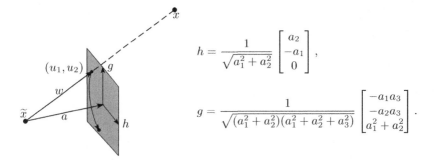

$$h = \frac{1}{\sqrt{a_1^2 + a_2^2}} \begin{bmatrix} a_2 \\ -a_1 \\ 0 \end{bmatrix},$$

$$g = \frac{1}{\sqrt{(a_1^2 + a_2^2)(a_1^2 + a_2^2 + a_3^2)}} \begin{bmatrix} -a_1 a_3 \\ -a_2 a_3 \\ a_1^2 + a_2^2 \end{bmatrix}.$$

Fig. 11.4. Analytic formulas for central projection

The parameter λ is determined by the fact that $w - a$ is orthogonal to a:

$$\langle w - a, a \rangle = 0 \qquad \Rightarrow \qquad \lambda = \frac{\langle a, a \rangle}{\langle x - \widetilde{x}, a \rangle}. \tag{11.2}$$

Finally, the scalar products $u_1 = w \cdot h$ and $u_2 = w \cdot g$ (see Theorem 9.5) are the desired coordinates. If the camera is not held horizontally, one can perform a rotation (10.5) through the angle α. In this way, a perspective image is determined by seven parameters: three for a, three for $\widetilde{x}$ and one for α.

Stereograms. The beautiful stereograms in this book were calculated with the same formulas, simply by replacing $\widetilde{x}$ once by $\widetilde{x} - 3h$ (left eye) and once by $\widetilde{x} + 3h$ (right eye). The number 3, measured in cm, is the half-distance between the eyes of the observer. This procedure gives *two* images, one for the left eye and one for the right.

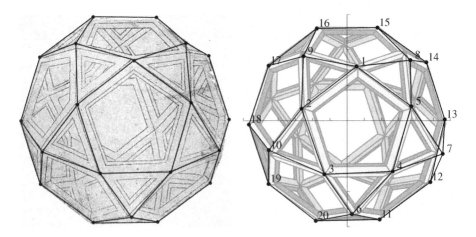

Fig. 11.5. Left: drawing by Leonardo da Vinci (1510) in grey, corrected vertices in black inserted by B. Gisin (2009), private communication; right: Leonardo's vertices in black, the corrected drawing in grey (Abdulle and Wanner, 2002)

'Correcting' Leonardo da Vinci. Renaissance artists made a great effort to study perspective. In contrast to today, errors in perspective were *true errors* ... We have taken the liberty of submitting a drawing of Leonardo to a rigorous scientific verification, for the first time in almost 500 years. As an example, we chose the drawing of an icosidodecahedron (Abdulle and Wanner, 2002). In order to perform the calculations, we measured the 20 visible vertices of the drawing and calculated their correct position in space. Then, by a numerical method called the "least squares method", we determined the best values for the above seven parameters. Once found, these values allow us to compare Leonardo da Vinci's drawing with the vertices of the closest possible *true* icosidodecahedron (see Fig. 11.3, left) and the drawing of the closest icosidodecahedron with the vertices of the original drawing (see Fig. 11.3, right). These calculations reveal several errors in the original drawing.

11.2 Poncelet's Principle of Central Projection

> "La doctrine est neuve, piquante et d'une vérité incontestable [the doctrine is new, surprising and of unquestionable truth]"
> (M. Brianchon 1819, quoted by Poncelet, 1862, vol. 2, p. 541)

After studying at the École Polytechnique, where he especially appreciated the lectures of Monge, Poncelet started on a military career. As *lieutenant du génie* he took part in Napoleon's disastrous invasion of Russia, where he survived several battles before being captured and imprisoned for two years in a camp on the river Volga. During this period, without access to any books or literature, he benefited from his recollection of Monge's lectures, and started to lay the foundations of projective geometry. This led to the *Traité des propriétés projectives des figures* in 1822. The original "cahiers" from Russia were finally published in 1862 (*Applications d'analyse et de géométrie*).

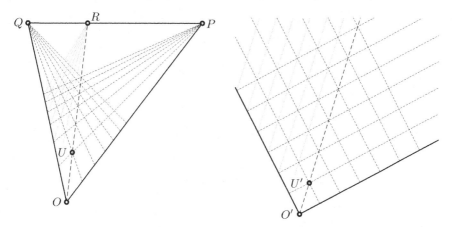

Fig. 11.6. Trying to transform a triangle into a quarter-plane

Poncelet's principle consists in transforming the *figure* of a theorem by a well-chosen central projection into an almost trivial form. With this method, Poncelet proved in a surprising and elegant manner a number of old and new theorems. We start with two lemmas followed by four theorems which demonstrate the power and elegance of this approach.

Poncelet's principal lemma. Given a perspective image in a Cartesian plane, we look for a central projection that maps the plane to the given image.

Lemma 11.1 (principal lemma on perspective). *Let OPQ be an arbitrary triangle and U be an arbitrary "unit point" inside OPQ (see Fig. 11.6, left). Then there exists a central projection which maps the line PQ to infinity and for which P and Q are the vanishing points of a pair of orthogonal axes centred at O', the image of O. The image U' of U is a unit point, i.e. O'U' is the diagonal of a square with sides on the axes O'P' and O'Q' (see Fig. 11.6, right).*

Proof. We produce the segment OU to find the point R on the line PQ. This point will be the vanishing point of the diagonal $O'U'$. We then place the triangle OPQ in a vertical plane so that PQ is horizontal, and choose the

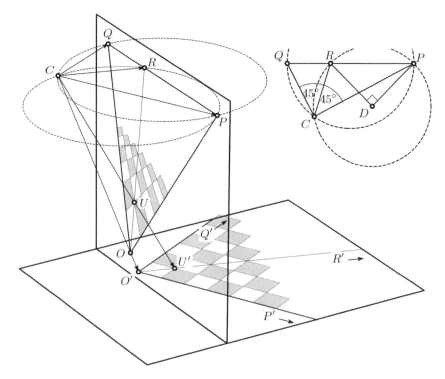

Fig. 11.7. Construction of a central perspective, see also Fig. 11.6

centre of projection C in the horizontal plane through P and Q. If we now project the figure $OPQR$ from C onto a plane parallel to CPQ (see Fig. 11.7), then the projected points P', Q' and R' will move to infinity. The projected region $P'O'Q'$ will be a quarter-plane, if the angle PCQ between the centre of projection and the vanishing points is a right angle. The projected point U' will have the same distance from each axis, i.e. be a unit point in an orthogonal Cartesian grid, if the angle PCR is $45°$. These two conditions allow us, by Eucl. III.20, to find the point C at the intersection of two circles, with central angles $180°$ and $90°$ respectively (see the inserted picture in Fig. 11.7).   □

Lemma 11.2 (second principal lemma, Poncelet 1814, IIIe Cahier, Princ. IV). *Consider a quadratic curve and a straight line d, both in the same plane (see Fig. 11.8, left). Then there exists a central projection that maps the curve onto a circle and the line d to infinity (see Fig. 11.8, right).*

Proof. We choose an arbitrary point P on d (see Fig. 11.8, middle). The polar of P cuts d at a point Q. The polar of Q meets that of P at the point O, and the respective tangents meet at U. It is thus sufficient to apply Lemma 11.1. The image of the curve under this central projection has tangents orthogonal to the axes at the points $x = \pm 1$, $y = 0$ and $y = \pm 1$, $x = 0$. It is therefore a circle.   □

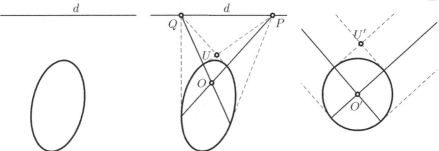

Fig. 11.8. Central projection transforming an ellipse into a circle

Remark. The light rays from the centre of projection C to a circle form a cone, if C is precisely above the circle's centre. So we understand immediately from the definitions of Chap. 3 that a central projection of this circle will become a conic section. In extension of this, Poncelet accepted in the above proof, without any further discussion, that the central projection of *any conic is again a conic*. Only the analytic treatment in Sect. 11.6 will confirm this intuition.

Theorem 11.3 (Pappus, *Collection*, Book VII, Props. 139, 143). *Let A, B and C be three points on a line, and let A', B' and C' be three points on another line in the same plane (see Fig. 11.9, left). Then the intersection points N, M, L of the pairs of lines AB' and BA', AC' and CA', BC' and CB' are collinear.*

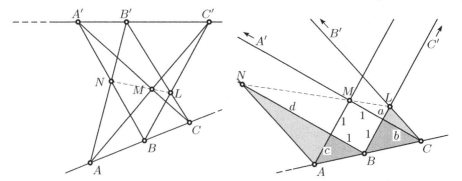

Fig. 11.9. Pappus' theorem and its proof by the "principle of central projection"

Proof. (Poncelet 1814, Cahier VII, 2e Partie, art. V.) We apply Lemma 11.1 to the triangle $BC'A'$ with unit point M. We obtain as image a square BM, whose side length we take to be 1 (see Fig. 11.9, right), and two parallel lines AN and CL. Thus the grey triangles are similar in pairs and we have $c = \frac{1}{b}$ and $\frac{d}{c} = \frac{b}{a}$ by Thales' theorem. This leads to $d = \frac{1}{a}$, which means that the two white triangles are also similar. Therefore N, M, L are collinear.   □

Theorem 11.4 (Desargues 1636). *For two given triangles ABC and $A'B'C'$, assume that the lines AA', BB', CC' are concurrent (see Fig. 11.10, left). Then the intersections $N = AB \cap A'B'$, $M = AC \cap A'C'$ and $L = BC \cap B'C'$ are collinear.*

In other words: If two triangles are perspective from a point, then they are perspective from a line.

Proof. By a central projection, we map two of these points, say N and L, to infinity (see Fig. 11.10, right). Consequently, the lines $AB, A'B'$ and $BC, B'C'$ become parallel. By Thales and Euclid I.4, the images of the triangles ABC and $A'B'C'$ are similar. Hence AC and $A'C'$ are also parallel. Therefore, the point M is collinear with N and L.   □

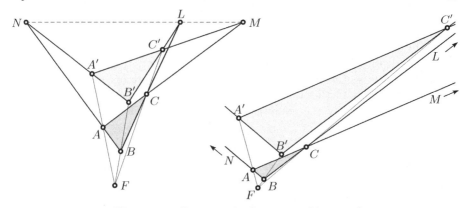

Fig. 11.10. Desargues's theorem and its proof

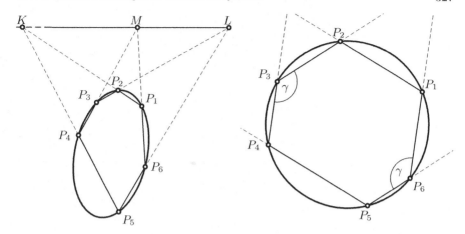

Fig. 11.11. Pascal's theorem and its proof

Theorem 11.5 (Pascal 1640[1]). *Let $P_1P_2P_3P_4P_5P_6$ be a hexagon inscribed in a conic (see Fig. 11.11, left; see also Fig. 11.13). Then the intersection points $K = P_1P_2 \cap P_4P_5$, $L = P_2P_3 \cap P_5P_6$ and $M = P_3P_4 \cap P_6P_1$ of pairs of opposite sides are collinear.*

The line through the points K, L and M is called the *Pascal line* (of the hexagon, with respect to the conic).

Proof. We apply Lemma 11.2 by mapping the line through two of these points, say K and L, to infinity (see Fig. 11.11, right). After this projection, the ellipse becomes a circle and two pairs of opposite sides are parallel. Therefore the arcs $P_2P_3P_4$ and $P_5P_6P_1$ have the same length. By Eucl. III.21, the two angles denoted by γ are thus equal and consequently, the third pair of opposite sides is also parallel. Therefore, the image of M lies on the line at infinity. □

Theorem 11.6 (Brianchon 1806). *Let $Q_1Q_2Q_3Q_4Q_5Q_6$ be a hexagon circumscribing a conic (see Fig. 11.12, left). Then the three diagonals joining pairs of opposite vertices are concurrent.*

The intersection point O is called the *Brianchon* point (of the hexagon, with respect to the conic).

Proof. The points $P_1, \ldots, P_6$ at which the hexagon touches the conic are the vertices of an *inscribed* hexagon. We apply to this hexagon the same projection as in the proof of Pascal's theorem. The images of the triangles $P_iQ_iP_{i+1}$ are isosceles, and the bases of opposite triangles are parallel. Consequently, their altitudes are concurrent, and all pass through the centre of the circle. □

---

[1]Pascal discovered this theorem at the age of 16.

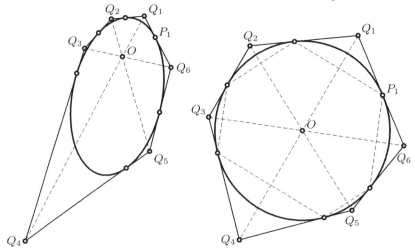

Fig. 11.12. Brianchon's theorem and its proof

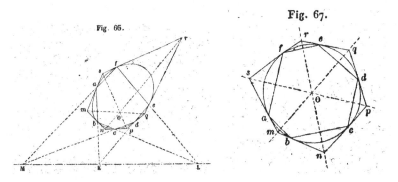

Fig. 11.13. Drawings by Poncelet (1814, publ. 1862) illustrating the theorems of
Pascal and Brianchon

Poncelet's continuity principle. Certain proofs are not always valid, for
example that of Pascal's theorem for the case when the line containing K, L
and M passes through the conic. Poncelet therefore formulated his *continuity
principle* which states that such a theorem persists. Poncelet's principle was
strongly criticised, in particular by Cauchy. The "principle of analytic contin-
uation" delivers us from all these problems and supports Poncelet's point of
view.

Poncelet's porism. One of the most spectacular visions of Poncelet was
his "Grand Théorème", also called Poncelet's porism or Poncelet's closure
theorem. Its rigorous proof challenged famous mathematicians, like Jacobi,
Cayley and Lebesgue for one century. For modern proofs of this result, we
refer to Griffiths and Harris (1978) and Tabachnikov (1993).

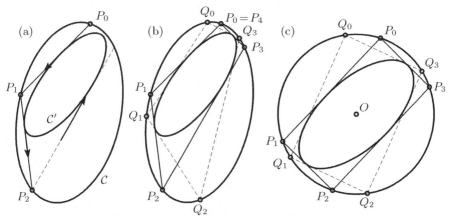

Fig. 11.14. Poncelet's porism and its proof

Theorem 11.7 (Poncelet 1813, VIe Cahier, Sect. III; Poncelet, 1822, Pl. XII, Fig. 97). *Let $\mathcal{C}$ and $\mathcal{C}'$ be two conics, $\mathcal{C}'$ lying inside $\mathcal{C}$, and let P_0 be a point on $\mathcal{C}$. We construct a polygonal line $P_0 P_1 P_2 \ldots$ that is both inscribed in $\mathcal{C}$ and circumscribed to $\mathcal{C}'$ (see Fig. 11.14 (a)). If the polygonal line closes for a certain integer n with $P_n = P_0$, then it closes independently of the choice of the point P_0 on $\mathcal{C}$ (see Fig. 11.14 (b)).*

Proof. The material that we developed in Chap. 7 allows elegant proofs for the particular cases $n = 3$ and $n = 4$. In the spirit of Poncelet, we start by reducing the theorem to a simpler configuration. Let P be a point whose polars with respect to the two conics coincide, see Fig. 11.15. (Algebraically, this leads to a generalised eigenvalue problem for a 3×3 matrix, see Exercise 14 below.) Next we choose a point Q on

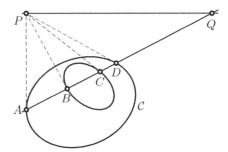

Fig. 11.15. Simultaneous transformation of two ellipses to a simpler form

this polar and apply Lemma 11.2, so that after projection the line PQ is at infinity and $\mathcal{C}$ is a circle.

For a good choice of Q on the polar we have two possibilities: (a) we move it in such a way that after projection $AB = CD$ is satisfied.[2] In this case the inner ellipse will be concentric with the circle (see Fig. 11.14 (c)). This is the situation of *Monge's circle* (see Fig. 7.5 on page 192). If (7.15) is satisfied, the polygonal line forms a rectangle for any initial point and Poncelet's theorem is proved for $n = 4$.

---

[2]This means, in virtue of (11.11) below, that the fourth harmonic points of ADQ and BCQ must be the same. Applying (11.12) twice, this leads to a quadratic equation for the coordinate of Q.

(b) We move Q such that the inner ellipse also becomes a circle, which however has another centre. In this case Poncelet's theorem is, for $n = 3$, a consequence of the Chapple–Euler–Lhuilier theorem (Theorem 7.22).   □

11.3 The Projective Line

"Die durch den Ponceletschen Traité eingeleitete Bewegung pflanzte sich nach Deutschland fort und ward einerseits von den Analytikern *Moebius* (1790–1868) und *Plücker* (1801–1868) und andererseits von den Synthetikern *Steiner* (1796–1863) und *von Staudt* (1798–1867) weitergeführt. [The movement initiated by Poncelet's Traité spread to Germany and was continued, on the one hand, by the analysts Moebius and Plücker and, on the other hand, by the geometers Steiner and von Staudt.]"   (F. Klein, 1928, p. 11)

Projective transformations. Following on Poncelet's treatise, the analytic theory of projective geometry was founded in Germany by Möbius and Plücker (see the quotation). The theory starts with analytic expressions for projective transformations in *one* dimension.

Consider two lines with origins O and O', respectively, and a central projection between them with centre C, see the figure below. According to Thales, we have

$$\frac{h}{x+g} = \frac{x'+f}{x+e},$$

$$x' = \frac{hx+he}{x+g} - f,$$

which is of the form

$$x' = \frac{ax+b}{cx+d}.$$

(11.3)

Consequently, we call the map (for $ad - bc \neq 0$)

$$x \mapsto x' = \frac{ax+b}{cx+d} \qquad \Leftrightarrow \qquad x' \mapsto x = \frac{dx'-b}{-cx'+a} \tag{11.4}$$

a *projective transformation* or a *Möbius transformation*. We have already encountered this transformation when discussing Carnot's solution of the Cramer–Castillon problem (6.31). Recall that the Möbius transformations form a *group* (the composition of Möbius transformations and the inverse of a Möbius transformation are again Möbius transformations).

The projective line. A projective transformation maps a "point at infinity" to an ordinary point (the point $x' = \frac{a}{c}$), and an ordinary one (the point $x = -\frac{d}{c}$) "to infinity". In order to include these particular cases, we declare that the *projective line* $\mathbb{P}$ consists of the real line $\mathbb{R}$ plus *one* point at infinity.

Homogeneous coordinates. In order to capture the point at infinity in a clear way, and to simplify several other concepts of analytic geometry, Plücker introduced *homogeneous coordinates* in 1830, see Plücker (1830a) and (1830b): each point x on the line is represented by a *pair* of numbers (x_1, x_2) by setting

$$x = \frac{x_1}{x_2} = \frac{\rho x_1}{\rho x_2}. \tag{11.5}$$

These coordinates are far from being unique, since one can multiply both by an arbitrary factor ρ. By setting $x_2 = 0$, one obtains the point at infinity. The formula (11.4) for projective transformations

$$\frac{x_1'}{x_2'} = \frac{a\frac{x_1}{x_2} + b}{c\frac{x_1}{x_2} + d}$$

becomes particularly elegant on multiplying numerator and denominator by x_2; this gives

$$\begin{bmatrix} x_1' \\ x_2' \end{bmatrix} = \begin{bmatrix} a & b \\ c & d \end{bmatrix} \begin{bmatrix} x_1 \\ x_2 \end{bmatrix} \quad \text{and} \quad \begin{bmatrix} x_1 \\ x_2 \end{bmatrix} = \begin{bmatrix} d & -b \\ -c & a \end{bmatrix} \begin{bmatrix} x_1' \\ x_2' \end{bmatrix}. \tag{11.6}$$

Note that for the inverse map, the usual division by the determinant is unnecessary since we are dealing with homogeneous coordinates.

There are several ways of interpreting the projective line $\mathbb{P}$ (see Fig. 11.16). We can consider it as

(a) the set of all straight lines in the plane that pass through the origin (i.e. all one dimensional subspaces of the Euclidean plane). The homogeneous coordinates are interpreted as a direction vector;

(b) the circle S^1 where antipodal points x' are $-x'$ are identified;

(c) the circle S^1 of points $\tilde{x}$ under a stereographic projection (see Chap. 5). In topology this is called a one-point compactification of $\mathbb{R}$.

The cross-ratio. In the preceding chapters, we saw that Thales's theorem is the central pillar of Euclidean geometry: the ratio of the lengths of two segments is unchanged by a *parallel* projection. Unfortunately, this nice property is destroyed by *perspective* projections. However, there will be a substitute in projective geometry: the *cross-ratio*.

Definition 11.8. *Let P_1, P_2, P_3, P_4 be four points on a line with (affine) coordinates x_1, x_2, x_3, x_4. Then, the number*

$$\mathrm{XR}\,(P_1, P_2, P_3, P_4) = \frac{P_1 P_3}{P_2 P_3} : \frac{P_1 P_4}{P_2 P_4} = \frac{x_3 - x_1}{x_3 - x_2} : \frac{x_4 - x_1}{x_4 - x_2} \tag{11.7}$$

is called the cross-ratio of the four points.

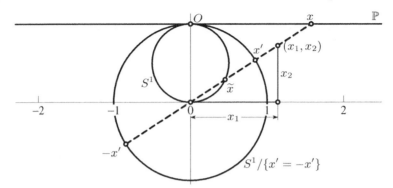

Fig. 11.16. Various interpretations of the projective line

Theorem 11.9 (Pappus, *Collection*, Book VII, Prop. 129). *The cross-ratio of four points is invariant under projective transformations, i.e.*

$$\mathrm{XR}\,(P_1', P_2', P_3', P_4') = \mathrm{XR}\,(P_1, P_2, P_3, P_4).$$

Algebraic proof (from lecture notes by W. Gröbner 1962). We compute the difference of two points after projection (11.4):

$$x_i' - x_k' = \frac{ax_i + b}{cx_i + d} - \frac{ax_k + b}{cx_k + d} = \frac{\Delta \cdot (x_i - x_k)}{(cx_i + d)(cx_k + d)},$$

where $\Delta = ad - bc$. Inserting this four times into (11.7), all Δ's and all denominators $(cx_i + d)$ drop out and we obtain

$$\frac{x_3' - x_1'}{x_3' - x_2'} : \frac{x_4' - x_1'}{x_4' - x_2'} = \ldots = \frac{x_3 - x_1}{x_3 - x_2} : \frac{x_4 - x_1}{x_4 - x_2}. \qquad \square$$

Geometric proof (from Steiner, 1832). From the law of sines (see Chap. 5), applied to the triangles $P_1 P_3 C$ and $P_2 P_3 C$ (see Fig. 11.17 (a)), we get

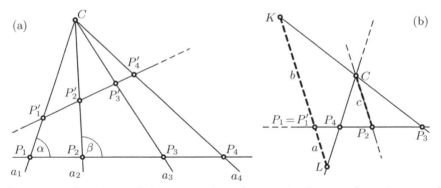

Fig. 11.17. Invariance of the cross-ratio under a projective transformation; proof by sine rule (a), Pappus' proof (b)

$$\left.\begin{array}{l}\dfrac{P_1P_3}{CP_3}=\dfrac{\sin a_1a_3}{\sin \alpha}\\[3mm]\dfrac{P_2P_3}{CP_3}=\dfrac{\sin a_2a_3}{\sin \beta}\end{array}\right\}\quad\Rightarrow\quad \dfrac{P_1P_3}{P_2P_3}=\dfrac{\sin a_1a_3}{\sin a_2a_3}\cdot\dfrac{\sin \beta}{\sin \alpha}\,,$$

where a_1a_3 denotes the angle between the lines a_1 and a_3, etc.

Similarly we get for the triangles P_1P_4C and P_2P_4C

$$\frac{P_1P_4}{P_2P_4}=\frac{\sin a_1a_4}{\sin a_2a_4}\cdot\frac{\sin \beta}{\sin \alpha}\,.$$

By taking the ratio of these expressions, the common factor $\frac{\sin \beta}{\sin \alpha}$ cancels and the cross-ratio becomes

$$\mathrm{XR}\,(P_1,P_2,P_3,P_4)=\frac{\sin a_1a_3}{\sin a_2a_3}:\frac{\sin a_1a_4}{\sin a_2a_4}\,. \tag{11.8}$$

This expression depends only on the four concurrent lines a_1, a_2, a_3, a_4 and is therefore also called the *cross-ratio of four lines*. □

Pappus' proof. Pappus' original proof extends over $1\frac{1}{2}$ pages with 8 figures. Since we are no longer afraid of negative quantities, we can present it, by keeping the same idea, in a much shorter way (see also Heath, 1921, vol. II, p. 420). The first idea is to suppose that P_1 and P_1' coincide. This can be achieved by a parallel displacement of the line $P_1'P_2'P_3'P_4'$, which preserves ratios, hence also the cross-ratio. Then we place the remaining points in the order $P_1 \to P_4 \to P_2 \to P_3$ on a line through P_1. The crucial idea is to draw a line parallel to CP_2 through $P_1 = P_1'$ whose intersections with CP_3 and CP_4 determine the points K and L, respectively (see Fig. 11.17 (b)). This creates two pairs of similar triangles: P_1LP_4 is similar to P_2CP_4 and P_1KP_3 is similar to P_2CP_3. Thales' theorem gives

$$\frac{x_3-x_1}{x_3-x_2}:\frac{x_4-x_1}{x_4-x_2}=\frac{b}{c}:\frac{a}{-c}=-\frac{b}{a}\,.$$

This last ratio is independent of the position of the line $P_4P_2P_3$ through P_1, as long as the lines CP_3, CP_2 and CP_4 are fixed. □

Harmonic points. The four points P_1, P_2, P_3, P_4 are called *harmonic*, if

$$\mathrm{XR}\,(P_1,P_2,P_3,P_4)=-1\,. \tag{11.9}$$

By putting P_1 and P_2 at ∓1 (see Fig. 11.18 (a)), this condition becomes

$$\mathrm{XR}=\frac{x_3+1}{x_3-1}:\frac{x_4+1}{x_4-1}=-1$$
$$\Leftrightarrow\quad (x_3+1)(x_4-1)=(1-x_3)(x_4+1)\quad\Leftrightarrow\quad x_3x_4=1\,, \tag{11.10}$$

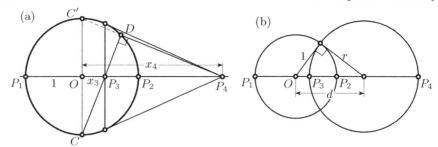

Fig. 11.18. Harmonic points

a particularly simple relation. If we set $x_3 = 0$, we see that

the points   and the lines

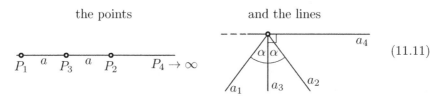

$$(11.11)$$

are harmonic. Formula (11.10) expresses the fact that (see Fig. 11.18)

(a) P_3 lies on the polar of P_4;

(b) the segment CD passing through P_3 (with $C = (0, -1)$ and D on the circle) is orthogonal to DP_4. This is seen as follows: by Eucl. III.20 the segment CD, which has slope $\frac{1}{x_3}$, is orthogonal to $C'D$, which by (11.10) has slope $-\frac{1}{x_4}$. Since $C'P_4$ has the same slope, the points C', D, P_4 are collinear and DP_4 also is orthogonal to CD;

(c) the distance P_1P_2 is the *harmonic mean*[3] of the distances P_1P_3 and P_1P_4,

$$\frac{1}{P_1P_2} = \frac{1}{2}\left(\frac{1}{P_1P_3} + \frac{1}{P_1P_4}\right) \quad \text{since} \quad \frac{1}{2} = \frac{1}{2}\left(\frac{1}{1+x_3} + \frac{1}{1+x_4}\right); \quad (11.12)$$

(d) the circles with diameters P_1P_2 and P_3P_4 intersect at right angles (see Fig. 11.18 (b)), because (by Eucl. III.36)

$$x_3x_4 = (d-r)(d+r) = 1. \qquad (11.13)$$

The complete quadrilateral. A *complete quadrilateral* consists of four lines in general position (i.e. no three are concurrent). Taken in pairs, these four lines intersect in six points A, B, C, D, E, F. The three additional lines AC, DB and EF are called the *diagonals of the quadrilateral*. Each diagonal cuts the other two, see Fig. 11.19 (left).

---

[3]The expression comes from music. The wave lengths $\frac{\lambda}{2}$, $\frac{\lambda}{3}$, $\frac{\lambda}{4}$, ... are the "harmonics" of the fundamental wave λ; each is the harmonic mean of its neighbours.

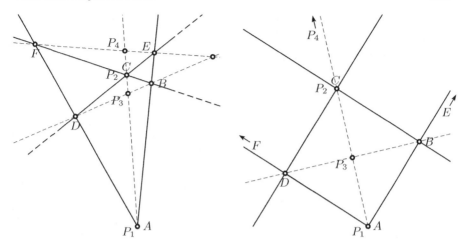

Fig. 11.19. Harmonic points from the complete quadrilateral

Theorem 11.10. *The four intersection points on each diagonal of a complete quadrilateral are harmonic.*

Proof. We apply Lemma 11.1 to the triangle AEF with C as unit point (see Fig. 11.19, right). Then the quadrilateral $ABCD$ becomes a square and the diagonal FE together with two intersection points move to infinity. The assertion is now clear from (11.11). □

11.4 The Projective Plane

Plücker coordinates. We begin with the equation $y = px + q$ of a line in $\mathbb{R}^2$ (see Fig. 7.2). The values of p and q are considered as fixed. By varying x and y according to this equation, one obtains *the set of points lying on this line* (see Fig. 11.20, left).

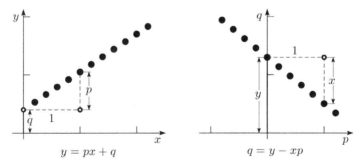

$$y = px + q \qquad q = y - xp$$

Fig. 11.20. Coordinates of points lying on the line (p, q) (left); coordinates of lines passing through the point (x, y) (right)

The idea of Plücker (1830b) consists in reversing the roles of the variables by considering (p, q) as *coordinates of this line*.[4] If we now choose (x, y) to be fixed and solve $y = px + q$ for q, we obtain $q = y - xp$ as equation for the *set of lines passing through this point* (see Fig. 11.20, right).

This equation becomes neater if one replaces (p, q) by a triple (u_1, u_2, u_3) satisfying

$$p = -\frac{u_1}{u_2}, \quad q = -\frac{u_3}{u_2},$$

which results in

$$u_1 x + u_2 y + u_3 = 0. \tag{11.14}$$

The numbers (u_1, u_2, u_3) are called the *Plücker coordinates* of the line. The last step towards perfect harmony is achieved by using, in a similar way as in (11.5), two-dimensional homogeneous coordinates (x_1, x_2, x_3) for the point (x, y):

$$x = \frac{x_1}{x_3}, \quad y = \frac{x_2}{x_3}.$$

Then equation (11.14) finally obtains the symmetric form

$$u_1 x_1 + u_2 x_2 + u_3 x_3 = 0. \tag{11.15}$$

A point thus lies on the line (11.15) if its homogeneous coordinates (x_1, x_2, x_3) are (formally) orthogonal to the Plücker coordinates of this line.

The feet of the altitudes of a triangle. Gut and Waldvogel (2008) gave a simple algorithm to construct the feet of the altitudes of an n-simplex with the help of Plücker coordinates. We illustrate here their idea for $n = 2$, i.e. for triangles.

Let $A_1 A_2 A_3$ be a triangle in $\mathbb{R}^2$ with homogeneous coordinates $A_k = (a_{1k}, a_{2k}, 1)$ and let

$$A = \begin{bmatrix} a_{11} & a_{12} & a_{13} \\ a_{21} & a_{22} & a_{23} \\ 1 & 1 & 1 \end{bmatrix}.$$

By construction, the columns of $S = (A^{-1})^{\mathsf{T}}$ are orthogonal to those of A. Therefore, the k-th column $S_k = (s_{1k}, s_{2k}, s_{3k})$ of S contains the Plücker coordinates of the line through the points A_ℓ for $\ell \neq k$.

The foot F_k of the altitude through A_k is thus given by

$$F_k = A_k + \lambda S_k^0$$

with $S_k^0 = (s_{1k}, s_{2k}, 0)$ and λ determined by $S_k^{\mathsf{T}} F_k = 0$. This finally yields

$$\begin{bmatrix} f_{1k} \\ f_{2k} \end{bmatrix} = \begin{bmatrix} a_{1k} \\ a_{2k} \end{bmatrix} - \frac{1}{s_{1k}^2 + s_{2k}^2} \begin{bmatrix} s_{1k} \\ s_{2k} \end{bmatrix}.$$

---

[4]This was the first time in history that "coordinates" were something else than coordinates of a point.

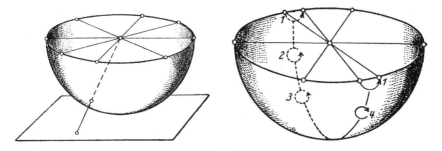

Fig. 11.21. Projective plane, drawing by F. Klein (1928), p. 15

It is worth noting that the same approach works in any dimension, see Gut and Waldvogel (2008). For tetrahedra, we refer to Exercise 12 below.

The projective plane. One "line" is particular among all those of (11.15), namely the line

$$x_3 = 0 , \qquad \text{i.e. the coordinates } u_1 \text{ and } u_2 \text{ are zero}, \tag{11.16}$$

since it represents *the points at infinity*. Thus, we can state that the *projective plane* $\mathbb{P}^2$ consists of the plane $\mathbb{R}^2$ and *the projective line* $\mathbb{P}$ *at infinity*.

We can consider the projective plane $\mathbb{P}^2$ as (see Fig. 11.21, left)

(a) the set of all lines in $\mathbb{R}^3$ passing through the origin (that is, all one dimensional subspaces of the Euclidean space);

(b) the sphere S^2 where antipodal points x' and $-x'$ are identified;

(c) the half-sphere bordered by a circle representing $\mathbb{P}^1$, the line at infinity.

On the other hand, the interpretation as a *stereographic projection* is no longer possible, since infinity would then be represented by *a point* and not by a line.

Theorem 11.11 (F. Klein, *Math. Ann.* vol. 7, 1874, p. 550, and L. Schläfli[5]). *The projective plane* $\mathbb{P}^2$ *is a non-orientable manifold*.

Proof. We choose a figure drawn by the discoverer himself (see Fig. 11.21, right). Let a small circle in position "1" cross the line at infinity. By moving it to the positions "2", "3", and finally "4", we see that its orientation has changed. □

---

[5]Ludwig Schläfli (1812–1895) taught mathematics in a school at Thun (Switzerland) and trained himself in higher mathematics. In 1843, when Steiner went to Rome with Jacobi, Dirichlet and Borchardt, Schläfli was chosen as their interpreter. Dirichlet gave him daily lessons (Dieudonné, *Abrégé d'histoire des mathématiques* II, p. 453).

Projective transformations of the plane. In homogeneous coordinates the analogue of the transformation (11.6) for the plane becomes (Chasles 1837, von Staudt 1847)

$$
\begin{bmatrix} x_1' \\ x_2' \\ x_3' \end{bmatrix} = \begin{bmatrix} a_{11} & a_{12} & a_{13} \\ a_{21} & a_{22} & a_{23} \\ a_{31} & a_{32} & a_{33} \end{bmatrix} \begin{bmatrix} x_1 \\ x_2 \\ x_3 \end{bmatrix} \qquad \text{or simply} \quad x' = Ax, \tag{11.17}
$$

with $\det A \neq 0$. The set of projective transformations (in terms of homogeneous coordinates) is hence given by the general linear group (the invertible matrices) modulo invertible multiples of the identity. It was Klein (1872) who, in his *Erlanger Programm*, emphasised the role of group theory in geometry.

Contragredient transformation. In order to understand how the coordinates of a line are transformed by a projective transformation (11.17) of points, we set $u' = Bu$. The condition $u^\mathsf{T} x = 0$ (the point x lies on the line u) is equivalent to $(u')^\mathsf{T} x' = 0$. Since

$$
(u')^\mathsf{T} x' = u^\mathsf{T} B^\mathsf{T} A x ,
$$

we obtain

$$
u' = (A^\mathsf{T})^{-1} u = (A^{-1})^\mathsf{T} u . \tag{11.18}
$$

Such a transformation is called *contragredient* with respect to (11.17).

11.5 The Principle of Duality

> "The principle of duality may properly be ascribed to Gergonne (1771–1859). Poncelet protested that it was nothing but his method of reciprocation with respect to a conic (polarity), and Gergonne replied that the conic is irrelevant ... It is sad that such a beautiful discovery was marred by bitter controversy over the question of priority."
>
> (H.S.M. Coxeter, *The real projective plane* (1949), pp. 13–14)

> "Die Auffindung des Dualitätsprinzips, das von unserem heutigen Standpunkt aus nicht allzu tiefliegend erscheint, stellte eine wesentliche wissenschaftliche Leistung dar. Man erkennt dies am besten daran, dass rund 150 Jahre nach der Auffindung des Pascalschen Satzes vergangen sind, ehe der Satz des Brianchon gefunden wurde ..."
>
> (F. Klein, 1928, p. 38)

Polar reciprocation (Poncelet, 1817). On comparing Figs. 11.11 and 11.12 we observe that the sides $P_i P_{i+1}$ are the *polars* of the points Q_i. By Theorem 7.3, the intersections K, L and M are the poles of the diagonals $Q_i Q_{i+3}$. The fact that K, L and M are collinear (Pascal's theorem) is equivalent to the fact that the polars pass through one point, the corresponding pole. In this way, Brianchon's theorem can be seen as a "dual" version of Pascal's theorem. It took 150 years for this "triviality" to be discovered (see the quotation).

Axiomatic duality (Gergonne, 1824/27). Gergonne discovered that the above duality follows directly from the axioms characterising lines and points, on which he based projective geometry. His concept of duality relies on the observation that points and lines are interchangeable objects, and that conics are no longer required. For more details on this axiomatic approach, we refer to the book by Coxeter (1961).

Duality by coordinates (Plücker, 1830b). A third approach to duality relies on the perfect symmetry of the expressions (11.15). In fact, there is a symmetry between the two problems:

A point x is the intersection of two lines u and v	A line u connects two points x and y
$u_1 x_1 + u_2 x_2 + u_3 x_3 = 0$	$x_1 u_1 + x_2 u_2 + x_3 u_3 = 0$
$v_1 x_1 + v_2 x_2 + v_3 x_3 = 0$	$y_1 u_1 + y_2 u_2 + y_3 u_3 = 0$

The solution is each time given by the cross product (9.23), in one case by $x = u \times v$, and in the other by $u = x \times y$. In contrast to Euclidean geometry, two lines *always* meet. The intersection point, however, sometimes lies at infinity.

The principle of duality. *To each theorem in projective geometry, there corresponds a "dual theorem" in which each expression in one column*

$$
\begin{array}{rcl}
line & \leftrightarrow & point \\
pass\ through & \leftrightarrow & lie\ on \\
intersection\ point\ of\ two\ lines & \leftrightarrow & line\ connecting\ two\ points \\
concurrent & \leftrightarrow & collinear \\
polar & \leftrightarrow & pole
\end{array}
$$

is replaced by the corresponding expression in the other column.

Having stated this principle, we must add a small caveat: The hope of effortlessly finding many new theorems by applying this principle turns out to be illusory. For example, the theorems of Pappus and Desargues simply reproduce each other under duality (see Fig. 11.22).

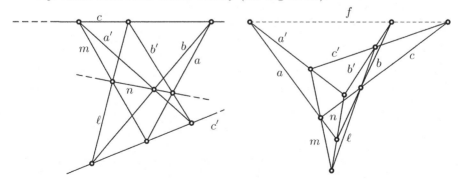

Fig. 11.22. The theorems of Pappus and Desargues are self-dual

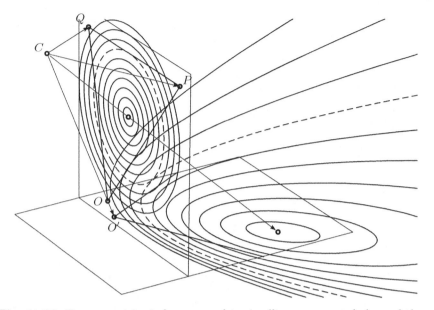

Fig. 11.23. Ten concentric circles, mapped to six ellipses, one parabola, and three hyperbolas

11.6 The Projective Theory of Conics

Here again, homogeneous coordinates show their strength. Consider a general conic (7.5) in the Euclidean plane. We introduce homogeneous coordinates (11.15) and multiply by x_3^2 to obtain

$$ax_1^2 + 2bx_1x_2 + cx_2^2 + 2dx_1x_3 + 2ex_2x_3 + fx_3^2 = \begin{bmatrix} x_1 & x_2 & x_3 \end{bmatrix} \begin{bmatrix} a & b & d \\ b & c & e \\ d & e & f \end{bmatrix} \begin{bmatrix} x_1 \\ x_2 \\ x_3 \end{bmatrix} = 0$$

or in matrix notation

$$x^\mathsf{T} A x = 0 \,, \tag{11.19}$$

where A is a symmetric 3×3 matrix. A projective transformation (11.17) $x = Tx'$ transforms this equation into

$$(x')^\mathsf{T} A' x' = 0 \,, \qquad \text{where} \qquad A' = T^\mathsf{T} A T \,. \tag{11.20}$$

Such a transformation no longer preserves ellipses, parabolas and hyperbolas (see Fig. 11.23). The curves differ only by their position with respect to the line at infinity $x_3 = 0$. A parabola possesses this line as tangent (the seventh circle in Fig. 11.23), the hyperbolas meet it in two points, and ellipses do not meet it at all.

The projective classification of conics. Any real *symmetric* matrix A has real eigenvalues and a basis of orthogonal eigenvectors. Therefore, there exists a non-singular matrix T such that

$$T^{\mathsf{T}} A T = \operatorname{diag}(\lambda_1, \lambda_2, \lambda_3), \qquad \lambda_i \in \{0, \pm 1\}, \tag{11.21}$$

see Exercise 13 below. From the projective point of view, conics can be classified as follows:

$(\lambda_1, \lambda_2, \lambda_3)$	equation	conic
$(0,0,0)$	$0 = 0$	projective plane
$(1,0,0)$	$x_1^2 = 0$	(double)line
$(1,1,0)$	$x_1^2 + x_2^2 = 0$	point
$(1,-1,0)$	$x_1^2 - x_2^2 = 0$	two crossing lines
$(1,1,1)$	$x_1^2 + x_2^2 + x_3^2 = 0$	empty set
$(1,1,-1)$	$x_1^2 + x_2^2 - x_3^2 = 0$	circle

From this table we see once again that ellipses, parabolas and hyperbolas can not be distinguished from a projective point of view.

Polars. Let x_1 be a point on the conic defined by A, i.e. $x_1^{\mathsf{T}} A x_1 = 0$, and let x be a second point satisfying

$$x_1^{\mathsf{T}} A x = 0 \qquad \text{or equivalently} \qquad x^{\mathsf{T}} A x_1 = 0. \tag{11.22}$$

If x were also on the conic, then for each real λ, we would have

$$\left(x_1 + \lambda(x - x_1)\right)^{\mathsf{T}} A \left(x_1 + \lambda(x - x_1)\right) = 0,$$

i.e. the entire line connecting x_1 and x would lie on the conic. If the conic is not degenerate, this is impossible and we obtain the following result.

If $x_1^{\mathsf{T}} A x_1 = 0$, then formula (11.22) is the equation of the tangent at x_1.

For an arbitrary point x_0 in the projective plane, as in Section 7.3,

$$u_0^{\mathsf{T}} x = 0 \qquad \text{with} \qquad u_0^{\mathsf{T}} = x_0^{\mathsf{T}} A \tag{11.23}$$

is the equation of the polar of x_0. All the beautiful properties that we know from Section 7.3 remain valid. In particular, we have the following:

(a) A polar, given by (11.23), is tangent if

$$x_0^{\mathsf{T}} A x_0 = 0 \qquad \Leftrightarrow \qquad u_0^{\mathsf{T}} A^{-1} u_0 = 0. \tag{11.24}$$

The latter is the *condition on the coordinates u_0 of a line for it to be tangent.*

(b) The centre of the conic is the pole of the line at infinity.

(c) A diameter of the conic is the polar of a pole at infinity.

(d) Two diameters are conjugate if the pole of the first lies on the polar of the second, and conversely.

(e) Let a point P_3 lie on the polar of a point P_4. If the line connecting P_4 with P_3 cuts the conic in two points P_2 and P_1, then the points P_1, P_2, P_3, P_4 are harmonic (see Fig. 11.18).

11.7 Exercises

1. Give an alternative proof of Pappus' theorem 11.3 by moving the line ML to infinity. The points A', B' and C' then exchange their positions and we obtain the picture of Fig. 11.24 (left). It remains to prove that if $A'C$ is parallel to AC' and $B'C$ to BC', then $A'B$ is parallel to AB'.

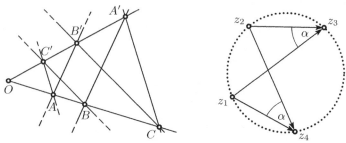

Fig. 11.24. Alternative proof of Pappus' theorem (left); solution for the relation between a circle and the cross-ratio (right)

2. Let five points of a conic be given. Construct other points on the conic with the help of Pascal's theorem. An additional challenge is to find, in the case of an ellipse, from these five points the centre of the ellipse, a pair of conjugate diameters and finally its axes (Pappus, *Collection*, Book VIII, Prop. 13, "Cum autem quæ situm sit circa quinque data puncta $HKLMN$ ellipsim describere").

3. (see Hurwitz and Courant 1922, p. 274) Let z_1, z_2, z_3, z_4 be four numbers in the complex plane (see Fig. 11.24, right). Show that they lie on a circle if and only if their cross ratio $\mathrm{XR}\,(z_1, z_2, z_3, z_4)$ is a real number. Use this result to show that the map $z \mapsto \frac{1}{z}$ (or more generally any Möbius transformation) transforms circles to circles (if lines are considered as degenerate circles).

4. Given three collinear points P_1, P_2 and P_3, show that there exists a unique harmonic point P_4. Construct this point using only a ruler.

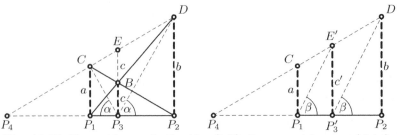

Fig. 11.25. Two cableways in the Alps (left); the geometric mean (right)

5. (Professor Stewart's *Mathematical Curiosity* no. 94, see Stewart (2008)) In a mountain village called *Après-le-Ski*, located between two mountain

peaks of height a and b respectively, two cableways run, one from the foot of each mountain to the opposite peak. At which height c do the two cables meet? You are first allowed to use Thales in order to show that

(a) $\quad \dfrac{1}{c} = \dfrac{1}{a} + \dfrac{1}{b}\,;$ (11.25)

(b) $P_3 B$ is the bisector of angle $CP_3 D$.

Then use higher projective education to clarify the apparent similarity of the first formula with the harmonic mean and to give an elegant explanation of the second result. In formula (5.21), we have already seen result (a) in a slightly different form.

6. Compare the two pictures in Fig. 11.25: while in the picture on the left $P_3 P_4$ is the *harmonic mean* of $P_1 P_4$ and $P_2 P_4$, in the picture on the right $P_3' P_4$ is their *geometric mean* (by similar figures $\frac{b}{c'} = \frac{c'}{a}$, hence $c' = \sqrt{ab}$). Conclude that the harmonic mean is smaller than the geometric mean (for $a \neq b$), i.e. give another geometric proof of the inequality in (7.59) on page 223.

7. Show, (a) by an analytic calculation, (b) purely geometrically, that the circumcentre O, the nine-point centre N, the centroid G and the orthocentre H of a triangle on the Euler line are harmonic.

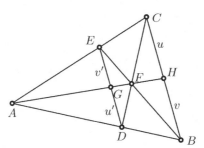

Fig. 11.26. Bisecting a side of a triangle

8. Given a triangle ABC (see Fig. 11.26), draw a segment DE parallel to BC, intersect CD and BE to find F, draw AF to find H on BC. Show, first using Thales, then using projectivity, that H is the midpoint of BC.

9. As an extension of (11.13), show that the four intersection points P_1, P_2, P_3, P_4 of a line with two circles intersecting at right angles lie in harmonic position, if the line $P_1 P_2 P_3 P_4$ is diameter of *at least one* of the circles (see Fig. 11.27 (a)).

10. Prove the following result of Steiner (1846a) (see Fig. 11.27 (b)): *Draw the circle of curvature of an ellipse at a point B. Then produce the diameter through B of this circle to the point A outside the ellipse, such that $AB = r$, the radius of curvature. The circle with diameter AB intersects Monge's circle of radius $\sqrt{a^2 + b^2}$ at right angles.*

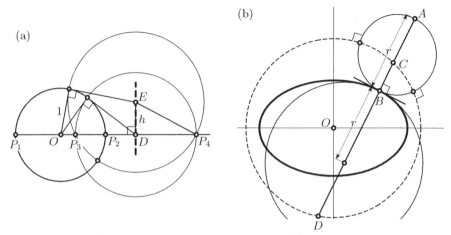

Fig. 11.27. Harmonic points and orthogonal circles (a); Steiner's theorem on the radius of curvature of a ellipse and its relation to harmonic points (b)

Together with the previous Exercise 9 this means that *the points A, B, C, D are harmonic*. Steiner found this result sufficiently interesting for a separate publication, but nevertheless did not include any indication about its discovery or any hint of a proof. Find an elegant proof.

11. Formulate the dual version of Theorem 11.10. The resulting configuration is called a *complete quadrangle*.

12. Plücker coordinates can also be defined in higher dimensions. For example, the plane $ax + by + cz + d = 0$ in $\mathbb{R}^3$ is determined by the homogeneous coordinates (a, b, c, d).

    Let $A_1 A_2 A_3 A_4$ be a given tetrahedron in $\mathbb{R}^3$ with homogeneous coordinates $A_k = (a_{1k}, a_{2k}, a_{3k}, 1)$, and consider the matrix A with columns A_k. Show that the ℓ-th column of $S = (A^{-1})^\mathsf{T}$ contains the Plücker coordinates of the face opposite to A_ℓ. Find an explicit formula for the feet of the altitudes of the tetrahedron.

13. Find an explicit formula for the transformation T in (11.21).

14. Given two conics, find a point P for which its polars with respect to the conics coincide.

12

Solutions to the Exercises

At the age of thirteen or fourteen, L. Euler used to work alone on difficult mathematical problems. Every Saturday afternoon he was admitted at the home of Johann Bernoulli, then the world-leading mathematician, to be fostered with unsolved problems and questions. Later in his life, Euler mentioned frequently[1] that throughout the week he worked hard to solve as many problems as he could and to have to ask Johann Bernoulli as few questions as possible. For Euler this was the best method for making rapid progress in mathematics.

Now also, 300 years later, the efforts for solving exercises without looking up a solution are of extreme importance. For those of our readers, however, who don't have—in case of difficulties—a Johann Bernoulli at hand, we present below our solutions to the exercises. This allows the readers to compare with their own solutions and to find the one or other idea for improvements. Whenever a reader finds a shorter or more elegant solution, the authors will be glad to include it in a future edition of this book.

12.1 Solutions for Chapter 1

1. He is. A modern computer gives the value $1, 43\ 55\ 22\ 58\ 27\ 57\ 56\ldots$, because

$$\sqrt{3} = 1.73205080756888$$

$$
\begin{aligned}
\text{remainder} &= 0.73205080756888 & \times 60 &= 43.92304845413 \\
\text{remainder} &= 0.92304845413 & \times 60 &= 55.382907248 \\
\text{remainder} &= 0.382907248 & \times 60 &= 22.9744349 \\
\text{remainder} &= 0.9744349 & \times 60 &= 58.46609 \\
\text{remainder} &= 0.46609 & \times 60 &= 27.9656 \text{ etc.}
\end{aligned}
$$

---

[1] reported by Nicolaus Fuss 1783 in his *Lobrede auf Herrn Leonhard Euler*, *Opera Omnia*, vol. I, p. LII

A. Ostermann and G. Wanner, *Geometry by Its History,* 345
Undergraduate Texts in Mathematics, DOI: 10.1007/978-3-642-29163-0_12,
© Springer-Verlag Berlin Heidelberg 2012

2. Call the angle $DAB = \beta$, so that $DOB = 2\beta$. Further call the angle $DAC = \gamma$, so that $DOC = 2\gamma$. The result follows by subtracting β, respectively 2β.

3. Because both inscribed angles correspond to the same central angle.

4. One obtains the *Fibonacci numbers* $1, 1, 2, 3, 5, 8, 13, 21, 34, \ldots$, where each is the sum of its two immediate predecessors. The fractions r_k tend to the golden ratio. If one does not simplify these fractions, one obtains them in the form

$$1 + \cfrac{1}{1+1}, \qquad 1 + \cfrac{1}{1+\cfrac{1}{1+1}}, \qquad 1 + \cfrac{1}{1+\cfrac{1}{1+\cfrac{1}{1+1}}}, \qquad \ldots$$

as *continued fractions*.

5. Since $\Phi = 1 + \frac{1}{\Phi}$, the rectangles are similar.

6. Observe that all dimensions of these pieces are Fibonacci numbers. Their slopes are close, but not equal, to Φ. Therefore the diagonal in the picture on the left is not a straight line.

7. The two shaded rectangles have area $2 \cdot (1, 25) \cdot \delta$ and, if we neglect the tiny square in the upper right corner, this must be $0, 00\ 25$. Hence we have to subtract from $1, 25$ the value of

$$\delta = \frac{0, 00\ 25}{2, 50} = 8.823529411765/60^2 = 0, 00\ 08\ 49\ 24 \ldots$$

which leads to $1, 24\ 51\ 10\ 35 \ldots$

8. Inspired by an idea which Gauss had as schoolboy, we add a second triangle upside down:

$$\Rightarrow \quad 2t_n = n(n+1) \quad \text{or} \quad t_n = \frac{n(n+1)}{2}.$$

9. We cut the pentagon into three triangles and correct with one column of n dots

$$\Rightarrow \quad p_n = 3t_{n-1} + n = 3\frac{n(n-1)}{2} + n = \frac{n(3n-1)}{2}.$$

The same proof also applies to higher "polygonal numbers" (see Heath, 1921, p. 79).

10. Adding up the dots along the zig-zag lines in any of the three directions, we obtain $1 + 3 + 5 + \ldots$, so the result is the same as in (1.7).

11. The theorem of Pythagoras, what else?

12. Just use the formula for the area of a parallelogram.

13. If you want to do this in an elegant way, apply Theorem 4.2 of Chap. 4 to the triangle $B\Gamma\Pi$. One can also apply Thales' theorem. For Heron's original proof, which relies only on the principles of Euclid's first book, see Heath (1926, vol. I, p. 366).

14. Draw the circle with diameter OP, and rotate the picture so that OC and PD are vertical. Then *look*.

15. This is the identity $(p + q)^2 - p^2 - q^2 = 2pq = 2h^2$ (using the altitude theorem in (1.10) and the notations of Fig. 1.20), multiplied by $\frac{\pi}{8}$.

16. Denote by h the altitude of the tent, and by ℓ the length of the projection of EB onto the base. Then we have two right-angled triangles with sides $h, \ell, 1$ and $\frac{\Phi}{2}, \frac{\Phi-1}{2}, \ell$ respectively. Applying Pythagoras' theorem to both and eliminating ℓ^2 leads to $h = \frac{1}{2}$ by using $\Phi^2 - \Phi = 1$. The slopes of the triangles are $\frac{2h}{\Phi-1}$ and the slopes of the quadrilaterals are $\frac{2h}{\Phi}$. The product of these slopes is 1, so one is the inverse of the other.

17. (Pythagorean triples.) There are many such triples on the Babylonian tablet "Plimpton 322", which O. Neugebauer and A. Sachs deciphered (see R.C. Buck, 1980, for an excellent account). But no tablet is known which explains how these triples were actually discovered. A student whose childhood was spent learning algebraic identities would see at once that

$$a = u^2 - v^2, \qquad b = 2uv, \qquad c = u^2 + v^2 \qquad (12.1)$$

are, for u and v arbitrary integers with $u > v > 0$, Pythagorean triples. We will see in the next exercise that in fact *all* such triples have this form. But someone who instead spent his youth playing with marbles might prefer to represent square numbers by arranging marbles in a square array (see Fig.12.1, left). He would then see that the difference of the two square numbers $(n+1)^2 - (n-1)^2$ is $4n$. Since 4 is itself a square, it simply suffices

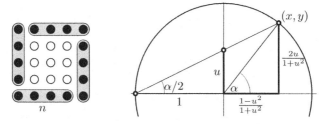

Fig. 12.1. Pythagorean triples and rational points on the unit circle

to put $n = u^2$ to make the third number a square. This gives (12.1) for $v = 1$. (This last solution is attributed to Plato, the general procedure is given in Eucl. X.28; see Heath, 1926, vol. I, p. 356, and vol. III, p. 63.)

18. We see in Fig. 12.1 (right) two similar triangles with sides $1, u$ and $1+x, y$. Thus $y = \lambda u$ and $x = \lambda - 1$. From $x^2 + y^2 = 1$ we have $\lambda = \frac{2}{1+u^2}$. This gives for x and y the values indicated in Fig. 12.1. These are rational values, if u is rational. These values are clearly related to the above Pythagorean triples. Conversely, if x and y are rational, so are λ and u.

19. Connect the centres of the circles to the intersection points of the line through P with the respective circle. You obtain a sequence of similar triangles. Prof. Michel Mayor, explaining his discovery of the first Exoplanet, once called this effect the "Doppler dragon-fly".

20. Each quarter of this set has the same area as the triangle with the same vertices. This is seen in the same way as in Hippocrates' squaring of the lunes (1.12). Thus the total area $\mathcal{A}$ is the area of the square with side length $\sqrt{2}$, hence $\mathcal{A} = 2$.

21. The two grey quadrilaterals have the same angles and sides, hence have the same area. One represents $\frac{1}{2}(a^2 + b^2)$ plus the area of the triangle; the other represents $\frac{1}{2}c^2$ plus the area of the triangle.

22. Because of the identities $(r + a)(r - a) = r^2 - a^2$ (Eucl. II.5 in the next chapter) and $r^2 - a^2 = b^2$ (Pythagoras).[2]

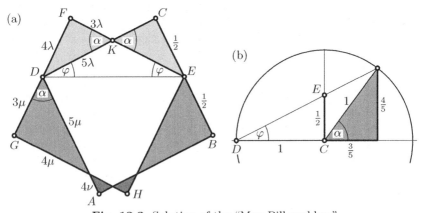

Fig. 12.2. Solution of the "Max-Bill-problem"

23. We denote by φ the angle by which the square $ABCD$ is rotated to the left (with respect to the line DE) and the square $EFGH$ is rotated to the right (see Fig. 12.2 (a)). The angle EKC is an exterior angle of DEK and by (1.2) equal to $\alpha = 2\varphi$. We place the triangle DEC into the unit circle

as in Fig. 12.2 (b), which is the same as in Fig. 12.1 (right) for $u = \frac{1}{2}$. By Eucl. III.20 we again find $\alpha = 2\varphi$ and conclude that all six triangles in Fig. 12.2 (a), two-by-two of the same size, are similar to the triangle with sides 3, 4, 5. Hence there are three constants λ, μ and ν such the sides are as indicated in this figure. From the condition $FD = \frac{1}{2}$ we obtain $\lambda = \frac{1}{8}$, from $DG = \frac{1}{2}$ we obtain $\mu = \frac{1}{6}$, and from $DA = 1$ we get $5\mu + 4\nu = 1$ or $\nu = \frac{1}{24}$. Thus the three areas are $6\lambda^2 = \frac{3}{32}$, $6\mu^2 = \frac{1}{6}$ and $6\nu^2 = \frac{1}{144}$. The value of the angle $\alpha = \arctan \frac{4}{3}$ is not particularly interesting; a numerical calculation gives $\alpha = 53°7'48''22'''6''''31'''''$.

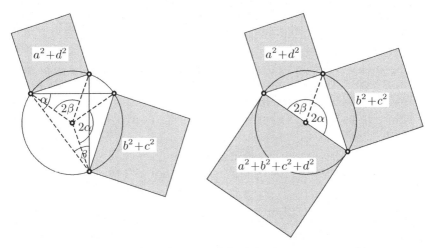

Fig. 12.3. Proof "without words" of the four-squares problem

24. The proof "without words" is given in Fig. 12.3. If words are allowed, we would say that Pythagoras' theorem and Eucl. III.20 are used repeatedly, and that $\alpha + \beta = \llcorner$ by Eucl. I.32, hence $2\alpha + 2\beta = 2\llcorner$.

25. A square with diagonal $10d$, i.e. area $50d^2$, is replaced by a circle of diameter $8d$, i.e. area $16d^2\pi$. If both areas were the same, we would have $\pi = \frac{25}{8} = 3.125$, slightly better than the Egyptian value of $\pi = 3.1605$.

12.2 Solutions for Chapter 2

1. Cut the polygon, say by the lines NB and NC, into $n - 2$ triangles and apply Eucl. I.32 to each of these. If the polygon is *convex*, i.e. if all angles are $< 2\llcorner$, this dissection requires no precautions. In the case of the above drawing, however, one could not cut along the line DA.

2. Denote by γ the two angles in the "X", which are equal by Eucl. I.15. Then by Eucl. I.32 we have $\alpha + \gamma + \llcorner = \beta + \gamma + \llcorner$ and the result follows by subtracting $\gamma + \llcorner$.

3. Let $AB = 1$, so that AF must be $\frac{1}{\Phi}$ and $MF = \frac{1}{\Phi} - \frac{1}{2} = \Phi - \frac{3}{2}$. Then, since $ME = 1$ and $GE = \frac{1}{4}$, we have $MG = \frac{\sqrt{15}}{4}$. Thus by Thales applied to the similar triangles FMD and EGD we obtain $MF = \frac{1}{4+2\sqrt{5}} = \frac{2\sqrt{5}-4}{4} = \Phi - \frac{3}{2}$.

4. Suppose, instead, that C is *not* on the circle, for example outside it (see the picture). Then let D be the point of intersection of line AC with the circle. Now the angle ADB is right by Eucl. III.20, the angle ACB is right by hypothesis; a contradiction with Eucl. I.16.

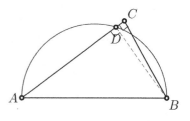

5. The angles BCE and CBD are parallel angles to β and γ respectively (Eucl. I.29). Also by construction $BD = b$ and $CE = c$. Hence, by Eucl. I.4, the triangles CBD and CEB have the same sides. The result follows from Eucl. I.7 (what we have actually proved here, is the converse of Eucl. I.33).

6. A still more clever student will discover that one foot of the perpendiculars through E lies *inside* the triangle and one lies *outside*.

7. The triangles BDA and CDA have the same sides, thus by Eucl. I.22 the angles BDA and CDA are the same. By Post. 4, they are right angles.

8. The triangles PC_1Q and PC_2Q are isosceles. Thus the result follows from Exercise 7.

9. We have $t = \sqrt{3(d+3)}$ by Eucl. III.36 and $\frac{d+3}{9} = \frac{t}{r}$ by Thales. Inserting t from the first equation and using $d = 2r$ leads to

$$(d+3) \cdot d^2 = 12 \cdot 9^2 = (9+3) \cdot 9^2 \,.$$

Comparing the first and last expression we see that $d = 9 \,\text{li}$.

10. By Eucl. I.29, the angles α and β in the picture are the same, respectively. Since the sides a are the same too, we have by Eucl. I.26 that $AE = CE$ and $BE = DE$. The result for the rhombus then follows from Exercise 7.

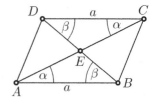

11. By Eucl. I.16 we have $\gamma < \delta$ and by Eucl. I.5 we have $\delta = \varepsilon < \beta$ (see Fig. 12.4 (a)); thus, as Euclid says, β "is much greater" than γ.

12. By Eucl. I.5 we have $\delta = \varepsilon$ (see Fig. 12.4 (b)), which itself is smaller than η. Hence, by Eucl. I.19 (which is the converse of Eucl. I.18 of the previous

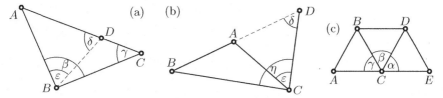

Fig. 12.4. Proofs for Eucl. I.18, Eucl. I.20 and Eucl. IV.15.

exercise and can be proved by contradiction), BD (which is $BA + AC$) is greater than BC.

13. By Eucl. I.5 all angles of an equilateral triangle are the same; by Eucl. I.32 they are all $2 \llcorner /3$. Therefore $\alpha + \beta + \gamma = 2 \llcorner$ (see Fig. 12.4 (c)), and the result follows from Eucl. I.14.

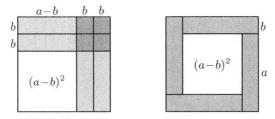

Fig. 12.5. Proofs for Eucl. II.8; Euclid's proof (left), Clavius' proof (right).

14. Euclid's original proof (picture on the left in Fig. 12.5) represents the difference of the two squares of sides $a + b$ and $a - b$ as four rectangles $b \times (a - b)$ and four squares $b \times b$, which together fill four rectangles $b \times a$. The picture on the right, attributed by Heath to "Clavius and others", shows the result immediately without any calculation.

15. By construction and Pythagoras we obtain $(s + \frac{1}{2})^2 = 1 + \frac{1}{4}$, which on the one hand gives $s^2 + s = 1$ using Eucl. II.4, the desired equation, and on the other hand by taking square roots $s + \frac{1}{2} = \frac{\sqrt{5}}{2}$.

16. If C is not the orthogonal projection (Eucl. I.12) of F onto this line, let another point G be this projection and denote the angle FCG by α. Both angles at G, the exterior and the interior, are right angles. By Eucl. I.16, $\alpha < \llcorner$. Then, by Eucl. I.18 (see Exercise 11), FG is shorter than FC. On the other hand, G is outside the circle and hence FG is longer than FC. This is a contradiction.

17. Euclid's original proof is very long and based on Eucl. III.18 (see Exercise 16). Today, we are less scrupulous about passing to the limit, and simply let the point C move towards B, always making the same angle α, and in the limit the line CB will become the tangent.

18. We attach the three squares a^2, b^2 and c^2 to the three sides of the triangle ABC (see Fig. 12.6, left). By producing the three altitudes of this triangle we cut each of these squares into two rectangles. For precisely the same reason as in Euclid's proof, pairs of these rectangles have the same area. The square c^2 is $\mathcal{A}_1 + \mathcal{A}_2$, which is the same as $a^2 + b^2$, if the two rectangles of area $\mathcal{A}_3 = av$ are subtracted.

   This elegant proof, found by Grégoire de Saint-Vincent in 1647, has since then been rediscovered again and again (see Heath, 1926, vol. 1, p. 404 and Steiner and Arrigo, 2010, p. 109).

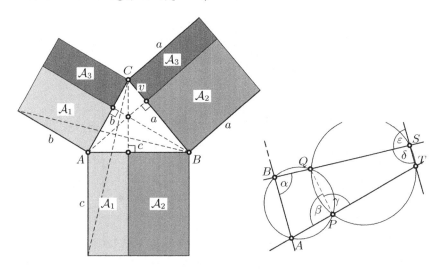

Fig. 12.6. Direct proof of Eucl. II.13 (left); proof of the two-circles problem (right)

19. As with Clavius' corollary of Eucl. III.36, a more general result is easier to prove: the two lines TS and AB in Fig. 2.43 (b) are parallel. This result follows from Eucl. III.22 applied to the two circles. The remainder of the proof is displayed in Fig. 12.6 (right): Subtract, add, and subtract $\alpha + \beta = 2 \llcorner$, $\beta + \gamma = 2 \llcorner$, $\gamma + \delta = 2 \llcorner$, $\delta + \varepsilon = 2 \llcorner$ and you obtain $\alpha = \varepsilon$.

20. The authors were unable to find a precise reference to work of J. Steiner for this theorem. However, a clear proof was given by A. Miquel (1838b) as follows: apply Eucl. III.22 (as drawn in Fig. 2.17 (c)) to the four quadrilaterals inscribed in the four circles, to conclude that the angles denoted by α, β, γ and δ in Fig. 2.44 (right) are the same, respectively. Then, again by Eucl. III.22 (and its converse) the points A, B, C, D and A', B', C', D' are concyclic if and only if $\alpha + \beta + \gamma + \delta = 2 \llcorner$.

21. (a) We compute, using Eucl. III.36, $BE^2 = AE \cdot CE = \frac{9b}{5} \frac{4b}{5}$, so that $BE = \frac{6b}{5}$. By Eucl. III.32 (see Exercise 17), the two angles marked α are the same. Together with the common angle β, we see that BCE is similar to ABE, thus by Thales $\frac{BC}{BE} = \frac{AB}{AE}$, which gives $BC = \frac{2a}{3}$. If we extend

BC to D such that $BD = BA$, i.e. ADB is an equilateral triangle, we see that $\frac{BC}{CD} = 2 = \frac{FC}{CA}$. Hence the triangles FCB and ACD are similar with similarity factor 2. As a consequence $BF = 2 \cdot AD = 2a$ and the angle CBF is the angle $CDA = 60°$, which gives the desired result.

(b) With C as before, draw a second $60°$-circle based on CF and let B be the intersection of the two circles (Fig. 2.45 (c)). Then BC is the angle bisector of the triangle ABF and $BF = 2 \cdot BA$ follows from Eucl. VI.3.

22. (a) We take $A\Gamma = 1$. By Pythagoras, $b = \sqrt{1 - a^2}$ (see Fig. 2.46, right). We see by Eucl. III.20 that the triangles in question are similar. Hence, by Thales, $ZA = \frac{b}{y}$ and $HZ = \frac{x^2}{y}$. This shows that $y = HA = HZ + ZA = \frac{x^2}{y} + \frac{b}{y}$. We obtain $y^2 - x^2 = b$ or, by Pythagoras, $1 - 2x^2 = b$ and $2y^2 - 1 = b$. This gives the two formulas (which will be encountered later, see (5.9))

$$y = \sqrt{\frac{1+b}{2}}, \qquad x = \sqrt{\frac{1-b}{2}}.$$

Archimedes obtained in this way the estimates

$$B\Gamma > \frac{780}{1560}, \; H\Gamma > \frac{780}{3013\frac{3}{4}}, \; \Theta\Gamma > \frac{240}{1838\frac{9}{11}}, \; K\Gamma > \frac{66}{1009\frac{1}{6}}, \; A\Gamma > \frac{66}{2017\frac{1}{4}}.$$

Try to do this without a computer and without floating point calculations, computing all the roots by rational approximations.

(b) Again we set $\Gamma E = 1$. By Eucl. VI.3 we have (see Fig. 2.47, right)

$$\frac{s-t}{\sqrt{1+s^2}} = \frac{t}{1} \quad \text{or} \quad t = \frac{s}{\sqrt{1+s^2}+1}.$$

With this Archimedes obtained the estimates

$$\Gamma Z < \frac{153}{256}, \; \Gamma H < \frac{153}{571}, \; \Gamma\Theta < \frac{153}{1162\frac{1}{8}}, \; \Gamma K < \frac{153}{2334\frac{1}{4}}, \; \Gamma\Lambda < \frac{153}{4673\frac{1}{2}}.$$

For more details, see the footnotes of Ver Eecke in Archimedes' *Opera* (1921), or the article of Miel (1983).

23. We discover that for all these polyhedra the relation

$$s_0 - s_1 + s_2 = 2 \qquad (\text{in Euler's notation} \quad E + F = K + 2)$$

is true. As simple as this relation is, if you want a general proof, Euler will tell you about the *demonstrationis difficultatem* ... and, in fact, simple proofs were only given a century later (Cauchy, Steiner (1826d), von Staudt (1847); see Pont (1974) for details).

12.3 Solutions for Chapter 3

1. Denote by x_1 and x_2 the abscissae of B, Q and P, A (see Fig. 3.8, left); by y_1 and y_2 the ordinates of B, P and Q, A. Then by Pythagoras

$$OP^2 + OP'^2 = OP^2 + OQ^2 = x_2^2 + y_1^2 + x_1^2 + y_2^2 = x_2^2 + y_2^2 + x_1^2 + y_1^2 = a^2 + b^2,$$

which means, of course, that "summa quadratorum binarum diametrorum coniugatorum semper est constans".

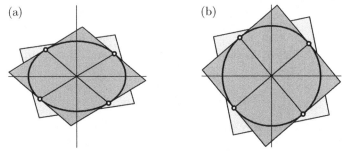

Fig. 12.7. Parallelograms about the conjugate diameters of an ellipse

2. Well, if we stretch the ellipse of Fig. 12.7 (a) into a circle (Fig. 12.7 (b)), the conjugate diameters become orthogonal and the parallelograms turn into squares rotating about this circle. The theorem can thus *really* be seen "almost by Intuition, even without Demonstration". Newton himself just says that the result "Constat ex Conicis". It is actually Apoll. VII.31, Apollonius' proof, however, extends over four pages.

3. This is equivalent to Pappus' definition of the parabola (see the first picture in Fig. 3.1 and also Exercise 4 below): we have $PS = PV$ exactly if $PU = PQ$, under the condition that $VU = SQ$, the radius of the circle. The centre of the circle is the focus of the parabola. If the line intersects the circle, the parabola passes through the intersection points, because here both distances are zero.

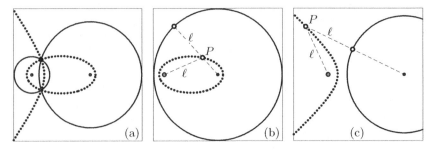

Fig. 12.8. Points with same distance from two circles

4. If the two circles intersect, the answer is quite tricky (see Fig. 12.8 (a)). The question is easier if the two circles are disjoint, either one inside the other, or one next to the other. One can then reduce the smaller of the two circles to a point by subtracting (or adding) a constant to both radii. We thus arrive at the problem: *Given a circle and a point, find the set of points at the same distance from the circle and the point.* The answer is an easy consequence of (3.5) or (3.12) and leads to a *third* characterisation of the conics (see Fig. 12.8 (b) and (c)). The centres of the circles are located at the foci and the radius of the remaining circle is $2a$.

5. They are perpendicular. For a proof, look at the angles denoted by α in Fig. 3.4 and denote the corresponding angles of Fig. 3.11 by β. One then sees that $2\alpha + 2\beta = 2 \llcorner$ and that the angle between the two tangents is $\alpha + \beta$, which must therefore be $\llcorner$. Another way of seeing this result is Fig. 12.8 (a), because the points on the two confocal conics have the same distance from the two circles, hence the tangents to the conics are the angle bisectors of the tangents to the circles.

6. Draw the circle with diameter AB. Call its centre M and its radius r. Then draw the diameter PQ passing through C. Think of all three objects as being fixed to the triangle during its movement (see Fig. 3.15 (b)). Draw the circle ED with centre O and radius $2r$. Because the angle AMD is twice the angle EOD, while the radii are inversely proportional, the arc lengths AD and ED are the same. *Hence we can imagine that the circle $PAQB$ rolls inside the circle DE while the diameter AB moves on the rectangular axes.* By Proclus' construction of Fig. 3.8 (page 68) any point on this diameter moves on an ellipse. But PQ, which carries the point C, is just another diameter which will touch the outer circle somewhere else. Therefore by the same result, C will move on an ellipse which is rotated by half the angle AMQ.

   Remark. If we replace $2r$ by $3r$, the point A moves on a much more interesting curve, Steiner's deltoid (see Fig. 7.17 on page 205).

7. Since the point E stays on the diagonal of the parallelogram $OIPG$, we have $IE = EG$. The point G moves on a circle centred at H, hence E has the same distance from this circle and from H, i.e. it moves on an ellipse by the third characterisation mentioned in the solution to Exercise 4.

8. Draw the line $F'P$ and produce it to the point B as in the picture on the right of Fig. 3.4. Then the point R lies halfway between the points B and F (the triangle BPF is isosceles). Since O is halfway between F' and F, the triangles FOR and $FF'B$ are similar with ratio $1 : 2$. Since $F'B = 2a$ by (3.5), we have $OR = a$.

9. Since the tangent at the vertex is halfway between F and the directrix d, the result follows, with the same argument as in the previous exercise (see Fig. 3.1), from Thales' theorem.

10. They are parabolas with the midpoint F as focus. In order to see this, note that the midpoint M between A and B, which is the orthogonal projection of F onto AB, moves on a straight $45°$ line. The result thus follows from the previous exercise (see Fig. 3.17, right).

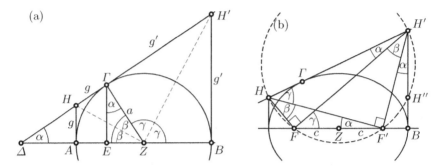

Fig. 12.9. Proofs of Apoll. I.34, I.36 and III.42 (a); Proof of Apoll. III.46 (b)

11. As suggested, we stretch the ellipse into a circle (see Fig. 12.9 (a)). Then the triangles $\Gamma H'B$ and ΓHA are isosceles, since the sides g' and g are tangents to the circle from the same point and thus have the same length. So by Thales for the "horizontal" g, g' we have $\frac{g'}{g} = \frac{u'}{u}$ (see Fig. 3.19 (a)), and for the the "vertical" g, g' we have $\frac{g'}{g} = \frac{h'}{h}$. This proves Apoll. I.34.

    The angles α are right angles. By Thales' theorem $ZE/Z\Gamma = Z\Gamma/Z\Delta$, which proves Apoll. I.36. Since $2\beta + 2\gamma = 2\llcorner$, the angle $H'ZH$ is a right angle and $gg' = a^2$ is the altitude theorem (Eucl. II.14). This proves Apoll. III.42, because $g = \frac{a}{b}h$ and $g' = \frac{a}{b}h'$.

    We have kept the same names for the points as in Apollonius' drawing for I.36, but his proofs are much longer.

12. See 13.

13. We draw the Thales circle with diameter HH' (see Fig. 12.9 (b)). Since its centre lies directly above Z, this circle cuts the axis AB in two points F and F' which have the same distance c from Z. Applying Clavius' corollary of Eucl. III.36 (see (2.6)) to the point B, we obtain $hh' = (a - c)(a + c)$, hence by Apoll. III.42 and Eucl. II.5, $a^2 - c^2 = b^2$ or $c^2 = a^2 - b^2$ (you might argue that perhaps this Thales circle does *not* intersect AB; we would then raise AB upwards until it becomes tangent and Eucl. III.36 would lead to the contradiction $a^2 < b^2$). Since the arc HF is equal to the arc $F'H''$, all three angles marked α are equal by Eucl. III.21 (and similarly for the others).

14. If $\Theta\Gamma$ *were* perpendicular, then from the triangles $\Theta\Gamma H$ and $\Theta\Gamma H'$ we would have $\tan\alpha / \tan\gamma = u/u'$ (anticipating a notation from Chap. 5; see Fig. 3.19 (a)). On the other hand, from the triangles $F'BH'$ and FAH we

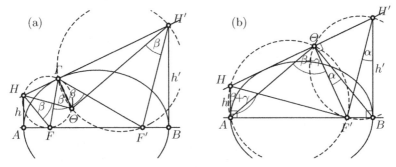

Fig. 12.10. Proof of Apoll. III.48 (a); Proof of Apoll. III.49 (b)

have $\tan\alpha/\tan\gamma = h/h'$, since $F'B = AF$. This is in accordance with Apoll. I.34. Since this "backwards" argument does not meet the rigour established by Euclid, Apollonius turns this into a lengthy proof showing that the opposite hypothesis leads to a contradiction.

15. The quadrilateral $\Gamma H'F'\Theta$ of Fig. 3.20 (b) has two opposite angles which are right. Hence Thales' circle with diameter $\Theta H'$ will pass through Γ and F'. By Eucl. III.21 we see that the angles $F'\Gamma\Theta$ and $F'H'\Theta$ are equal to β (see Fig. 12.10 (a)). Similarly we prove that the angle $\Theta\Gamma F$ is β.

16. As in the previous exercise we draw circles with diameters $F'H'$ and $F'H$. Two applications of Eucl. III.21 (see Fig. 12.10 (b)) then show that the angle $A\Theta B$ is $\alpha + \beta + \gamma$. But we see, from Fig. 3.20 (a) and Euclid's Postulate 5, that $2\alpha + 2\beta + 2\gamma = 2 \, \llcorner$.

17. The proof is contained in Fig. 3.21, left: by applying Eucl. I.32 to the triangle SFS', we see that $\gamma = \beta - \alpha$. We then project F orthogonally onto the three tangents, to obtain N', n and N, which lie on the circumcircle of the ellipse (this is Apoll. III.50, which Poncelet attributes to Maclaurin). We also draw M and M' as the points on that circle, opposite to N and N' with respect to F. Because of the right angles, the points S', N', n and F are concyclic. By Eucl. III.21 applied to this circle, we move α to N', and then, by Eucl. III.20 applied to the circumcircle, we see that the angle nOM' is 2α. Similarly, the angle nOM is 2β. Thus 2γ is the angle $M'OM$, which is independent of the position of the third tangent. If we move either S or S' to P, we obtain the second result as an easy corollary.

18. By Apoll. I.36 (see Exercise 11) we have $xw = a^2$ in the notation of Fig. 12.11 (a). By stretching (see the "Hint" for that same exercise) we also have the analogous formula $yv = b^2$ and by Thales $\frac{z}{y} = \frac{v}{w}$, i.e. $zw = yv = b^2$. Hence, $PO \cdot QO = (x - z)w = xw - zw = a^2 - b^2$.

19. If we stretch the ellipse to a circle (see Fig. 12.11 (b)), then ZAE and EAH are similar, hence $\frac{u}{w} = \frac{w}{v}$ or $uv = w^2$. Because ΔE and ZH are parallel, the stretching preserves the ratios on these lines.

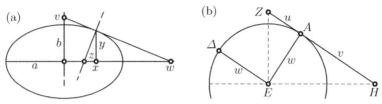

(a)

Fig. 12.11. Solutions of Exercises 18 (a) and 19 (b)

20. By Apoll. III.50 the points R and R' lie on the Thales circle with diameter AB. The triangles $H'BD$, $F'R'D$, FRD and HAD are all similar. Hence by Thales there is a constant k (which is the cotangent of the angle ADH) such that

$$d \cdot d' = k^2 \cdot RD \cdot R'D = k^2 \cdot AD \cdot BD = h \cdot h' = b^2 \,.$$

The second relation is Clavius' corollary to Eucl. III.36, the last one is Apoll. III.42.

12.4 Solutions for Chapter 4

1. Indeed, we have with the notation of Fig. 4.29 (b),

$$z = \frac{v}{y} = \frac{vw}{x^2} = \frac{vwv^2}{w^4} = \frac{v^3}{w^3} = u^3 \,.$$

Here, the first and last identity follows from Thales' theorem, the second and third identities follow from Eucl. II.14 applied to the right triangles $\Gamma\Lambda A$ and ΛAH, which gives $yw = x^2$ and $xv = w^2$ respectively.

2. We weigh the parabola strip by strip and compare these weights to those of a triangle. A strip of length x placed on a lever arm of length $1 - x$ is in equilibrium with a strip of the parabola of length $x(1-x)$ placed on a lever arm of length 1. We sum up all areas and concentrate the entire triangle (which has area $4\mathcal{T}$) at its centre of gravity. Since its distance is $\frac{1}{3}$ (by Theorem 4.1), we have equilibrium with the entire parabola concentrated on the lever arm of length 1 if $\mathcal{P} = \frac{4}{3}\mathcal{T}$.

3. We have to show that the Miquel point M lies on all three altitudes; by symmetry it suffices to show this for one altitude, say for AD. The angles EBA and ACF in Fig. 4.15 on page 95 are then orthogonal, hence equal. Thus $\frac{AE}{AF} = \frac{AB}{AC}$ or $AF \cdot AB = AE \cdot AC$, which means that the point A has the same power with respect to both circles FDB and EDC. Consequently, A lies on the radical axis of these two circles, i.e. the points A, M, D are collinear.

4. The angles marked δ in Fig. 4.31 (a) are orthogonal, hence equal. Therefore CFB and AFH are similar triangles. Hence $\frac{b}{c} = \frac{y}{a}$ or $y = \frac{ab}{c}$. The second computation exchanges $a \leftrightarrow b$ and leads to exactly the same result.

5. If $y = \frac{ab}{c}$ from the solution of the preceding Exercise 4, then also $c = \frac{ab}{y}$.
One can also see the result geometrically from Fig. 4.31 (a): If C and H are exchanged, then the sides AC, BC and the altitudes BE, AD exchange their roles by remaining mutually perpendicular. See also Fig. 12.25 on page 394.

6. The triangles APF and BVF are similar, as are APU and BVP. Thus $\frac{PU}{PV} = \frac{PA}{BV} = \frac{c_1}{c_2}$. Now insert into this $\frac{PU}{PC} = \frac{b_2}{b_1}$ divided by $\frac{PV}{PC} = \frac{a_1}{a_2}$, which is precisely Eucl. VI.2, and conclude.

7. Denote by $\mathcal{C}_1, \mathcal{C}_2, \mathcal{A}_1, \mathcal{A}_2, \mathcal{B}_1, \mathcal{B}_2$ the areas of these triangles and let h_1 resp. h_2 be the distance of A resp. B from the line CP. These are the altitudes of the triangles APC and BPC respectively. Hence

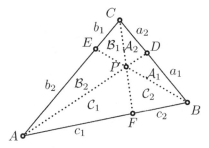

$$\frac{c_2}{c_1} = \frac{h_2}{h_1} = \frac{\mathcal{A}_1 + \mathcal{A}_2}{\mathcal{B}_1 + \mathcal{B}_2}.$$

Multiplying all three corresponding factors $\frac{a_2}{a_1} \cdot \frac{b_2}{b_1} \cdot \frac{c_2}{c_1}$ clearly gives 1. This proof (from 1816) is due to A.L. Crelle, see Baptist (1992, p. 61).

8. (a) We learned from Joh. Bernoulli (see Exercise 6) that, when everything is oblique, we have to draw some parallel line, in order to be able to apply other theorems. We choose to draw DG parallel to BF. This creates two pairs of similar triangles: CGD is similar to CFB and GDH is similar to FAH. From the first pair we obtain by Thales $DG = \frac{BF}{2} = \frac{c}{4}$ and $CG = GF$, and then from the second pair $GH = \frac{HF}{2}$. This leads to $CH = 2 \cdot HF$, another proof for the position of the barycentre known since Archimedes. Now Eucl. I.41 gives $\mathcal{D} = 2 \cdot \mathcal{A}$ and also $\mathcal{A} + \mathcal{D} = \frac{\Delta}{2}$. Thus $\mathcal{A} = \frac{\Delta}{6}$ and $\mathcal{D} = \frac{\Delta}{3}$. Similarly, $AH = 2 \cdot HD$, so that also, again by Eucl. I.41, $\mathcal{D} = 2 \cdot \mathcal{B}$, i.e. $\mathcal{B} = \mathcal{A}$ and $\mathcal{C} = \mathcal{D}$.

(b) We follow precisely the proof of Jakob Steiner, who solved the analogous problem for more general ratios. We draw EG parallel to BF, which again gives us two pairs of similar triangles: EGC is similar to AFC and EGH is similar to BFH. Thales gives $EG = \frac{AF}{3} = \frac{c}{9}$ and $HB = EH \cdot \frac{FB}{EG} = 6 \cdot EH$, i.e. by Eucl. I.41, $\mathcal{B} = 6\mathcal{A}$. Again by Eucl. I.41, $\mathcal{B} + \mathcal{A} = \frac{\Delta}{3}$, hence $\mathcal{A} = \frac{\Delta}{21}$ and $\mathcal{B} = \frac{6\Delta}{21}$. By a symmetric argument, the triangles ABL and CAK have area $\frac{6\Delta}{21}$; all three, together with $\mathcal{T}$, make up the entire triangle. Hence $\mathcal{T} = (1 - \frac{3 \cdot 6}{21})\Delta = \frac{\Delta}{7}$.

9. Let P be the intersection of the circles, say BCD and ACE, inside the triangle. Then, by Eucl. III.22, the angles BPC and CPA are $120°$, because they are inscribed angles corresponding to central angles of $240°$. Hence the angle APB is also $120°$, and, again by Eucl. III.22, P must lie on the third circle, which proves (b). Statement (c) then follows from

Eucl. III.21, because, for example, the angles DBC and DPC are inscribed angles on the same arc. Consequently, A, P, D are collinear, as are B, P, E and C, P, F, which proves (a). For (d) use a property of the radical axis of two circles (see Exercise 8 in Chap. 2). For (e), define a point H on the arc between B and F such that the arc AP is equal to the arc BH. Join AP, AH, PF and HF and obtain two equilateral triangles, which lead to the desired conclusion.

10. (a) Since O and H lie symmetrically with respect to the circle centre N, and since $A'O$ and HA are parallel, we have $LO = HK$ (see Fig. 4.35). From the third property of the nine-point circle, we know that K is the midpoint between H and A. By definition of A', L is the midpoint between O and A'. Therefore $A'OAH$ is a parallelogram with N as midpoint and ANA' a diagonal.

(b) By construction, $A'COB$ is also a parallelogram, more precisely a rhombus (Eucl. Def. 22) with side length R, the circumradius. From part (a) we already know that OA and $A'H$ have the same length. All mentioned circumcircles have the same radius R.

(c) The lines $A'B$, $A'C$ and $A'H$ are parallel to BO, CO and OA, respectively; thus the result is the same as statement (c) of Theorem 4.3.

11. An elegant solution (see the right picture of Fig. 4.36) is obtained by increasing the radius of this circle by d, which gives the dashed circle tangent to the axis CO in O. Then

by Eucl. III.36   and by Eucl. II.14
$$a^2 = (R + d)(R - e) \qquad \text{the construction}$$

12. *Motivation:* Since R should be on the bisector of BAW, and on that of WBA, it must be the incentre of the triangle ABW and also lie on the angle bisector of AWB, i.e. also of QWP (Eucl. IV.4). If we assume,

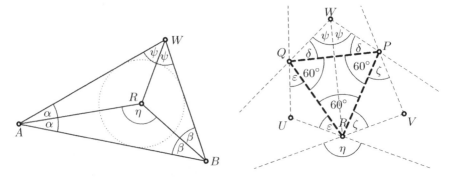

Fig. 12.12. Roger Penrose's backwards proof of Morley's theorem

proceeding *backwards*, that QPR is an equilateral triangle, QWP must therefore be an isosceles triangle.

So we choose, first arbitrarily, three angles δ, ε and ζ and attach to PQR three isosceles triangles with δ, ε and ζ as basis angles (see Fig. 12.12, right). We then produce their sides which will eventually meet in three points A, B and C (outside the figure).

Let ψ and η be as in Fig. 12.12. We have by Eucl. I.32 and Eucl. I.15, respectively,

$$\psi = 90° - \delta \qquad \text{and} \qquad \eta = \varepsilon + \zeta + 60° . \qquad (12.2)$$

By construction, R lies on the angle bisector AWB. Under *one* additional condition, R will be the incentre of the triangle ABW. Applying Eucl. I.32 to the triangles ARB and AWB of the picture on the left of Fig. 12.12, this condition becomes

$$\eta + \alpha + \beta = 180° \text{ and } 2\psi + 2\alpha + 2\beta = 180° ; \quad \text{i.e. } \eta - \psi = 90° . \text{ (12.3)}$$

Inserting ψ and η from (12.2), this becomes

$$\delta + \varepsilon + \zeta = 120° . \qquad (12.4)$$

With this condition, the two *lower* angles α and β at A and B are equal. The perfect symmetry of this condition allows one to apply the same argument around the triangle and finally *all three* angles α, β and γ at all three points A, B and C will be equal.

We further have from (12.3)

$$\alpha + \beta = \delta , \quad \text{and similarly} \quad \beta + \gamma = \varepsilon , \qquad \gamma + \alpha = \zeta \qquad (12.5)$$

which allows us, with α, β and γ given and satisfying (4.19), to determine δ, ε and ζ in accordance with (12.4).

13. The triangles CLQ and CKP are similar. Hence (by Eucl. I.15 and Eucl. I.5) the triangle QPO is isoceles, where, by Eucl. IV.5, O is the centre of the circumcircle. Hence P and Q have the same power with respect to this circle, and by Eucl. III.35 we have $RQ \cdot QC = RP \cdot PC$. This proves the result by Thales and Eucl. I.41, because QC and PC are in the same ratio as QL and PK, hence in the same ratio as the altitudes of the triangles LQR and KPR.

14. With the notations of Fig. 12.13 we have from Lemma 5.1

$$(w + v)(m + n) = ac + bd .$$

Inserting $v = \frac{mn}{w}$ (from Eucl. III.35), $c = \frac{am}{w}$ (from Eucl. III.21 and Thales) and $d = \frac{bn}{w}$ gives the desired result after multiplication by w.

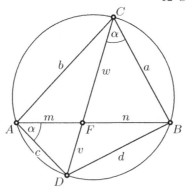

Fig. 12.13. Another proof of Stewart's theorem

12.5 Solutions for Chapter 5

1.

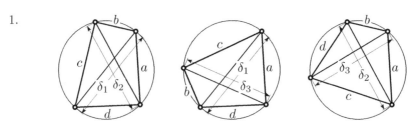

Permute the sides, apply three times Ptolemy, divide — and conquer.

2. Simplifying $6\cos\frac{\pi}{6} = 3\sqrt{3} = \sqrt{27}$ to $\sqrt{25} = 5$, we obtain the numbers 6, 5, 3, 0, -3, -5, -6 and get *la règle des douzièmes* [the rule of twelfths]: hour per hour the sea level falls by $\frac{1}{12}, \frac{2}{12}, \frac{3}{12}, \frac{3}{12}, \frac{2}{12}, \frac{1}{12}$ of the total difference.

3. By Eucl. III.20, ICB is similar to BDA, hence $IC = \sin\alpha\tan\beta$, $IB = \frac{\sin\alpha}{\cos\beta}$. So $AI = \cos\alpha - \sin\alpha\tan\beta$. Also IEA is similar to BDA, hence $AE = AI \cdot \cos\beta$ which leads directly to the second of our formulas. The first one is slightly trickier: by Thales we compute $EI = AI \cdot \sin\beta = \cos\alpha\sin\beta - \sin\alpha\frac{\sin^2\beta}{\cos\beta}$ and obtain EB as $EI + IB = \sin\alpha(\frac{1}{\cos\beta} - \frac{\sin^2\beta}{\cos\beta}) + \cos\alpha\sin\beta$, which gives the desired result by using Pythagoras, $1 - \sin^2\beta = \cos^2\beta$.

4. The value $\sin 18° = \frac{1}{2\Phi}$ is seen in Fig. 1.10 by drawing the altitude of the isosceles triangle DCF; from Fig. 1.22 (right) we see that $\cos 36° = \frac{\Phi}{2}$. The values for $30°, 45°$ and $60°$ follow from the equilateral triangle (Fig. 1.22, left) and the square. The remaining values are obtained by the addition formulas (5.6) and (5.7) and Pythagoras' theorem (5.2).

5. This follows at once by inserting (5.10) into (5.9). The only additional idea you need for this exercise is to replace expressions like $u^2 - v^2$ by $(u+v)(u-v)$.

6. Insert the formula for $\sin\frac{\alpha}{2}$ of Exercise 5 and the analogous relations for $\sin\frac{\beta}{2}$, $\sin\frac{\gamma}{2}$ and use $s = \frac{a+b+c}{2}$. All square roots disappear and the identity drops out by straightforward algebraic simplifications.

7. The result is obtained by adding and subtracting the formulas (5.6) and (5.7) to or from each other.

8. For the analytic proof, simply replace a and b on the left by $2R\sin\alpha$ and $2R\sin\beta$ respectively (the law of sines (5.12)) and simplify with the help of the addition formulas (5.62). Then the result drops out.

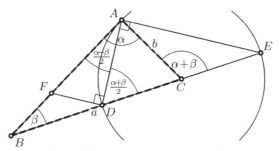

Fig. 12.14. Viète's proof of the law of tangents

Viète himself deduced this identity by geometric arguments (cf. Fig. 12.14): Draw the circle with centre C and radius b, which intersects the line BC at the points D and E with $BE = a+b$ and $BD = a-b$. Then the angles marked $\alpha+\beta$, $\frac{\alpha+\beta}{2}$ and $\frac{\alpha-\beta}{2}$ are determined, in this order, by Eucl. I.32, Eucl. III.20, and again Eucl. I.32. By Thales' circle, DAE is a right angle. We draw FD perpendicular to DA and have $FD = AD \cdot \tan\frac{\alpha-\beta}{2}$ and $AE = AD \cdot \tan\frac{\alpha+\beta}{2}$. But $\frac{FD}{AE} = \frac{a-b}{a+b}$ by Thales, which gives the stated result.

9. The tireless Euler arrived at these results several times in his work (in particular, see Euler (1748), Caput XIV, §237, and Euler (1783)). One can, of course, verify them by simply repeatedly applying the formulas (5.6) and (5.62). Euler himself used complex analysis (his formula (8.9) in Chap. 8 below). Experts in Numerical Analysis will recognise in (5.65) the factorisation of the Chebyshev polynomials based on their roots.

10. Applying three times the first formula of (5.8) in the form

$$2\sin\alpha = \frac{\sin 2\alpha}{\cos\alpha}$$

we obtain for the quarter of the perimeter of the 16-gon

$$8\sin\frac{\pi}{16} = \frac{4\sin\frac{\pi}{8}}{\cos\frac{\pi}{16}} = \frac{2\sin\frac{\pi}{4}}{\cos\frac{\pi}{16}\cos\frac{\pi}{8}} = \frac{1}{\cos\frac{\pi}{16}\cos\frac{\pi}{8}\cos\frac{\pi}{4}}\,,$$

since $\sin \frac{\pi}{2} = 1$. We now iterate the second formula of (5.9) to obtain successively

$$\cos \frac{\pi}{4} = \sqrt{\frac{1}{2}}, \quad \cos \frac{\pi}{8} = \sqrt{\frac{1}{2} + \frac{1}{2}\sqrt{\frac{1}{2}}}, \quad \cos \frac{\pi}{16} = \sqrt{\frac{1}{2} + \frac{1}{2}\sqrt{\frac{1}{2} + \frac{1}{2}\sqrt{\frac{1}{2}}}},$$

and so on.

11. They are. The correct values are $\sin 38°20' = 0.620235491268260$ and $\sin 51°40' = 0.784415664919576$.

12. We denote the unknown distance BC by x. We then have $x = \frac{h}{\tan \beta}$ and $a + x = \frac{h}{\tan \alpha}$. This leads to

$$h = \frac{a}{\cot \alpha - \cot \beta},$$

a famous formula in practical geodesy.

13. We know that the barycentre G divides the median in the ratio $2 : 1$. If the Euler line, which contains HG, is parallel to AB, then by Thales the orthocentre H divides the altitude in the same ratio. Since the angle BAD is $90° - \beta$, this means that $\tan \alpha = 3 \cdot \tan(90° - \beta)$. We conclude with $\tan(90° - \beta) = 1/\tan \beta$.

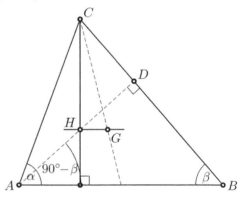

14. The similarity factor between the similar triangles AEF and ABC is $\frac{AF}{AC} = \cos \alpha$; analogously for the other triangles.

15. Apply the law of sines (formula (5.12)) to the triangle ABC of Fig. 4.27 to obtain $AB = 2r \sin 3\gamma$, and then again the law of sines to the triangle ABR, by using the fact that by Eucl. I.32 the angle $ARB = 180° - \alpha - \beta = 120° + \gamma$ and using Euler's formula (5.64) for $\sin 3\gamma$. This gives

$$AR = 8r \sin(60° + \gamma) \sin \beta \sin \gamma, \quad \text{and similarly}$$
$$AQ = 8r \sin(60° + \beta) \sin \beta \sin \gamma.$$

Now the triangle ARQ is determined by Eucl. I.4 (SAS), and one could compute QR by the law of cosines. But, more elegantly, we see that $(60° + \gamma) + (60° + \beta) + \alpha = 180°$. Hence, since the angle $QAR = \alpha$, the only possibility to verify all three equations of the law of sines for this triangle is to have the angles $AQR = 60° + \gamma$, $ARQ = 60° + \beta$, and the side length

$$QR = 8r \sin \alpha \sin \beta \sin \gamma$$

(a formula first given in Taylor and Marr 1913). The symmetry of this formula with respect to α, β and γ shows that the triangle PQR is equilateral.

16. Imagine the sides of the triangle in Fig. 5.8 (right) to be slightly rounded. *Law of cosines.* We apply (5.23) to the two right-angled triangles of the picture:

$$\cos b = \cos u \cdot \cos h , \qquad \cos c = \cos h \cdot \cos(a - u) = \frac{\cos b}{\cos u} \cdot \cos(a - u) .$$

We then use the addition theorem (5.6) for $\cos(a - u)$ and (5.26) in the form $\tan u = \cos \gamma \cdot \tan b$. This at once gives the equation (5.35).

Law of sines. We apply (5.25) to the two right-angled triangles of Fig. 5.8 and have $\sin \gamma = \frac{\sin h}{\sin b}$ and $\sin \beta = \frac{\sin h}{\sin c}$. Dividing one equation by the other gives the sine theorem.

17. (a) We project the triangle horizontally onto the circumscribed cylinder, which replaces the latitudes by the sine values, and apply Theorem 5.6. This produces a rectangle of height $1 - \sin \varphi$ and width γ, with area

$$\mathcal{A}_1 = \gamma(1 - \sin \varphi) .$$

(b) We compute the angle α at A of the spherical triangle by dividing it into two right-angled triangles with hypotenuse $90° - \varphi$ and angle $\frac{\gamma}{2}$ at N. Then the angle α at A is determined by (5.29), which gives

$$\tan \alpha = \frac{1}{\tan \frac{\gamma}{2} \cdot \cos(90° - \varphi)} \quad \text{or} \quad \tan(90° - \alpha) = \tan \frac{\gamma}{2} \cdot \sin \varphi .$$

With Girard's formula (5.46) we get

$$\mathcal{A}_2 = \gamma + 2\alpha - \pi .$$

(c) We set $\beta = 90° - \alpha$ and obtain for the difference

$$\mathcal{A}_1 - \mathcal{A}_2 = 2\beta - \gamma \sin \varphi \qquad \text{where} \qquad \beta = \arctan \left(\tan \frac{\gamma}{2} \cdot \sin \varphi \right) . \quad (12.6)$$

For small values of γ we are allowed to use series expansions (e.g. Hairer and Wanner, 1997, pp. 47 and 51) to obtain

$$\mathcal{A}_1 - \mathcal{A}_2 \approx \frac{\gamma^3}{12}(\sin \varphi - \sin^3 \varphi) .$$

18. As in the preceding exercise, we use Theorem 5.6 and replace the latitudes by the sine values. This gives a region which is the sum of two rectangles, with area $ab + cd$ where $a = (12\frac{50}{60} - 11\frac{50}{60}) \cdot \frac{\pi}{12} = \frac{\pi}{12}$, $c = (12\frac{35}{60} - 11\frac{50}{60}) \cdot \frac{\pi}{12} = \frac{\pi}{16}$, $b = \sin 22° - \sin 11°$, $d = \sin 24°30' - \sin 22°$. The result is 0.1782π steradians or approximately 0.45% of the total area of the sky.

19. The required angle β is given by (12.6), where γ is the difference in longitude and φ is the latitude. For our example, we get $\beta = 55°58'32''$. So don't be astonished to see a lot of ice.

20. We denote Trondheim by A, Tromsø by B, the unknown position of the ship by C and the North Pole by N. We then consider the spherical triangles ABN, ABC and CAN. We use standard notation for the sides and angles of ABC. The sides and angles of ABN, in contrast, are denoted by a', b', c and α', β', ν'. From the data we know the angles $a' = 90° - 69°39'$, $b' = 90° - 63°26'$, $\nu' = 18°59' - 10°24' = 8°35'$ and the angles $NBC = \varphi = 107°17'$ and $NAC = \psi = 74°13'$. To find the ship, we have to compute d and ε. The algorithm for their computation is displayed in Fig. 12.15. The numerical values obtained by this sequence of formulas are

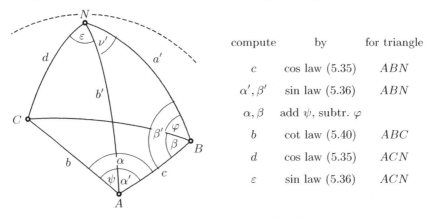

compute	by	for triangle
c	cos law (5.35)	ABN
α', β'	sin law (5.36)	ABN
α, β	add ψ, subtr. φ	
b	cot law (5.40)	ABC
d	cos law (5.35)	ACN
ε	sin law (5.36)	ACN

Fig. 12.15. Solution of the Norway ship rescue problem (caution: the drawing does not correspond to the actual data)

$c = 0.12354876086763$ $\alpha' = 0.434723616997564$ $\beta' = 2.5692006160581$
$\alpha = 1.7300488101860$ $\beta = 0.69675321687682$ $b = 0.11993261707074$
$d = 0.44515373520827$ $\varepsilon = 0.27067640710951$

The only "dangerous bend" was the value for β', which is greater than 90°, and for which a careless use of arcsin would have given a wrong value. If we transform the last results to degrees, subtract the first from 90°, subtract the second from the longitude of Trondheim, we find that the ship's position is

$$64°29'40''27'''4''''N, \quad 5°6'31''1'''0''''W.$$

Since the original data are precise only to a minute, all these sexagesimal fractions are totally useless, but they help you to check your (and our) computer.

21. As in Example 3 on page 137, we compute the angles of the sundial by

$$\cot z = \frac{\cos \sigma \cot \alpha - \sin \sigma \sin \varphi}{\cos \varphi},$$

where φ denotes the latitude, σ the angle by which the wall deviates from the east-west direction, and α the reduced time ($\alpha = 0$ at noon). With the values $\varphi = 47° 17'$ and $\sigma = -79°$, we obtain the following result:

XII	I	II	III	IV	V
0°0'	25°20'	32°49'	36°39'	39°13'	41°18'

22. Following the hints, the computations for $\sin \frac{\alpha}{2}$ and $\cos \frac{\alpha}{2}$ are *really* straightforward, the only thing to use are identities like $\frac{b-c+a}{2} = s - c$ or $\frac{a-b+c}{2} = s - b$. The result for $\tan \frac{\alpha}{2}$ is obtained by dividing the formula for sin by that for cos. The second set of formulas is obtained by inserting (5.41) instead of (5.35).

23. The first statement is seen in the same way as in the proof of Eucl. IV.4 on page 83. Let F be the intersection of the side a with the perpendicular great circle through I. To prove the second statement, we apply (5.27) to the right-angled triangle formed by the sides AF (with length $s - a$), FI (with length ρ) and IA as hypotenuse. This gives

$$\tan \rho = \tan \frac{\alpha}{2} \cdot \sin(s - a),$$

and the stated result follows after inserting (5.68) for $\tan \frac{\alpha}{2}$.

24. The first statement is seen in the same way as in the proof of Eucl. IV.5. For the computation of r we observe that the angles which the great circles AO, BO and CO (which are all of length r) make with the sides a, b and c of the triangle are given by $\sigma - \alpha$, $\sigma - \beta$ and $\sigma - \gamma$ (similar to the calculations in the remark on page 83). We apply (5.26) to the right-angled triangle formed by the first half of the side BC, the perpendicular side bisector and the side BO as hypotenuse:

$$\cos(\sigma - \alpha) = \frac{\tan \frac{a}{2}}{\tan r} \qquad \text{or} \qquad \cot r = \frac{\cos(\sigma - \alpha)}{\tan \frac{a}{2}}.$$

This gives the stated result after inserting (5.69) for $\tan \frac{a}{2}$.

25. This follows at once by applying (5.27) to the right-angled triangles ACF and BCF, which gives

$$\tan p = \sin h \tan \delta, \qquad \tan q = \sin h \tan \varepsilon.$$

Since $\delta + \varepsilon = \frac{\pi}{2}$, the product $\tan \delta \tan \varepsilon$ is 1.

26. Since the movement of rotation is uniform, we have the ratios

$$\frac{QP}{2a\pi} = \frac{\Delta t}{T} \qquad \Rightarrow \qquad QU \approx \frac{2a\pi\Delta t}{T} .$$

As in Newton's proof, we use Eucl. II.14:

$$RQ \approx UP \approx \frac{QU^2}{2a} .$$

Inserting these last two relations into (5.71) leads to

$$f = \frac{2a\pi^2}{T^2} = \frac{Const}{a^2}$$

(the last identity by Kepler 3). Unfortunately for Newton, this simple access to the inverse-square law had also been found independently by his archenemy Robert Hooke. Edmund Halley, who paid for the publication of the *Principia*, urged him to include a kind acknowledgment, without success. In Newton's manuscript (1684) this result on the "circumferentiis circulorum", with precisely the same proof, is called "Theorem 2".

12.6 Solutions for Chapter 6

1. By solving the second equation of (6.2) we obtain $y = 1 + \sqrt{1 + c^2/a^2}$, i.e. $ay = a + \sqrt{a^2 + c^2}$, which, by construction, is the length of BG in Fig. 6.3. If we multiply the first equation of (6.2) by a, we have $x + \frac{a^2}{x} = ay = BG$. By Thales, $\frac{a^2}{x} = AE$. If we project the point E vertically down to a point H on the line BG, we obtain a right-angled triangle EHG which is similar to BDF. Because $EH = BD$ we have $HG = DF = x$. Hence this first equation corresponds to $BH + HG = BG$ and the circle in Pappus' construction leads to its solution.

Pappus' original solution is based on his Prop. VII.71: *If in Fig. 12.16 (a) $AB\Delta\Gamma$ is a square and BEZ a right angle, then*

$$\Delta Z^2 = \Gamma\Delta^2 + HE^2 . \tag{12.7}$$

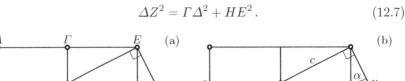
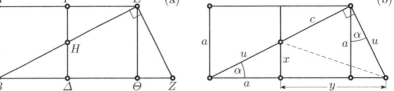

Fig. 12.16. Pappus' Proposition VII.71 (a) and it's proof (b)

Pappus' proof of (12.7) fills, together with Ver Eecke's explanations in footnotes, two pages. After having seen by orthogonal angles that the triangles $B\Delta H$ and $E\Theta Z$ are identical, you get the result in one line by computing twice the distance HZ^2 by Pythagoras as (see Fig. 12.16 (b))

$$c^2 + u^2 = x^2 + y^2 \quad \Rightarrow \quad c^2 + a^2 = y^2 \quad \text{(because } u^2 = a^2 + x^2\text{).}$$

2. We set $A\Lambda = 1$, $MA = w$, $A\Gamma = u$, $\Gamma K = v$ so that by Thales $u = vw$ and by construction $A\Delta = \Gamma Z = \Theta K = \frac{u}{2}$. Since HZ is parallel to $\Gamma\Theta$, we have $Z\Theta : 2 = \frac{u}{2} : v$ which gives $Z\Theta = \frac{u}{v} = w$. We finally apply Eucl. II.12 to the triangle $K\Gamma Z$ which gives

$$v + v^2 = (w + \tfrac{u}{2})^2 - (\tfrac{u}{2})^2 = w^2 + wu = w^2(1 + v) \quad \Rightarrow \quad v = w^2, \; u = w^3 \,.$$

3. Folding the Geisha fan as before, we obtain the relations

$$x = 1 + \frac{y}{w}, \qquad y = w + \frac{z}{x}, \qquad z = x + \frac{x}{y}, \qquad z = y + \frac{1}{z} \,.$$

Since we are mainly interested in the longest diagonal, we turn this into an equation for z by solving $y = z - \frac{1}{z}$, then $x = \frac{yz}{y+1}$, and $w = y - \frac{z}{x}$. Inserting everything into the first equation then leads to an equation in z which, after simplification, becomes

$$z^5 - 3z^4 - 3z^3 + 4z^2 + z - 1 = 0 \,,$$

an equation of degree 5 with no hope for a closed-form solution. We can, however, obtain numerical solutions to any desired precision:

$$z = 3.51333709166613518878217159629798184\ldots$$
$$y = 3.22870741511956490789458625906524250\ldots$$
$$x = 2.68250706566236233772362329783873543\ldots$$
$$w = 1.91898594722899477978073611413265539\ldots$$

4. This is precisely the identity (6.21) of the proof of Fermat's result, which here relates Ramanujan's formula to Pythagoras' theorem. The next entry in Ramanujan's notebook is the formula

$$\left(a + b - \sqrt{a^2 + b^2}\right)^2 = 2\left(\sqrt{a^2 + b^2} - a\right)\left(\sqrt{a^2 + b^2} - b\right)$$

which is an algebraic counterpart to the above geometric result.

5. The right-angled triangles in Fig. 6.17 yield the relations $\sin\frac{\alpha}{2} = \rho$, $\sin\frac{\beta}{2} = 2\rho$, $\sin\frac{\gamma}{2} = 3\rho$. This, inserted into (5.61), leads to the cubic equation $12\rho^3 + 14\rho^2 - 1 = 0$. For $x = \frac{1}{\rho}$ we obtain $x^3 - 14x - 12 = 0$. This polynomial is negative for $x = 0$ and for $x \to -\infty$, and positive for $x = -1$ and for $x \to \infty$. Therefore, it has one positive and two negative roots. Einstein, computing with pen, paper and a four-digit table of logarithms, obtained with formula (6.13) the result $\rho = 0.243$ for the positive root.

6. Fermat presented these porisms with much high-flown Latin and Greek as a sort of substitute for lost porisms of Euclid, but with no hint of a proof. We apply an idea similar to that in Euler's proof of (6.20), i.e. project the four points onto the line connecting N and M.

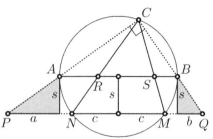

We choose the radius equal to 1. The assertion then follows from

$$\frac{PN \cdot MQ}{PM \cdot NQ} = \frac{(a+1-c)(b+1-c)}{(a+1+c)(b+1+c)} = \frac{(1-c)(2+a+b)}{(1+c)(2+a+b)} = \frac{(1-c)}{(1+c)},$$

where the fraction has been simplified using Pythagoras, $s^2 + c^2 = 1$, and Thales, $a/s = s/b$, i.e. $ab = s^2$.

7. The simplest proof is a Stone Age one (see Fig. 6.22, right): The first equation is, for the values of this figure and by Thales, equivalent to $\frac{1}{6} + \frac{2}{6} + \frac{3}{6} = 1$, which is evidently true. We also see from the bold segments in this figure that the sum of the numerators, here $1 + 2 + 3 = 6$, is always equal to the number of subdivisions.

The proof via areas is just as easy, using the picture for the solution of Exercise 7 of Chap. 4, page 359: the triangles BPC and BAC have the same base, and their altitudes are in the ratio $\frac{PD}{AD}$. Hence the ratio of their areas, i.e. $\frac{A_1 + A_2}{A_1 + A_2 + B_1 + B_2 + C_1 + C_2}$, is equal to $\frac{PD}{AD}$. If we add up the corresponding ratios for all three sides, we clearly obtain 1.

For the identities of the second and third lines we go over to the column on the right: if we add the first two, we obtain ("orietur ista aequatio identica") $1 + 1 + 1 = 3$ and see that they are equivalent; if we multiply the first equation by the common denominator and simplify, we obtain the third relation.

Remark. It is interesting to see Euler's original approach (1815) to these discoveries by obscure and complicated trigonometric calculations ("quod initio calculos satis abstrusos et molestos ..."), which he afterwards obtained so simply ("ad solutionem simplicissimam aeque ac elegantissimam ..."). For explanations in English see Sandifer (2006).

8. As good pupils of Descartes, we name the known distances: $AC = a$, $BD = b$, $AB = \ell$, and the unknown distance: $AF = x$, so that $FB = \ell - x$. We must have $CF = FD$, or $CF^2 = FD^2$, thus by Pythagoras $x^2 + a^2 = (\ell - x)^2 + b^2$, which is, after simplification, $x = \frac{\ell^2 + b^2 - a^2}{2\ell} = 32$. Leonardo himself applied Thales' theorem (see Fig. 12.17, left): The point F must lie on the perpendicular bisector of CD. We have two similar triangles and obtain $\delta = \frac{(b-a)(a+b)}{2\ell} = \frac{b^2 - a^2}{2\ell} = 7$, the same result.

Var. 1. Here we have to add another distance to CF respectively FD, therefore the square roots are more difficult to eliminate. We look for a

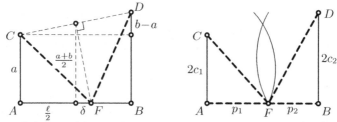

Fig. 12.17. Solution of Leonardo Pisano's problem using Thales (left); of the first variant (right)

more elegant solution. The theorems of Apollonius in Chap. 3 will help. We consider an ellipse with foci A and C, at distance $2c_1 = a$, with semi-major axis a_1, semi-minor axis b_1 and *latus rectum* p_1 (see Fig. 12.17, right). Then for F on this ellipse we have $CF + FA = 2a_1$ and $AF = p_1$ (see Apoll. III.52 and Fig. 3.4, left). We further have $p_1 = \frac{b_1^2}{a_1}$ (equation (3.7)) and $b_1^2 = a_1^2 - c_1^2$ (either Pythagoras or solution of Exercise 13 of Chap. 3). We place an analogous ellipse with foci B and D and with corresponding lengths a_2, b_2, c_2 and p_2. The equations to satisfy are thus

$$a_1 = a_2 =: u, \qquad p_1 + p_2 = \ell. \tag{12.8}$$

We insert the formulas we have into this and obtain

$$\frac{u^2 - c_1^2}{u} + \frac{u^2 - c_2^2}{u} = \ell \quad \Rightarrow \quad u = \frac{\ell}{4} + \sqrt{\frac{\ell^2}{16} + \frac{c_1^2 + c_2^2}{2}}.$$

Var. 2. For the perimeters to be equal, we must have

$$a_1 + c_1 = a_2 + c_2 =: u \quad \Rightarrow \quad a_1 = u - c_1, \quad a_2 = u - c_2 \tag{12.9}$$

and a calculation as above leads to the condition

$$2u = \ell + c_1 \frac{u}{u - c_1} + c_2 \frac{u}{u - c_2}.$$

When multiplied out, this yields an equation of degree 3 for u, whose largest root gives positive values for a_1 and a_2. We find

$$u = 51.83621385758, \quad p_1 = 30.72809425360, \quad p_2 = 19.27190574640.$$

Var. 3. At last something easier. The areas are equal if $ax = b(\ell - x)$, i.e. if $x = \frac{b\ell}{a+b}$.

9. Flip the Pythagorean triangle with sides 5, 12 and 13 over to the right and obtain the triangle with sides $9 - 5 = 4$, 15 and 13 and area $4 \cdot 6 = 24$. There are only a finite number of oblique triangles with integer sides and smaller area, none with integer area. In size, the next oblique triangles with integer sides and area have sides 9, 10 and 17, and 3, 25 and 26, both of area 36.

10. $AK^2 = AK \cdot A'K = DK \cdot EK$ by Eucl. III.35, hence $\mathcal{A}^2 = (2B_1C_1 \cdot DK) \cdot$ $(2B_1C_1 \cdot EK)$. We now use for a moment the notations of Fig. 4.20 (right) and formula (4.13) and obtain for the product $2B_1C_1 \cdot DK = 2a \cdot (r_1 + d_1) = 2ar_1 + a^2 + r_1^2 - r_2^2 = (r_1 + a)^2 - r_2^2 = DF \cdot DL$, i.e. the power of D with respect to the right circle. The point E lies on the same circle, so we have similarly $2B_1C_1 \cdot EK = EF \cdot EL$ and both factors of the above formula give together $\mathcal{A}^2 = DF \cdot DL \cdot EF \cdot EL$. Since $DC_1 = C_1E = \frac{c}{2}$, $C_1B_1 = \frac{a}{2}$, and $LB_1 = B_1F = \frac{b}{2}$, each of these four factors is one of the factors in (6.22).

11. One way of proving the first assertion is by trigonometry: $\mathcal{A} = bc\frac{1}{2}\sin\alpha = c\sin\frac{\alpha}{2} \cdot b\cos\frac{\alpha}{2} = b\sin\frac{\alpha}{2} \cdot c\cos\frac{\alpha}{2}$. Thébault divides the triangle into two parts by drawing the horizontal line through B; the first part has area $AB'' \cdot AB'$, the second has area $BB' \cdot B'C'$. The square of the area is then $\mathcal{A}^2 = CC'' \cdot BB' \cdot BB'' \cdot CC'$, which we cleverly group as $(BB'' \cdot CC'') \cdot (BB' \cdot CC')$. The first pair of factors can be considered as the product of the distances of the foci of an ellipse with foci B and C, passing through A, to the tangent at this point A. The second pair of factors is the same quantity for the corresponding hyperbola. The reason for this is that the tangents at A are known to be the angle bisectors (see Fig. 3.4 and Fig. 3.11). By (3.18) — and a similar formula for the hyperbola — these products are equal to the squares of the semi-minor axes of these two conics. Since we know for both conics the values of ℓ and ℓ' (here b and c respectively) as well as the distance of the foci (here a), we can find these semi-minor axes from (3.6) and (3.14) (with another meaning of the letters a, b, c). The result for the above product will then be $\frac{1}{16}((b+c)^2 - a^2)(a^2 - (b-c)^2)$. As in the last line of (6.25), one application of Eucl. II.5 will lead to the desired result.

12. Imagine that the lines in Fig. 4.6 on page 83 are slightly rounded and think of it as if it were a drawing of spherical triangles. The points which have the same distance from two great circles again lie on a great circle, the angle bisector. Indeed, these two great circles are the intersection of the sphere with two planes through the origin, and the angle-bisecting plane also passes through the origin. The same argument as in the proof of Eucl. IV.4 applies and the segments ID, IE and IF enclose right angles with the respective sides. We thus obtain two right-angled spherical triangles AIF and AEI which have the common side AI, and $EI = FI$ of the same length. Thus by (5.23) the third sides $AF = AE$ also have the same length. Applying the same argument to the other triangles, we obtain $AF = AE = s - a$ as before. Now, as in Remark (ii) on page 173, we consider the right-angled triangle AFI and obtain from (5.27) $\tan\rho = \tan\frac{\alpha}{2} \cdot \sin(s-a)$. This gives the desired result by inserting the expression for $\tan\frac{\alpha}{2}$ into (5.68).

13. Denote the length of the diagonal AC in Fig. 6.18 by g and compute g^2 twice by the law of cosines (5.10), applied once to the triangle ACD, and once to ACB. Subtract one formula from the other to obtain

$$2(ad + bc)\cos\alpha = a^2 + d^2 - b^2 - c^2. \tag{12.10}$$

Express the area of the quadrilateral by that of the two triangles as

$$\mathcal{A}_q = (ad + bc)\frac{\sin\alpha}{2}. \tag{12.11}$$

One thus obtains from (12.11) and (12.10):

$$16\mathcal{A}_q^2 = 4(ad + bc)^2(1 - \cos^2\alpha) = 4(ad + bc)^2 - (a^2 + d^2 - b^2 - c^2)^2.$$

This can be compared to the third formula of (6.25). Proceeding exactly as in this formula, applying Eucl. II.5, then Eucl. II.4, and finally Eucl. II.5, leads to the desired result. The last step is the same as the last step in the above proof of Theorem 6.4.

14. Thales' theorem requires that

$$\frac{\frac{2u_1}{1+u_1^2} - b_1}{\frac{1-u_1^2}{1+u_1^2} - a_1} - \frac{\frac{2u_2}{1+u_2^2} - b_1}{\frac{1-u_2^2}{1+u_2^2} - a_1} = 0.$$

After simplification, and factoring by $(u_2 - u_1)$, this leads to

$$-(1 + a_1)u_1u_2 + b_1u_2 + b_1u_1 + a_1 - 1 = 0.$$

From this u_2 can be computed, and this gives formula (6.39).

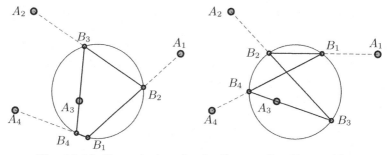

Fig. 12.18. Two solutions for the Cramer–Castillon problem

15. By multiplying the four matrices corresponding to (6.39), we obtain for (6.35)

$$u_1 = \frac{8.576u_1 - 5.248}{-8.24u_1 + 2.56}, \quad \text{or} \quad 8.24u_1^2 + 6.016u_1 - 5.248 = 0,$$

with solutions $u_1 = -1.2426325$ and $u_1 = 0.51253545$ (see Fig. 12.18). The second solution does not have the shape one might expect.

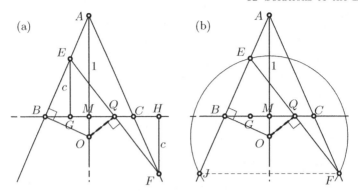

Fig. 12.19. Solution of the Armenian-Australian problem

16. We set $AM = 1$ and $BM = MC = a$, a given constant, so that by Thales $MO = a^2$ (see Fig. 12.19 (a)). Suppose first that $EQ = QF$. Then by Thales $EG = HF$; as good pupils of Descartes, we denote this length by c. Again by Thales $GQ = QH$ and $BG = CH = ac$ and hence, by Descartes' dictionary, $GH = 2a$, $GQ = a$ and $MQ = ac$. Therefore, QGE and OMQ are similar triangles and the angle GEQ is equal to the angle MQO. This shows that OQ is perpendicular to EQ. If, conversely, we suppose that these two lines are perpendicular, the steps of the above proof, taken in reverse order, lead to $EQ = QF$.

The following purely geometric argument is, however, much more elegant (see Fig. 12.19 (b)): we draw the circle with centre O passing through E. By orthogonality of OB to BA, $EB = BJ$. Because O is on the angle bisector, JF is parallel to BC. Thus $EQ = QF$ by Thales.

17. The answer is $120°$. This is easily seen in the case where BAC is isosceles. In this case $B'A'C'$ is also isosceles and by hypothesis half of a square. Further, C, B', C', B lie on a circle and $CB' = B'C' = C'B$ by Eucl. III.21. We conclude that the angles $B'CA'$ and $A'BC'$ are $30°$.

The general case requires algebraic calculations. Let a, b, c be the side lengths of our triangle. By Eucl. VI.3, the point B' divides the side AC in the ratio $a : c$. Therefore we have $B'A = \frac{bc}{a+c}$ and similar expressions for the five other segments. We next apply the law of cosines (5.10) to the triangles $A'BC'$ and CBA and eliminate $\cos \beta$. This gives for the square of the side $C'A'$ the expression

$$C'A'^2 = \frac{c^2 a^2}{(a+b)^2} + \frac{a^2 c^2}{(b+c)^2} + \frac{ac}{(a+b)(b+c)}(b^2 - a^2 - c^2).$$

By cyclic permutation we obtain $A'B'^2$ and $B'C'^2$. By hypothesis we know that $B'C'^2 - A'B'^2 - C'A'^2 = 0$. After inserting the above values, we obtain a rational expression in a, b, c which, when simplified, contains the non-trivial factor $a^2 - b^2 - c^2 - bc = 0$. But this means, again by the law of cosines, that $\cos \alpha = -\frac{1}{2}$.

18. Thales' theorem tells us that $C_1M/C_2M = r_1/r_2$. Another appeal to Thales then shows that the distances of C_1 and C_2 from the line KM must be in the same ratio. But if two pairs of circles satisfy this property, then so does the third one. Steiner also obtained similar statements for the *inner* similarity points (see Fig. 6.25, right).

19. One is required to find an s for which $s = \cos s$. For such an equation we need numerical methods. If we try, for example, $s = 40°$, we find $\frac{40\pi}{180} - \cos 40° < 0$, while $\frac{45\pi}{180} - \cos 45° > 0$. The solution, which by geometric intuition we know is unique, must therefore lie in the interval $[40°, 45°]$. Much in the spirit of Eucl. X.1, we next test the midpoint $42.5°$ and find the difference to be greater than 0. Thus, the solution is contained in $[40°, 42.5°]$. We continue this "bisection algorithm" and after every ten iterations obtain three additional decimal digits. After some 50 iterations, we have

$$s = 0.7390851332151606 = 42°20'47''15'''6''''29''''' .$$

Euler himself, who did all calculations by hand, used a more sophisticated method. Firstly, the use of logarithms turned the conversion factor $\frac{\pi}{180}$ into an addition. Secondly, the difference of the logarithms for $s = 40°$ was -0.0403166, while this difference for $s = 45°$ was 0.0456049. Hence the use of Thales' theorem allows one to find as a better value $40° + 5° \frac{0.0403166}{0.0403166 + 0.0456049}$. Repeating this "regula falsi" twice, first for the interval $[42°20', 43°]$, then for $[42°20', 42°21']$, allowed him to obtain three correct sexagesimal digits $42°20'47''15'''$ (see Euler's work (1748), §531 for details).

20. If we take AC as the base of the triangle, its altitude is $\sin 2s$ and its area $\frac{1}{2}\sin 2s$. The area of the sector $CBEA$ is s. Thus we have to solve $s = \sin 2s$. Using any of the above algorithms, we get the solution

$$s = 0.9477471335169904 = 54°18'6''52'''43''''55''''' .$$

Euler gave the value to this precision with a slight error in the last sexagesimal digit.

21. The piece of cake DAE has half the area of the segment above the line AD of Fig. 6.26 (IV) where s is replaced by $2s$. Therefore the solution is half the arc length of Exercise 22, i.e.

$$s = 1.154940730005029 = 66°10'23''37'''33''''15''''' .$$

22. As in Exercise 20 the triangle ACD has area $\frac{1}{2}\sin s$. The moon ADs hence has area $\frac{s}{2} - \frac{1}{2}\sin s$, which is required to be $\frac{\pi}{4}$. We obtain $s - \frac{\pi}{2} = \sin s = \cos(s - \frac{\pi}{2})$ which is the equation of Exercise 19, to whose answer we add $90°$:

$$s = 2.309881460010057 = 132°20'47''15'''6''''29''''' .$$

23. The moon BAs with area $\frac{s}{2} - \frac{1}{2}\sin s$ (see Exercise 22) must now be $\frac{\pi}{3}$. This time we set $u = s - 120°$ and obtain the equation $u = \sin(60° - u)$ with solution $u = 29°16'26''59'''43''''44'''''$ and

$$s = 2.605325674600903 = 149°16'26''59'''43''''44'''''\,.$$

24. Here one is required to solve $180° - s = 1 + \cos s + \sin s$, with the solution

$$s = 0.7295815096762676 = 41°48'6''59'''19''''27'''''\,.$$

   Since the above equation with its sums is not practical for logarithmic calculations, Euler transformed it, with the help of (5.9), (5.8) and (5.7) to the equivalent form $180° - s = 2\sqrt{2}\cos\frac{s}{2}\cos(45° - \frac{s}{2})$.

25. Since the area of the triangle CAE is $\frac{1}{2}\tan s$ and the area of the sector $\frac{s}{2}$, we have to solve $2s = \tan s$, with the solution

$$s = 1.165561185207211 = 66°46'54''15'''7''''20'''''\,.$$

26. Let G be the midpoint of the segment AE. Then triangles CGA and FCA are similar (because of the two right angles). Then by Thales $\frac{FA}{1} = \frac{1}{CG} = \frac{1}{\sin\frac{s}{2}}$. Thus our problem requires the solution of $s \cdot \sin\frac{s}{2} = 1$, with the result

$$s = 1.481681910190981 = 84°53'38''49'''55''''39'''''\,.$$

12.7 Solutions for Chapter 7

1. By (7.2c), the altitude through A is given by $y = \frac{b}{c}(x + a)$. Its intersection with the altitude $x = 0$ is $y_H = \frac{ab}{c}$, the same symmetric result as in Exercise 4 of Chap. 4.

   For the perpendicular bisector of BC we have the equation $y = \frac{c}{2} + \frac{b}{c}(x - \frac{b}{c})$. Its intersection with the perpendicular bisector of AB gives $x_O = \frac{b-a}{2}$, $y_O = (c - \frac{ab}{c})\frac{1}{2}$. The medians are obtained by (7.2 (d)) and lead to $x_G = \frac{b-a}{3}$, $y_G = \frac{c}{3}$. The property $G = \frac{2}{3}O + \frac{1}{3}H$ can now be seen.

2. If the centre of the unknown circle is (x, y) and its radius is r, then the tangency condition with the three given circles is

$$(x - x_i)^2 + (y - y_i)^2 = (r - r_i)^2, \qquad i = 1, 2, 3.$$

   If we take these equations by pairs, and subtract one equation from the other, we obtain two linear relations of the type

$$\begin{array}{ll} a_{11}x + a_{12}y = b_{11}r + b_{12} \\ a_{21}x + a_{22}y = b_{21}r + b_{22} \end{array} \quad \begin{array}{c} \text{solve} \\ \Rightarrow \end{array} \quad \begin{array}{l} x = c_{11}r + c_{12} \\ y = c_{21}r + c_{22}\,. \end{array}$$

This, inserted into one of the above equations, leads to a quadratic equation for r. Other sign combinations in the above formulas as $(r \pm r_i)^2$ lead to other solutions, with some circles touching from inside and others from outside (see Fig. 7.40, right).

3. If (x_i, y_i) are the coordinates of the vertices, then the left-hand expression is

$$(x_1 - x_2)^2 + (x_2 - x_3)^2 + (x_3 - x_4)^2 + (x_4 - x_1)^2 ,$$

plus analogous terms for the y-values. The right-hand side is, by $4 \cdot \frac{1}{2^2} = 1$,

$$(x_1 - x_3)^2 + (x_2 - x_4)^2 + ((x_1 + x_3) - (x_2 + x_4))^2 ,$$

etc. Multiplying out, the two expressions are seen to be identical. The verification of this result, which generalises the parallelogram law (4.6) to arbitrary quadrilaterals, may seem a trivial task, but it's another story to actually *discover* such a nice relation ("Why on earth didn't I think of that?").

Remark. J. Steiner gave in Steiner (1827), "64. Lehrsatz", without proof the following generalisation to *spherical quadrilaterals*:

$$\cos a + \cos b + \cos c + \cos d = 2 \cdot \cos \frac{\ell_1}{2} \cdot \cos \frac{\ell_2}{2} \cdot \cos e .$$

4. (a) By Pythagoras we have $u^2 + v^2 = (a + b)^2$ and by Thales $\frac{u}{a+b} = \frac{x}{a}$ and $\frac{v}{a+b} = \frac{y}{b}$. This leads to $\frac{x^2}{a^2} + \frac{y^2}{b^2} = 1$. For (b) use $\frac{x}{a} = \cos \alpha$, $\frac{y}{b} = \sin \alpha$ and (5.2). The computations for (c) are similar to those for (a), with $a + b$ replaced by $a - b$.

5. We apply Eucl. III.21: The points from which one sees the emperor under the same angle lie on circles passing through A and B (see the picture). The smaller the radius, the larger the angle α. Hence the largest α is obtained with the smallest circle which touches the eye-level line. We then have, by Eucl. III.36, $d^2 = s(s+h)$ and use Eucl. II.14 for a construction. The angle bisector is DF by Eucl. III.21 (arc AF = arc FB) and has a slope of $45°$ ($FC = CD$).

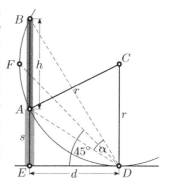

6. Here is a splendid opportunity for those who want to be cleverer than 17-year-old Einstein: let the major axis of the ellipse be $2a$, choose a point P with abscissa ax and ordinate $y = \sqrt{1 - x^2}$ (see Fig. 7.42 (b)). Since the slope of the tangent at P is $-\frac{x}{ay}$ (a-times smaller than the slope of the corresponding circle), the normal at P cuts the x-axis at the

point $(a - \frac{1}{a})x = \lambda x$. By Pythagoras we can compute r and the condition $r = \sqrt{1 - \lambda^2 x^2}$ leads to $a = \sqrt{2}$.

Remark. This result was stated without proof as "12. Lehrsatz" in Steiner (1828b).

7. We denote by $x_A, y_A, x_B, y_B, \ldots$ the coordinates of the bottle centres $A, B, \ldots$ The triangles FDA, ADB, BEC and CEH are isosceles with either vertical or horizontal base (see Fig. 7.43, right). Consequently

$$y_D = \frac{y_A + y_F}{2}, \quad y_E = \frac{y_C + y_H}{2}, \quad x_D = \frac{x_A + x_B}{2}, \quad x_E = \frac{x_B + x_C}{2}.$$

The quadrilateral $BEGD$ is a parallelogram (more precisely, a rhombus), hence $x_G = x_E + x_D - x_B$ (and similarly for the y's), which gives, after inserting the above formulas and $x_A = x_F$, $x_C = x_H$, $y_A = y_B = y_C$,

$$x_G = \frac{x_F + x_H}{2}, \quad y_G = \frac{y_F + y_H}{2}.$$

We see that F, G, H are collinear and equidistant. The positions of the following centres I, J, K, L, M are thus, by symmetry, just the centres E, D, C, B, A reflected through G. Since the first three bottles are on the same level, so are the last three. If w increases beyond $(2 + \sqrt{3})d$ and bottle B touches C, then bottle F will touch A, the symmetry will be lost and the result is no longer true.

8. Here and in the following four exercises we look for (x, y) such that

$$x^2 + y^2 = 1 \quad \text{hence} \quad x\,dx = -y\,dy, \quad (12.12)$$

which allows one to replace either dx or dy by the other. Here we have to solve $(x+1)y = \max$ which by inserting $x+dx$ and $y+dy$ and subtracting, leads to $(x + 1)\,dy + y\,dx = 0$ and, using (12.12), $x + 1 - \frac{y^2}{x} = 0$ and, with the first equation of (12.12), $2x^2 + x - 1 = 0$ with solution $x = \frac{1}{2}$.

9. Here we have $x + 2y = \max$ which gives as before $dx + 2dy = 0$ and $1 - \frac{2x}{y} = 0$, i.e. $y = 2x$ or $x = \frac{1}{\sqrt{5}}$, $y = \frac{2}{\sqrt{5}}$.

10. Adding the areas of a square and four rectangles, we obtain $4x^2 + 4 \cdot 2x(y - x) = \max$ or $2xy - x^2 = \max$. If we add to this $x^2 + y^2 = 1$ we find precisely the same maximum problem as in (7.19) with the same solution.

11. By dividing the volume formula by 2π, we obtain the problem $xy^2 = \max$, which leads to the condition $y^2\,dx + 2yx\,dy = 0$. Together with (12.12) this gives $y^2 - 2x^2 = 0$. Adding and subtracting $y^2 + x^2 = 1$ we find $x = \frac{\sqrt{3}}{3}$, $y = \frac{\sqrt{6}}{3}$.

12. By dividing the volume formula by $\frac{\pi}{3}$, we obtain the problem $(x + 1)y^2 = \max$, and, as above, $3x^2 + 2x - 1 = 0$ with the solution $x = \frac{1}{3}$.

13. The surface area of the cone is, following Archimedes (Prop. XV of *On the sphere and cylinder*, Ver Eecke (1921), vol. I, p. 35) $\mathcal{A} = y^2\pi + 2y\pi \cdot \frac{s}{2}$ (the factor $\frac{1}{2}$ comes from Eucl. I.41), which leads to the problem $y^2 + ys =$ max. With our choice of the unknown, we have by Thales $y = vs$ and by Pythagoras $s^2 = 1 - v^2$, where we have chosen the diameter of the sphere to be 1. This gives the problem $(1 - v^2)(v^2 + v) =$ max leading to $4v^3 + 3v^2 - 2v - 1 = 0$. Fermat, who had carefully studied Viète, saw that $v = -1$ is a solution, so that the equation can be divided by $v + 1$ and gives $4v^2 - v - 1 = 0$ with solution $v = \frac{1+\sqrt{17}}{8}$.

14. Theorem 4.3 and Fig. 4.9 (b) on page 86 immediately tell us that the orthic triangle solves this problem. But the solution might not be unique and we might not have the idea of looking at this theorem. Here is another idea:

    Fejér's solution. When H.A. Schwarz lectured on geometry at the University of Berlin towards 1900 and had just laboriously proved the above result, a young Hungarian student came up and said: "Sir, I have an easier solution." The student was Lipót Fejér, who was going to become famous for his ingeniously "simple" proofs of difficult results in analysis.[3] Fejér's solution is as follows (see Fig. 7.45 (b)): Choose the point F arbitrarily on AB. Then reflect F in AC to obtain F' and in BC to obtain F''. The perimeter of DEF is equal to the length of the line $F'EDF''$. This is as short as possible if F', E, D, F'' are collinear. If this is the case, the triangle $F'F''C$ is isosceles with angle at C equal to $2\delta + 2\varepsilon = 2(\delta + \varepsilon) = 2\gamma$. This angle is independent of the choice of F. So all these triangles are similar and the distance $F'F''$ is minimal if the distance $F'C = FC$ is minimal, i.e. if F is the foot of the altitude from C. By symmetry, the same property holds for E and D. The uniqueness of the solution also becomes obvious, and together with the minimal sum property (7.21), we have another proof of Theorem 4.3 (b).

15. (Solution by H. Egli, Zürich) If the point P moves along CB to Q, the area of $AHPG$ increases by that of the strip ⊞ and decreases by the area of the strip ⊟ (see Fig. 7.46 (b)). We have maximality if the areas of both strips are the same. By neglecting tiny triangles on both ends, ⊟ and ⊞ become parallelograms with common base PQ. Their respective altitudes are the distances of G and H from the line CB. Thus the line GH must be parallel to CB, i.e. the quadrilaterals $AHPG$ and $ABRC$ are similar with similarity centre A. Because of the right angles at B and C, AR is a diameter of the circumcircle of ABC (Thales circle). The point P lies thus at the intersection of AO and BC.

16. We see in Fig. 5.22 that $a = b$ if B is at the north pole, or A is at the south pole. It seems obvious (if not, use differentiation) that the maximal deviation $|a - b|_{\max}$, which we denote by δ, occurs if the "equator" bisects

---

[3] Ask one of your Hungarian friends to translate Fejér's motto "A tegnap bonyolult problémáját a holnap trivialitásává tenni."

the circular segment AB. The point B is then lower by $\delta/2$ than the position $\pi/4$, where it should be without error. The "meridian" through B, the equator and half of the segment AB then form a right spherical triangle with legs $\frac{\pi}{4} - \frac{\delta}{2}$, $\frac{\gamma}{2}$, and hypotenuse $\frac{\pi}{4}$. The cosine theorem (5.23) on page 128 then gives

$$\cos^2\left(\frac{\pi}{4} - \frac{\delta}{2}\right) \cdot \cos^2\frac{\gamma}{2} = \frac{1}{2} \quad \text{or} \quad \frac{1 + \cos(\frac{\pi}{2} - \delta)}{2} \cdot \cos^2\frac{\gamma}{2} = \frac{1}{2},$$

where (5.9) has been used. This leads, using $\cos(\frac{\pi}{2} - \delta) = \sin\delta$, to

$$\sin\delta = \frac{1}{\cos^2\frac{\gamma}{2}} - 1 = \frac{\sin^2\frac{\gamma}{2}}{\cos^2\frac{\gamma}{2}} = \tan^2\frac{\gamma}{2},$$

which is the required relation between γ and δ.

17. (a) If we express y as a function of x, i.e. if we cut the folium by parallel vertical lines, then the equation $x^3 + y^3 = 3xy$, for the unknown y with a given x, is of the third degree with a complicated solution. The integration requires ingenious substitutions, as for example the calculations of C. Huygens in manuscript XV of Huygens (1833, pp. 154/155), which fill one and a half pages. Another, very elegant substitution is due to Joh. Bernoulli (*Opera*, vol. 3, p. 403, Lectio Quarta, 1691/92): Introduce a new variable u by setting

$$y = \frac{3x^2}{u^2}; \quad \text{the equation of the folium becomes} \quad x^3 = \frac{1}{3}u^4 - \frac{1}{27}u^6.$$

If this is differentiated and the result divided by u^2 one obtains

$$\frac{3x^2}{u^2}\,dx = \left(\frac{4}{3}u - \frac{2}{9}u^3\right)du.$$

The left hand side is $y\,dx$ and the whole expression, when integrated, gives the required area.

(b) If, inspired by Huygens' drawing, we compute v of Fig. 12.20 (b) as a function of u, we set

$$x = \frac{u + v}{\sqrt{2}}, \qquad y = \frac{u - v}{\sqrt{2}}$$

and expect only two intersection points. Indeed, when inserted into $x^3 + y^3 = 3xy$, the terms containing v and v^3 cancel and the equation of the folium becomes $v = u \cdot \sqrt{\dfrac{3 - \sqrt{2}u}{3 + 3\sqrt{2}u}}$. This function can be integrated, from 0 to $\frac{3\sqrt{2}}{2}$, the value of u for the point E, by a standard substitution of Euler (see Hairer and Wanner 1997, Exercise II.5.2).

(c) We can do even better. If we cut the folium by a family of lines AM passing through A with various slopes, say by setting $x = t \cdot y$ with

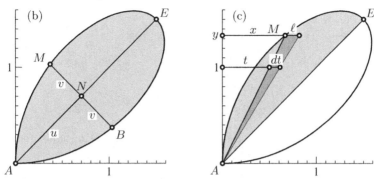

Fig. 12.20. Computing the area of Descartes' folium

$0 \le t \le 1$, these lines will cut the folium into small triangles whose areas fill its upper half (see Fig. 12.20 (c)). Inserted into $x^3 + y^3 = 3xy$ this gives $y^3(1 + t^3) = 3ty^2$, which can be divided by y^2 and yields a simple formula for the altitude of one of these triangles,

$$y = \frac{3t}{1 + t^3}, \quad \text{and for the base} \quad \ell = y\, dt$$

by Thales. By Eucl. I.41, the area of this slim triangle is $\frac{y\ell}{2}$, and adding up two of these gives an easy integral for the area of the entire folium:

$$\mathcal{A} = \int_0^1 y \cdot \ell = \int_0^1 y^2\, dt = 3 \int_0^1 \frac{3t^2\, dt}{(1 + t^3)^2} = 3 \int_1^2 \frac{dv}{v^2} = \frac{3}{2}. \qquad (12.13)$$

18. The two grey triangles in Fig. 7.47 (right) are similar; one angle is right, the other is equal by orthogonality (see Fig. 1.7). Hence by Thales $\frac{dz}{dx} = \frac{1}{\cos x}$. Also by Thales $\frac{dy}{dz} = \frac{1}{\cos x}$. Again by Pythagoras and Thales $\frac{1}{\cos x} = \sqrt{1 + y^2}$. All this leads to the formulas

$$\frac{dy}{dx} = \frac{1}{\cos^2 x} = 1 + y^2. \qquad (12.14)$$

The second term gives the derivative of arctan by interchanging $x \leftrightarrow y$.

Remark. With similar figures, Newton discovered in his manuscript *De Analysi* of 1669 the series expansions of many functions, his first great success, after which he was offered the Lucasian Chair at Cambridge (see Hairer and Wanner 1997, p. 52, Fig. 4.13).

19. If we give F_1 and F_2 the abscissa -1 and 1 respectively, our condition becomes $((x + 1)^2 + y^2) = C^2((x - 1)^2 + y^2)$, which is

$$x^2 + y^2 + 2\frac{1 + C^2}{1 - C^2}x + 1 = 0 \quad \text{or} \quad \left(x + \frac{1 + C^2}{1 - C^2}\right)^2 + y^2 = \left(\frac{1 + C^2}{1 - C^2}\right)^2 - 1,$$

the equation *of a circle!* The equation on the left exhibits a nice property: setting $y = 0$, we see that this circle intersects the x-axis at two points x_1

and x_2 with $x_1 x_2 = 1$ (Viète's identity), i.e. F_1, F_2, x_1, x_2 are "harmonic points" (see (11.10) of Chap. 11 and Apoll. I.36).

20. As in Exercise 17, the area is given by (here an easy integral)

$$\mathcal{A} = \int_{-\frac{\pi}{4}}^{\frac{\pi}{4}} \frac{r^2}{2}\, d\varphi = \frac{a^2}{2} \int_{-\frac{\pi}{4}}^{\frac{\pi}{4}} \cos 2\varphi\, d\varphi = \frac{a^2}{2}\,.$$

This result, which today appears so simple, was a great challenge for half a century, until G.C. Fagnano (1750) discovered it.

21. (a) The relation $x^2 - y^2 = 1$ becomes $\rho^2 \cos^2 \varphi - \rho^2 \sin^2 \varphi = 1$ or $\rho^2 \cos 2\varphi = 1$. Thus $\frac{1}{\rho}$ is (7.36) with $a = 1$. (b) Let the point P have coordinates $x_P = \cos\varphi/\sqrt{\cos 2\varphi}$ and $y_P = \sin\varphi/\sqrt{\cos 2\varphi}$. Then we obtain the equation of the tangent from (7.9) and the equation of the perpendicular through the origin from (7.2c). This gives

$$x \cos\varphi - y \sin\varphi = \sqrt{\cos 2\varphi}\,,$$
$$x \sin\varphi + y \cos\varphi = 0\,.$$

This linear system (with orthogonal matrix) has the solution $x_R = x_Q$, $y_R = -y_Q$.

22. By Pythagoras' theorem and the expressions from the proof of Thm. 7.18 (see Fig. 7.19 (b)), we have

$$ds^2 = dr^2 + r^2 d\varphi^2 = \left(1 + \frac{a^4 \cos^2 2\varphi}{a^4 \sin^2 2\varphi}\right) dr^2$$
$$= \frac{a^4}{a^4 \sin^2 2\varphi}\, dr^2 = \frac{a^4}{a^4 - a^4 \cos^2 2\varphi}\, dr^2 = \frac{a^4}{a^4 - r^4}\, dr^2\,.$$

23. R. Müller simply calls (a) and (b) "bekannte Formeln" [known formulas]; 100 years later, we have some difficulty in understanding them. Formula (b) is equivalent to y_H from Theorem 7.20, and (a) can be seen likewise by considering the similar triangles CHD and CBF. But it is simpler to use the extended triangle UVW of Gauss' proof in Fig. 4.9 (a), page 86, whose circumradius is $2R$ and whose circumcentre is H, then to apply Eucl. III.20 as in Fig. 5.9 (a) on page 120. The result (c) is Theorem 4.11 (d) on page 92. Finally, Eucl. III.35 tells us that $CH \cdot HM = R^2 - HO^2$, which leads to (7.67).

Hobson (1891, Art. 158) obtained this from the law of cosines (5.10) applied to the triangle OHC by using $OC = R$, $HC = 2R \cos\gamma$ and angle $OCH = \beta - \alpha$. After some formula manipulations, involving in particular (5.62), this leads to the same result.[4]

---

[4] We thank John Steinig for this reference.

24. If we denote the distance OC by d and the angle ACF of Fig. 7.48 (right) by φ we have, as in the proof of Theorem 7.32, $pp' = 1 - d^2$ (Eucl. III.35) and $p' - p = 2d \sin \varphi$. From this we obtain

$$p'^2 + p^2 = (p' - p)^2 + 2pp' = 4d^2 \sin^2 \varphi + 2 - 2d^2 = 2 - 2d^2 \cos 2\varphi$$

with (5.8). If we add up all three of these terms, we obtain 6, because

$$\cos 2\varphi + \cos(2\varphi + \frac{2\pi}{3}) + \cos(2\varphi + \frac{4\pi}{3}) = 0$$

(this is the real part of the barycentre of a rotating Mercedes star).

25. One possibility is to obtain, with Euler's formulas (cf. Theorem 7.20), after some simplifications,

$$AG^2 = x_G^2 + y_G^2 = \frac{1}{9}(2c^2 + 2b^2 - a^2). \qquad (12.15)$$

Adding to this the similar expressions for BG^2 and CG^2 gives $\frac{1}{3}(a^2 + b^2 + c^2)$, the desired result. More elegantly, one can obtain (12.15) from Pappus' formula (4.5) and the fact that $AG = \frac{2}{3} AD$ (see Fig. 4.8 (a)).

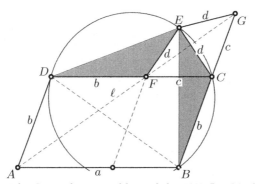

Fig. 12.21. The *Luxembourg* problem of the 2007 Int. Math. Olympiad

26. Once we are convinced that the stated conditions determine, for a given parallelogram $ABCD$, the point E uniquely[5], we give a *backwards* proof, i.e., *we suppose that the line ℓ is the angle bisector and show that E must lie on the circumcircle.* Under this assumption ADF is half of a rhombus, i.e. $DF = BC = b$ (see Fig. 12.21). Similarly, FCG is isosceles and the triangles FEC and CEG have the same exterior angles. Therefore the two shaded triangles are identical (by Eucl. I.4), which means that D and B

---

[5]If we move the point F from C to D, the intersection E of the perpendicular bisectors of FC and CG moves on a hyperbola through C (this can be checked by an analytical computation). Therefore there is at most one second intersection of this hyperbola with the circumcircle of BCD.

have the same inscribed angle on the arc EC. By the inverse application of Eucl. III.21, all four points $DBCE$ are concyclic.

Remark. The "official" IMO solution, which is direct, is more cumbersome.

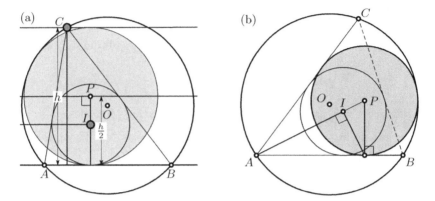

Fig. 12.22. The $h/2$ circle (a); the biggest chocolate egg (b)

27. We see this result by letting $p \to 0$; in this case D tends to $-\infty$ and the angle bisector tends to the horizontal line of ordinate $y_C/2$ (see Fig. 12.22 (a)). The projection $I \mapsto P$ becomes vertical.

28. We obtain the answer by running our machine so that D coincides with A (see Fig. 12.22 (b)). One obtains the centre P of the required circle by projecting the incentre I orthogonally to AI onto the side AB, and then orthogonally to AB back onto the angle bisector. The circle is thus by the factor $1 : \cos^2 \frac{\alpha}{2}$ larger than the incircle.

29. The curves are conics with the same foci as the reflecting ellipse. They are ellipses or hyperbolas, depending on the initial position and direction. The result follows from Poncelet's "second theorem" (Fig. 7.5, right, on page 192).

12.8 Solutions for Chapter 8

1. Denote by z_1, z_2, z_3, z_4 the complex numbers representing A_1, A_2, A_3, A_4 and by w_1, w_2, w_3, w_4 those representing B_1, B_2, B_3, B_4. Then we can go from A_1 to, say, B_1 by going half way along the segment $A_1 A_2$, turning 90° to the right (i.e. multiplying by $-i$) and going the same distance again. We thus have

$$
\begin{aligned}
w_1 &= z_1 + \tfrac{1-i}{2}(z_2 - z_1) \\
w_2 &= z_2 + \tfrac{1-i}{2}(z_3 - z_2) \\
w_3 &= z_3 + \tfrac{1-i}{2}(z_4 - z_3) \\
w_4 &= z_4 + \tfrac{1-i}{2}(z_1 - z_4)
\end{aligned}
\quad \Rightarrow \quad
\begin{aligned}
w_3 - w_1 &= \tfrac{1-i}{2}(z_4 - z_2) + \tfrac{1+i}{2}(z_3 - z_1), \\
w_4 - w_2 &= \tfrac{1-i}{2}(z_1 - z_3) + \tfrac{1+i}{2}(z_4 - z_2).
\end{aligned}
$$

We see that $w_4 - w_2 = i(w_3 - w_1)$.

2. The side we want is

$$s_{17} = 2 \cdot \sin \frac{\pi}{17} = \sqrt{(\varepsilon - 1)(\bar{\varepsilon} - 1)} = \sqrt{\varepsilon^1 \varepsilon^{16} - \varepsilon^1 - \varepsilon^{16} + 1} = \sqrt{2 - \beta_1}$$

from (8.34). If you insert all the further results from Gauss' proof, you obtain a square root containing square roots of square roots of ... of $\sqrt{17}$. To this you then apply algebraic simplifications as far as possible. Whenever you are afraid of making an error, check the results numerically to see whether both expressions give the value $s_{17} = 0.367499035633141$.

3. Be inspired by Eucl. IV.16 and construct a regular 17-gon and an equilateral triangle, both inscribed in the same circle with one common vertex. You will then find two vertices which have distance s_{51}.

4. (a) Book V of Apollonius' *Conica* is entirely devoted to such problems of minimal distances. Apoll. V.30 tells us that the segment $P_0 P_1$ must be orthogonal to the tangent at (x_1, y_1). The polar of this point is $xx_1 - yy_1 = 1$ and has slope $\frac{x_1}{y_1}$; the segment $P_0 P_1$ must thus have slope $-\frac{y_1}{x_1}$, hence we have

$$-\frac{y_1}{x_1} = \frac{y_1 - y_0}{x_1 - x_0} \quad \Rightarrow \quad \frac{y_0}{y_1} - 1 = 1 - \frac{x_0}{x_1}, \tag{12.16}$$

i.e. the point P_1 is obtained as the intersection of the given hyperbola with another hyperbola (Apoll. V.59).

(b) Fermat's method for

$$(x - x_0)^2 + (y - y_0)^2 = \min \quad \text{with} \quad x^2 - y^2 = 1$$

leads in Leibniz' notation to

$$(x - x_0)\, dx + (y - y_0)\, dy = 0 \quad \text{and} \quad x\, dx - y\, dy = 0.$$

Computing $\frac{dy}{dx}$ from both equations leads to the same equations as in (12.16). In what follows we set $(x_0, y_0) = (1, 1)$ and denote the expression on the right of (12.16) by λ. This gives $x_1 = \frac{1}{1-\lambda}$, $y_1 = \frac{1}{1+\lambda}$, which is on the hyperbola if

$$x_1^2 - y_1^2 = \frac{1}{(1 - \lambda)^2} - \frac{1}{(1 + \lambda)^2} = 1 \quad \Rightarrow \quad \lambda^4 - 2\lambda^2 - 4\lambda + 1 = 0.$$

This equation of degree 4 *might* hide two factors of degree 2 and thus *might* have roots which are constructible with ruler and compass. We thus set (with Euler, 1751, E170)

$$\lambda^4 - 2\lambda^2 - 4\lambda + 1 = (\lambda^2 + u\lambda + \alpha)(\lambda^2 - u\lambda + \beta),$$

where u, α and β are to be determined. Multiplying out we obtain

$$\alpha + \beta = -2 + u^2, \quad \alpha - \beta = \frac{4}{u}, \quad \alpha\beta = 1.$$

Adding and subtracting the first two equations gives 2α and 2β, whose product must be 4 from the last equation. This leads to

$$u^6 - 4u^4 - 16 = 0$$

or, with $u^2 = v$,

$$v^3 - 4v^2 - 16 = 0.$$

This equation now fits the proof we have seen and we conclude that neither v nor $u = \sqrt{v}$ nor λ are constructible with Euclid's instruments. A numerical calculation gives $\lambda = 0.22527042609892$.

5. By (7.2b) and Table 5.2, the segments AD resp. BF are

$$y = (x+1)\frac{\sqrt{2}}{2+\sqrt{2}}, \qquad \text{resp.} \qquad y = (x-1)\frac{-\sqrt{3}}{3}.$$

We write $(x-1) = (x+1) - 2$ in the second equation, subtract and obtain $(x+1)$ in a straightforward way. Then, by Pythagoras and the first equation, $AE = (x+1) \cdot \sqrt{1 + (\frac{\sqrt{2}}{2+\sqrt{2}})^2}$, which when simplified gives

$$AE = \frac{4\sqrt{2+\sqrt{2}}}{\sqrt{6}+\sqrt{2}+2} = 1.2604724 \ldots \quad \text{instead of} \quad \sqrt[3]{2} = 1.25992 \ldots$$

6. If we denote the coordinates of E by x, y and the distance EN by z, the three circles give by Pythagoras the equations

$$\begin{aligned} x^2 + y^2 &= 2 \\ (4-x)^2 + y^2 &= 4^2 \\ (8-x)^2 + y^2 &= z^2 \end{aligned} \qquad \Rightarrow \qquad x = \frac{1}{4}, \quad z^2 = 62.$$

This leads to $AF = 8 - \sqrt{62}$ and the approximation $10 \cdot AF = 1.25992126$ whereas $\sqrt[3]{2} = 1.25992104989487$. The approximation thus has an error smaller than $2 \cdot 10^{-7}$, which Finsler considers "sufficiently precise for constructional purposes".

7. All these approximations are easy to verify, but difficult to find!
(a) Because ADO is half of an equilateral triangle, we have by Thales $AD = \frac{1}{\sqrt{3}}$. By Pythagoras,

$$BC = \sqrt{(3 - \frac{\sqrt{3}}{3})^2 + 2^2} = \sqrt{\frac{40-6\sqrt{3}}{3}}.$$

(b) The triangles DOC and CFB are similar, hence $OD = \frac{1}{3}$ by Thales. By Pythagoras,

$$AE + ED = 3 + \sqrt{3^2 + (\tfrac{4}{3})^2} \quad \text{which gives} \quad \pi \approx \tfrac{9+\sqrt{97}}{6}.$$

(c) We compute DC by Pythagoras, add $AB = \tfrac{5}{2}$, divide by 2, and obtain $\pi \approx \tfrac{\sqrt{229}+10}{4}$. The precision of each of these values is displayed in Fig. 12.23. The precision of Ramanujan's approximations is extraordinary.

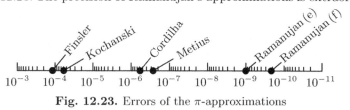

Fig. 12.23. Errors of the π-approximations

8. Ramanujan did not explain how he found this construction. He probably played around with numbers and saw that

$$355 = 4 \cdot 81 + 31, \qquad 113 = 4 \cdot 36 - 31, \qquad 31 = 4 \cdot 9 - 5.$$

So we start by finding a construction for $\sqrt{5}$ and do the rest with Pythagoras' theorem using triangles for which the square of one side is 4. Eucl. II.14: $QT = SR = \tfrac{\sqrt{5}}{3}$; Pyth.: $PS = \sqrt{4 - \tfrac{5}{9}} = \tfrac{\sqrt{31}}{3}$; Thales: $MN = \tfrac{\sqrt{31}}{3} \cdot \tfrac{2}{3} \cdot \tfrac{1}{2} = \tfrac{\sqrt{31}}{9} = PL$; Thales: $PM = \tfrac{\sqrt{31}}{6} = PK$; Pyth.: $KR = \sqrt{4 - \tfrac{31}{36}} = \tfrac{\sqrt{113}}{6}$; Pyth.: $LR = \sqrt{4 + \tfrac{31}{36}} = \tfrac{\sqrt{355}}{9}$; Thales: $RD = \tfrac{RC \cdot RL}{RK} = \tfrac{3}{2} \cdot \tfrac{\sqrt{355}}{9} \cdot \tfrac{6}{\sqrt{113}} = \sqrt{\tfrac{355}{113}}$.

9. Once again, the verification is easy, but our admiration is great for the man who found the construction: Since CB is the diagonal of a unit square, we have

$$AM = \sqrt{(1 + \tfrac{1}{3\sqrt{2}})^2 + (1 - \tfrac{1}{3\sqrt{2}})^2} = \sqrt{2 + \tfrac{1}{9}},$$

$$AN = \sqrt{(1 + \tfrac{\sqrt{2}}{3})^2 + (1 - \tfrac{\sqrt{2}}{3})^2} = \sqrt{2 + \tfrac{4}{9}}.$$

We then have by Thales and by construction

$$AQ = \frac{AM \cdot AP}{AN} = \frac{AM^2}{AN} = \frac{2 + \tfrac{1}{9}}{\sqrt{\tfrac{22}{9}}} \quad \Rightarrow \quad AS = AR = \frac{AQ}{3} = \frac{19}{9 \cdot \sqrt{22}}.$$

We compute SO by Pythagoras and obtain

$$SO = \sqrt{1 + AS^2} = \sqrt{1 + \frac{19^2}{9^2 \cdot 22}} = \sqrt{\left(9^2 + \frac{19^2}{22}\right)\frac{1}{3^4}}.$$

Taking square root of SO and dividing out the factor $\sqrt[4]{\tfrac{1}{3^4}} = \tfrac{1}{3}$ leads to the stated approximation.

12.9 Solutions for Chapter 9

1. We scale the coordinates (without changing ratios of volumes) so that the paraboloid has the equation

$$z = x^2 + y^2, \qquad 0 \le z \le 1, \quad 0 \le x^2 + y^2 \le 1 .$$

Then the surface of the liquid lies in the plane $z = -x$ and is bounded by $-x = x^2 + y^2$, hence $(x + \frac{1}{2})^2 + y^2 = \frac{1}{4}$ (see Fig. 9.30 (b)). If a guest has the kindness to look vertically into the glass, she will see the liquid as a circle with centre $-\frac{1}{2}$ and radius $\frac{1}{2}$. She is then asked to drink the rest with a straw, without tilting the glass any more. The surface would then sink to $z = x - d$, and she sees $(x + \frac{1}{2})^2 + y^2 = \frac{1}{4} - d$, a circle which diminishes in form of a parabola and disappears, together with the last drops of *Dom Pérignon*, for $d = \frac{1}{4}$. Thus the remaining champagne fills a paraboloid of revolution whose radius is one half of the original radius, and whose height is one fourth of the height of the glass. Its volume is thus

$$\frac{1}{4} \cdot \frac{1}{2} \cdot \frac{1}{2} = \frac{1}{16}, \qquad \text{i.e.} \quad 6.25\%$$

of the volume of the full glass.

2. Let a, b, c, p be the vector positions of the points A, B, C, P. Then the point A' is $\frac{p+b+c}{3} = q - \frac{a}{3}$, where $q = \frac{a+b+c+p}{3}$. Similarly B' is $q - \frac{b}{3}$ and C' is $q - \frac{c}{3}$. Thus the triangle $A'B'C'$ is similar to ABC with similarity factor $-\frac{1}{3}$. The same proof can easily be extended to, say, tetrahedra in space, where we would have the similarity factor $-\frac{1}{4}$.

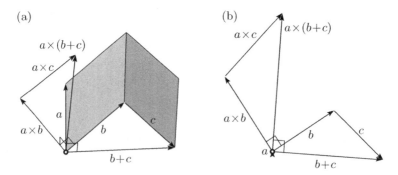

Fig. 12.24. Proof of $a \times (b + c) = a \times b + a \times c$ (left); projection onto plane perpendicular to a (right)

3. All three vectors $a \times (b+c)$, $a \times b$ and $a \times c$ lie in the plane perpendicular to a, their lengths being the areas of the indicated parallelograms (see Fig. 12.24 (a)). If we look at the picture in the direction of the vector a (Fig. 12.24 (b)), these lengths are the lengths of the projected vectors

multiplied by $|a|$. The formula is expressed by two similar triangles with similarity factor $|a|$, one rotated by $90°$ with respect to the other.

4. Proof of (9.51). The first factor $a \times b$ is perpendicular to a and b, therefore the exterior product by c, which is perpendicular to $a \times b$, must be of the form $\lambda a + \mu b$. Using the definition (9.23), we compute the first coefficient of $(a \times b) \times c$ and find $c_3 a_3 b_1 - c_3 a_1 b_3 - c_2 a_1 b_2 + c_2 a_2 b_1$. If we try to collect the coefficients which multiply the coordinates a_1 and b_1, we must find a symmetric expression. For this, we cleverly add $-a_1 b_1 c_1 + a_1 b_1 c_1$ and arrive at the stated expression, valid for all coefficients.

Proof of (9.52). If we write for the moment $u = a \times b$, we see that $u \cdot (c \times d)$ is a determinant (9.24), where we can permute the rows. So we obtain $(a \times b) \cdot (c \times d) = ((a \times b) \times c) \cdot d$. Here we use (9.51) and the scalar product by d.

5. Note that u, v, w are unit vectors. With (9.52) we obtain

$$(v \cdot w) - (u \cdot v)(u \cdot w) = |u \times v| \, |u \times w| \, \cos \alpha .$$

The required relations

$$|u \times v| = \sin c, \ |u \times w| = \sin b, \ v \cdot w = \cos a, \ u \cdot w = \cos b, \ u \cdot v = \cos c$$

follow at once from (9.27) and Theorem 9.4.

6. Following the hint, we obtain

$$\sin c \sin b \sin \alpha = |(u \times v) \times (u \times w)| = |\langle v, u \times w \rangle| ,$$

and in the same way

$$\sin c \sin a \sin \beta = |(v \times u) \times (v \times w)| = |\langle u, v \times w \rangle| .$$

We then conclude with (9.24).

7. (a) We imagine the lateral surface of the truncated cone to be composed of narrow isosceles triangles with C as vertex. They all have the same slope, determined by α. If we project them down onto the base AD (the lower part of Fig. 9.32 (a,b)), they fill an ellipse whose area is the lateral area $\mathcal{S}$ multiplied by $s = \sin \alpha$. We see from the figure that the major axis of this ellipse satisfies $2a = (v + w) \cdot s$. The semi-minor axis b is determined by the dashed circle, which is located midway between the levels of A and B, because the centre of the ellipse is at the midpoint between the projections of A and B. From Thales we have $DE = vs$ (half of the diameter of the circle at level B) and $EF = ws$ (half of the diameter of the circle at level A). By Eucl. II.14 we obtain

$$b = \sin \alpha \cdot \sqrt{vw} . \tag{12.17}$$

The desired result finally follows by inserting this into Archimedes' formula (3.10) on page 67 and then dividing the result by $s = \sin\alpha$.

(b) By Eucl. XII.10 (formula (2.13) on page 50) the desired volume is one third of the area of the ellipse in the plane AB, multiplied by the altitude h. Together with Archimedes, we obtain $V = \frac{h}{3} \cdot \frac{AB}{2} \cdot b \cdot \pi$. Comparing two formulas for the area of the triangle ABC we have

$$\frac{AB}{2} \cdot h = vw\,\frac{\sin 2\alpha}{2} = vw\sin\alpha\cos\alpha\,. \tag{12.18}$$

Inserting this and (12.17) into the expression for V gives the required formula (9.54).

(c) From Apollonius' (3.7) on page 66 we obtain Bernoulli's formula by inserting first (12.17), then (12.18); a straightforward calculation yields

$$p = \frac{b^2}{AB/2} = \frac{\sin^2\alpha \cdot vw}{AB/2} = \frac{\sin\alpha}{\cos\alpha} \cdot h\,.$$

This elegant proof by Johann Bernoulli (in a different notation) of his brother's discovery led to the very first mathematical article in Johann's Opera (vol. I, pp. 45–46).

Remark. The fact that the volume in formula (9.54), for a fixed cone, depends only on the product vw, and not on the individual values of v and w, leads, together with Apoll. III.43 of Fig. 10.18 on page 314, to the interesting conclusion that all planes, which cut a fixed volume from a cone, are the planes tangent to a hyperboloid of rotation for which the generatrices of the cone are asymptotes.

8. We follow the boundary, and count each boundary point as often as we meet it. For the Maltese cross of Fig. 9.32 (d) we thus have $b = 24$ and $i = 16$. Since the figure decomposes into m simple polygons, we have to subtract m times the correction -2 from b. Thus

$$\mathcal{A} = i + \frac{b - 2m}{2} = \text{(in our case)} \quad = 16 + \frac{24 - 2 \cdot 4}{2} = 24\,.$$

One can, of course, apply Pick's theorem four times, once to each of the four simple polygons or, for the Maltese cross, by just adding 8 times a triangle of base 3 and height 2 (Eucl. I.41).

9. We start by proving this property for a Platonic solid. In fact, any two adjacent faces of such a solid meet under the same angle, which can be determined by spherical trigonometry and was calculated earlier (see formulas (5.32) on page 131). Thus the perpendiculars through the centres of two adjacent faces meet in a point O which is at the same distance from both faces and, repeating the argument around the body, at the same distance from *all* faces. The distances between O and the vertices

are then determined by Pythagoras' theorem and all have the same value. The same argument extends to all Archimedean solids, because they all have vertices arranged symmetrically around the centres of the faces of the Platonic solids from which they have been constructed by one of the procedures described in Figures 9.22 through 9.28.

10. Suppose first that the solid is composed of k-gons of two types, say, of k_1-gons and k_2-gons, where ℓ_1 respectively ℓ_2 of these k-gons meet at each vertex. Projecting the body onto the sphere produces spherical k_1-gons and k_2-gons, whose isosceles triangles above each side again have as vertex angles $\alpha_1 = \frac{2\pi}{k_1}$, respectively $\alpha_2 = \frac{2\pi}{k_2}$. The angles β_i at the bases, however, are more difficult to compute. We write the cosine rule (5.41) for both triangles

$$\cos a = \frac{\cos \alpha_i + \cos^2 \beta_i}{\sin^2 \beta_i},$$

which must be equal for $i = 1, 2$. Writing $\cos^2 \beta_i = 1 - \sin^2 \beta_i$ in this formula and simplifying leads to the condition

$$\frac{\sin \beta_1}{\sqrt{\cos \alpha_1 + 1}} = \frac{\sin \beta_2}{\sqrt{\cos \alpha_2 + 1}}.$$

This, together with $\ell_1\beta_1 + \ell_2\beta_2 = \pi$, is a system of two equations, one nonlinear, one linear, for the computation of β_1 and β_2. We solve this system numerically and obtain the following values:

KepNo.	k_1	k_2	ℓ_1	ℓ_2	β_1	β_2	a	R	ρ/R
1	3	8	1	2	0.5480	1.2968	32°38′59″	1.778824	0.678598
2	3	6	1	2	0.5857	1.2780	50°28′43″	1.172604	0.522233
3	3	10	1	2	0.5320	1.3048	19°23′14″	2.969449	0.838505
4	5	6	1	2	0.9720	1.0848	23°16′53″	2.478019	0.914958
5	4	6	1	2	0.8411	1.1503	36°52′11″	1.581139	0.774597
8	3	4	2	2	0.6155	0.9553	60°0′0″	1.000000	0.707107
9	3	5	2	2	0.5536	1.0172	36°0′0″	1.618034	0.850651
10	3	4	1	3	0.5649	0.8589	41°52′55″	1.398966	0.862856
12	3	4	4	1	0.5689	0.8661	43°41′26″	1.343713	0.850340
13	3	5	4	1	0.5399	0.9822	26°49′16″	2.155837	0.918861

Only the solids No. 8 and 9 have simple expressions: indeed, for both the edges form "geodesic lines" which are regular hexagons and regular decagons, respectively.

For the solids No. 6, 7 and 11, there are three different types of faces and we obtain a system of three equations to solve, with the following results:

Kep.	k_1	k_2	k_3	ℓ_1	ℓ_2	ℓ_3	β_1	β_2	β_3	a	R	ρ/R
6	4	6	8	1	1	1	0.810	1.091	1.241	24°55′4″	2.317611	0.825943
7	4	6	10	1	1	1	0.794	1.063	1.285	15°6′44″	3.802394	0.904944
11	3	4	5	1	2	1	0.539	0.812	0.979	25°52′43″	2.232951	0.924594

11. The inscribed sphere touches the faces with the largest k-value. Therefore, we use formula (5.34) with the largest values of β from the above tables. This leads to the values of ρ/R of the last columns. We see that the rhombicosidodecahedron (No. 11) is the "roundest", followed by the snub dodecahedron (No. 13), while the FIFA-ball (No. 4) only occupies rank three, however, with a much smaller number of pieces of leather.

12.10 Solutions for Chapter 10

1. One gets this formula from (10.16) by inserting (9.19) for the scalar products, then factoring out from each column and each row the common factors $|a|$, $|b|$ and $|c|$, respectively, and finally expanding the remaining determinant according to formula (9.13).

2. One obtains (10.17) from (10.62) by expanding the determinant (10.62) with respect to its first column, then with respect to its first row. The sign changes are required, because in the second expansion the nonzero coefficient 1 is in the position $(1, 4)$.

   One obtains (10.18) from (10.62) by adding the last row multiplied by $|a|^2$, $|b|^2$, $|c|^2$, respectively, to the second, third, and fourth row; and then by adding the last column, again multiplied by $|a|^2$, $|b|^2$, $|c|^2$, respectively, to the second, third, and fourth column.

3. We rotate the coordinate system until the projection plane becomes the plane $x_3 = 0$, and the axonometric projection just removes the third coordinate from a vector. Then let our orthonormal vectors be the *columns* of the orthogonal matrix

$$Q = \begin{bmatrix} p_1 & q_1 & r_1 \\ p_2 & q_2 & r_2 \\ p_3 & q_3 & r_3 \end{bmatrix} \tag{12.19}$$

so that

$$z_p = p_1 + ip_2, \quad z_q = q_1 + iq_2, \quad z_r = r_1 + ir_2.$$

We use the characterisation (b) of Theorem 10.5, i.e. the first two *rows* of the above matrix must be orthonormal. Then $z_p^2 + z_q^2 + z_r^2$ becomes

$$(p_1^2 + q_1^2 + r_1^2) + 2i(p_1 p_2 + q_1 q_2 + r_1 r_2) + i^2(p_2^2 + q_2^2 + r_2^2) = 1 + 2i \cdot 0 - 1 = 0.$$

Conversely, if (10.63) is satisfied, then the first two rows of (12.19) are orthonormal. We extend them to an orthonormal basis and, after transposition, see that *any* triple of vectors satisfying (10.63) is the axonometric pre-image of an orthonormal basis in $\mathbb{R}^3$.

4. Any non-degenerate hyperbola can be transformed, by a translation and a rotation, to the equation $\frac{z_1^2}{a^2} - \frac{z_2^2}{b^2} = 1$ as in (10.51). We remove the denominators a^2 and b^2 by similarity transformations in the same way as we simplified the equation (3.9) for ellipses. Then the asymptotes will have slope ± 1. After a rotation through $45°$ the asymptotes coincide with the axes, and the equation of the hyperbola simplifies to $xy = \frac{1}{2}$ as in (3.1) (see the picture).

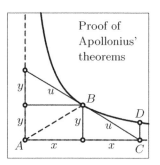

Proof of Apollonius' theorems

The key to Apoll. II.3 and Apoll. II.4 is the fact that the slope of the tangent at B is minus the slope of AB. This follows either from $(x + dx)(y + dy) = xy$, hence $x\,dy = -y\,dx$, or from the equation of the polar (7.9) at a given point (x_0, y_0), which here becomes $x_0 y + x y_0 = 1$. Then all four triangles in this picture are congruent and these results are true, since affine transformations preserve ratios of lengths of parallel segments.

Apoll. II.13 follows from $xy = \text{Const}$, Apoll. III.43 from $2x \cdot 2y = \text{Const}$, and Apoll. III.34 means that $2x \cdot \frac{y}{2} = xy$. A nice interpretation of Apoll. III.43 is the fact that all triangles formed by a tangent to a fixed hyperbola and its asymptotes have the same area.

Apollonius obtained Apoll. II.8 by drawing the tangent parallel to EZ and applying Apoll. II.3 and Thales' theorem to obtain $EM = MZ$. The result then follows from the fact that the conjugate diameter ΔM bisects $A\Gamma$.

5. Similarly to the preceding exercise we transform the hyperbola to $y = \frac{1}{x}$. Let the coordinates of the vertices of the triangle ABC be $(a, \frac{1}{a})$, $(b, \frac{1}{b})$ and $(c, \frac{1}{c})$. Then the slope of, say BC, is $\frac{\frac{1}{b} - \frac{1}{c}}{b - c} = -\frac{1}{bc}$. We now choose a point H on the hyperbola with coordinates $(h, \frac{1}{h})$. Then AH is perpendicular to BC if

$$-\frac{1}{bc} = ah, \qquad \text{i.e.} \quad abch = -1. \qquad (12.20)$$

The symmetry of this condition shows that the line connecting any two of the four points A, B, C and H is perpendicular to the line connecting the remaining two. Thus H is the orthocentre of ABC (see Fig. 12.25).

6. For each of these lines, in general, *one* of the vertices of the triangle lies on one side of the line, and the two other vertices lie on the other side. We look at the side on which there is only one vertex. On this side, for the case (a), the perimeter cut off from the triangle must have constant value s, which is half of the total perimeter. Thus one point of intersection of these lines approaches the vertex on one side with the same speed as the second point departs on the other side. We have precisely the situation already encountered in Exercise 10 of Chap. 3 (see Fig. 3.18) for the case of a right angle. After an affine transformation we see that the envelopes

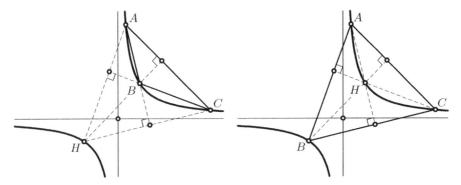

Fig. 12.25. Equilateral hyperbola and orthocentre

are parabolas. When the line crosses one of the vertices, the envelope
will jump from one parabola to another. For the case (b) we are in the
situation of constant areas cut off by a line from a fixed angle. This, by
Apoll. III.43 seen in the preceding exercise, produces hyperbolas.

7. Barrow's solution extends over two pages. We transform the angle ABC
to a right angle by an affine transformation, rotate, shift and scale the
axes until ABC lies on the lines $x = 1$, $y = 1$ and the point D is at the
origin (see Fig. 12.26). We let (x, y) be the coordinates of O, so that by
Thales the coordinates of N and M become $(\frac{x}{y}, 1)$ and $(1, \frac{y}{x})$, respectively.
For the condition $DO = MN$ we can use either the value of the abscissa
or of the ordinate, which lead to

$$x = 1 - \frac{x}{y} \qquad \text{or} \qquad y = \frac{y}{x} - 1.$$

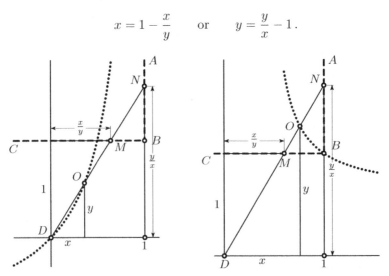

Fig. 12.26. Solution of Barrow's first problem (left) and Barrow's second problem
(for $q = 0.4$; right)

Both equations are equivalent to

$$xy + x - y = 0 \qquad \text{or} \qquad (x-1)(y+1) = -1 \,.$$

The curve is thus a hyperbola, with reversed orientation and the lines $x = 1$ and $y = -1$ as asymptotes.

8. Proceeding as before, the condition $MO = q \cdot MN$ becomes

$$x - \frac{x}{y} = q \cdot \left(1 - \frac{x}{y}\right) \qquad \text{or} \qquad y - 1 = q \cdot \left(\frac{y}{x} - 1\right) \,.$$

Again, both equations are equivalent to

$$xy - (1-q)x - qy = 0 \qquad \text{or} \qquad (x-q)(y-(1-q)) = q(1-q) \,.$$

This is a hyperbola with asymptotes $x = q$ and $y = 1 - q$.

9. The proof is a straightforward extension of the proof of Theorem 7.7 on page 191, by extending the definition of a polar (which is here a plane) in Definition 7.2 and comparing this equation with the equation of the plane in (10.64) together with the condition that the pole x_0 lies on the ellipsoid.

10. If we suppose $n_{i1}x_1 + n_{i2}x_2 + n_{i3}x_3 = d_i$ (with $i = 1, 2, 3$) to be the equations of the three planes, where (n_{i1}, n_{i2}, n_{i3}) are three mutually perpendicular unit vectors, then $d_1, d_2, , d_3$ are the distances of these planes from the origin. Conditions (10.65) for these three planes read:

$$n_{11}^2 a_1^2 + n_{12}^2 a_2^2 + n_{13}^2 a_3^2 = d_1^2$$
$$n_{21}^2 a_1^2 + n_{22}^2 a_2^2 + n_{23}^2 a_3^2 = d_2^2 \qquad (12.21)$$
$$n_{31}^2 a_1^2 + n_{32}^2 a_2^2 + n_{33}^2 a_3^2 = d_3^2 \,.$$

By hypothesis, the *rows* of the matrix $[n_{ij}]$ in (12.21) form an orthonormal basis. By Theorem 10.5, its *columns* must also be orthonormal. Consequently, the sum of the three equations in (12.21) simplifies to

$$d_1^2 + d_2^2 + d_3^2 = a_1^2 + a_2^2 + a_3^2 \,,$$

which, by Pythagoras' theorem, is the distance of the intersection point of the three planes from the origin.

11. The original derivation used differential calculus, based on the fact that the parabola and the circle have the same tangent and the same curvature at the point A, i.e. the same first and second derivatives (a consequence of (5.53)). We can also say that the circle and the parabola coincide at three infinitely close points around A and at the point B. Many decades later, two school teachers from the Canton de Vaud, Sylvie Conod (Gymnase de La Tour-de-Peilz) and Christoph Soland (Gymnase du Bugnon), independently discovered a more elegant geometric proof.

As often in mathematics, a problem becomes easier, if one makes it more complicated, i.e. if instead of looking at the two separate curves

$$y - a - bx - cx^2 = 0 \quad \text{(the parabola)} \qquad x^2 + y^2 - 1 = 0 \quad \text{(the circle)}$$

we consider the family of curves

$$\mu \cdot (y - a - bx - cx^2) + (1 - \mu)(x^2 + y^2 - 1) = 0 \quad (\mu \text{ a parameter}). \quad (12.22)$$

If μ varies from 0 to 1, these curves represent ellipses which transform the circle into the parabola; for $\mu > 1$ we obtain hyperbolas. All these conics pass through the same four points as the two "generators", i.e. the triple point at A and the point B (see Fig. 10.21 (b)). For a particular value of μ, this hyperbola will degenerate into its asymptotes, a pair of straight lines. Since they must recover the triple point at A as well as the point B, one of these asymptotes will be the tangent at A, the other one will join A to B. The crucial observation is that none of the equations (12.22) for these conics contains an xy-term, i.e. the value b in equation (10.50) is zero and the matrix T in (10.51) effects no rotation. Therefore all these conics have axes parallel or orthogonal to ON. As a consequence, the asymptotes of the degenerate hyperbola will have symmetric slopes with respect to ON (see Fig. 10.21 (c)). This means that the angle α to the left of A, which is an angle orthogonal to NOA, will reappear to the right of A, and, by Eucl. III.20, produces the angle 2α for BOA'. Hence $\beta = \alpha + 2\alpha = 3\alpha$.

12. The proof is precisely the same as for the preceding exercise. The degenerate hyperbola will, in the configuration of the figure, consist of two lines, one joining A_1 to A_2, the second joining A_3 to A_4. Then the vector $OA_1 + OA_2$, which makes an angle $\frac{\alpha_1 + \alpha_2}{2}$ with ON, will be orthogonal to the first asymptote, and the vector $OA_3 + OA_4$, which makes an angle $\frac{\alpha_3 + \alpha_4}{2}$, will be orthogonal to the second. Since both asymptotes are equally steep, the result is clear.

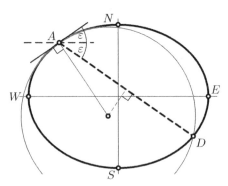

Fig. 12.27. Construction of the osculating circle for a point on an ellipse

13. We start by establishing the construction of an osculating circle for any point A on an ellipse displayed in Fig. 12.27. *The point D, at which this*

circle again cuts the ellipse, is obtained by drawing the segment AD whose slope is symmetric to that of the tangent. This is seen from Fig. 10.21 (b), because the circle drawn there is the osculating circle at the point A for *all* conics of the family (12.22).

We then forget the osculating circle and transform the ellipse into a circle by multiplying the abscissa by $\frac{a}{b}$. The two angles ε will become two equal angles, called α in Fig. 10.21 (c). Therefore, exactly as in (10.66), when A moves from W to N, the point D will move three times as fast through the arc $WSEN$. So for a fixed point D on this arc there will be a first solution A to Steiner's question (see Fig. 10.22, right). If our point moves into the next quadrant, the point D will travel through another three quadrants and we will have a second solution B, 120° behind the solution A. Another 120° later we will find the third solution C.

When transformed into the circle, ABC is an equilateral triangle, hence the tangent at B is parallel to AC. This property remains valid under the affine transformation back to the ellipse and we conclude by the above construction that the slope BD is symmetric to the slope AC. This allows us to show that A, B, C, D are concyclic by a modification of the idea employed in (12.22): the lines connecting AC and DB are of the form $x + cy + d = 0$ and $x - cy + e = 0$, hence these four points lie on the degenerate conic defined by $(x + cy + d)(x - cy + e) = x^2 - c^2y^2 + \ldots = 0$. By combining this with the equation of the ellipse, these four points lie on each conic of the family

$$\mu \cdot (x^2 - c^2y^2 + \ldots) + (1 - \mu)\left(\frac{x^2}{a^2} + \frac{y^2}{b^2} - 1\right) = 0 \qquad (12.23)$$

one of which is, by a judicious choice of μ, the equation of a circle.

12.11 Solutions for Chapter 11

1. By Thales, the hypothesis implies that $\frac{OA'}{OC} = \frac{OC'}{OA}$ and $\frac{OB'}{OC} = \frac{OC'}{OB}$. Dividing one equation by the other, we obtain $\frac{OA'}{OB'} = \frac{OB}{OA}$ or $\frac{OA'}{OB} = \frac{OB'}{OA}$, the desired result.

    Remark. The theorem in this form is a cornerstone of Hilbert's development of geometry from his system of axioms (see Sect. 2.7 and Hilbert, 1899). He proves it as his "Theorem 21" and deduces, the other way round, the validity of Thales' theorem as "Theorem 22".

2. Call the given points $P_1, \ldots, P_5$, construct the intersection point $K = P_1P_2 \cap P_4P_5$, and draw an arbitrary line ℓ through P_5 which will contain the required point P_6. This line cuts P_2P_3 at the point L. Next draw the Pascal line KL and denote its intersection with P_3P_4 by M. The

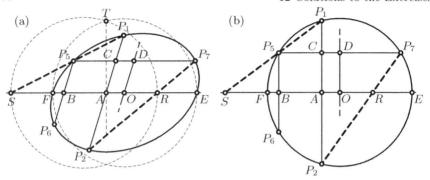

Fig. 12.28. Pappus' construction of an ellipse from 5 points (a); proof (b)

intersection of ℓ and MP_1 is then the required point P_6 on the conic.
More points of the conic can be obtained by repeating the procedure.

Pappus' construction (Collection, Book VIII, Prop. 13). As above, we
draw a line through P_5 in order to find a new point P_6. The main idea is
to choose the line ℓ *parallel* to another known segment, say P_1P_2. Let A
and B be the midpoints of the segments P_1P_2 and P_5P_6. Then the line
d through A and B determines the position of a diameter of the ellipse
(see Fig. 12.28 (a)). If we then choose ℓ, again through P_5, parallel to d,
we obtain a segment P_5P_7 whose midpoint D determines the position of
the second conjugate diameter (which is parallel to P_1P_2). We thus have
found the centre O of the ellipse.

We next determine $r = EO = OF$, where E and F denote the endpoints
of the diameter through AB. Pappus' idea is to cut the (extended) lines
P_2P_7 and P_1P_5 with d to find the points R and S, respectively. In order
to apply Euclid's propositions directly (instead of using the cross-ratio, as
Pappus did), we transform the ellipse into a circle, by keeping all points on
the line d fixed (see Fig. 12.28 (b)). Then, by applying Eucl. III.35 twice,
followed by Thales' theorem for the similar triangles P_1CP_5 and P_1AS,
as well as for CP_7P_2 and ARP_2, we obtain

$$\frac{FA \cdot AE}{P_2A \cdot AP_1} = 1 = \frac{P_5C \cdot CP_7}{CP_1 \cdot P_2C} = \frac{SA \cdot AR}{AP_1 \cdot P_2A} \quad \Rightarrow \quad FA \cdot AE = SA \cdot AR.$$

This last equation relates only to points on the line d and is thus valid
for *both* figures. From here on, Pappus' argumentation is cumbersome (see
also Heath, 1921, vol. II, p. 436). Taking square roots in the last expression
of the above formula, we see by Eucl. II.14 that the circles with diameters
SR and FE must meet in a point T on the perpendicular through A (see
Fig. 12.28 (a)). Since S, R and A are known, we can construct T and have
$r = OT$.

For the last step, after having found the second conjugate diameter in
the same way, Pappus states without proof a construction based on Exer-

cise 19 on page 78. The elegant construction of Rytz, given in Fig. 3.7 (a) on page 68, is simpler.

3. This is probably our last application of Eucl. III.21. Since the angles at z_3 and z_4 are both equal to α, we have by the property of complex division (see Fig. 8.2, right) that $\frac{z_3 - z_1}{z_3 - z_2} = C_1 \cdot e^{i\alpha}$ and $\frac{z_4 - z_1}{z_4 - z_2} = C_2 \cdot e^{i\alpha}$, so that their ratio is real. The invariance of this cross-ratio, and hence of circles under a Möbius transformation was shown in Theorem 11.9.

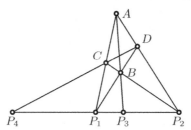

Fig. 12.29. Construction of fourth harmonic point

4. Apply Theorem 11.10 (see Fig. 12.29): Choose A arbitrarily not on the line $P_1 P_2$; join A to P_1, P_2, P_3; choose B arbitrarily on AP_3; then find C, D and P_4 as indicated in the figure.

5. (a) Set $P_1 P_2 = \lambda$. Then the two pairs of similar triangles give $P_1 P_3 = \frac{\lambda c}{b}$ and $P_3 P_2 = \frac{\lambda c}{a}$. The sum of these must be λ, whence $c \cdot (\frac{1}{b} + \frac{1}{a}) = 1$.

   (b) The angles marked α both have tangent $\frac{ab}{\lambda c}$. Higher education: complete the figure by drawing $DECP_4$ and obtain a figure which is precisely that in Fig. 12.29, where the point A has been moved to infinity orthogonally to $P_4 P_2$. Hence $2c$ is the harmonic mean of a and b, because the distances of P_3, P_1 and P_2 from P_4 have this property. The lines CP_3, DP_3, EP_3 and $P_4 P_3$ are also harmonic, so that the two angles marked α must be equal.

6. If the point P_3 is in the position of the harmonic mean (left picture), the angle $P_3 P_1 E$ is larger than α, because $EP_3 > CP_3$. Therefore, P_3 must be moved to the right in order to make the two angles marked β equal.

7. (a) If we place O at $x_1 = -1$ and N at $x_2 = 1$, then, because N is the midpoint between O and H, we have H at $x_4 = 3$. Since OG is one third of $OH = 4$ (see Theorem 4.10 on page 91), we have $x_3 = -1 + \frac{4}{3} = \frac{1}{3}$ and (11.10) is satisfied.

   (b) We see in Fig. 4.35 on page 111 that the points O, A', L, ∞ are harmonic by (11.11). The points O, N, G, H are the central projections from A of these points onto the Euler line HO and hence are also harmonic.

8. (a) If we set $\frac{GF}{FH} = p$, then $u' = pu$ and $v' = pv$, by similarity with centre F. Next, with $\frac{AH}{AG} = q$, we have $u = qv'$ and $v = qu'$ by similarity with

centre A. This gives $u = pqv = (pq)^2 u$. Because p and q are both > 0, we obtain $pq = 1$ and $u = v$.

(b) If we compare Fig. 11.26 with Fig. 12.29, we see that B, C, H together with the intersection point of DE with BC, which is at infinity, are harmonic. We also observe that pq of the foregoing proof, except for the sign, is a cross-ratio. The result $uv = 1$ is thus related to the fact that A, F, G, H are harmonic.

9. We consider the point P_4 as a circle of radius 0. Any circle intersecting this "circle" at right angles must pass through P_4. Moreover, the point D has equal powers with respect to the circles centred in O and P_4, and the perpendicular DE is the line of equal powers. Fig. 11.27 (a) is then a particular case of the right picture of Fig. 4.20 on page 99.

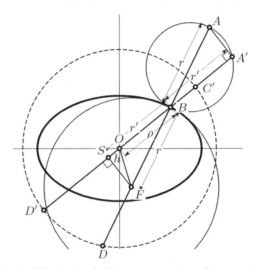

Fig. 12.30. Proof of Steiner's challenge concerning the curvature of an ellipse

10. The idea is to benefit from Exercise 9 and choose the line $A'BC'D'$, which is a diameter of Monge's circle (see Fig. 12.30). By Thales' circle, A' is the orthogonal projection of A onto this line, so if we project F orthogonally to S, the distances $SB = BA' = r'$ will be the same. We have simple expressions for the coordinates of $F = \left(c^3(a - \frac{b^2}{a}), s^3(b - \frac{a^2}{b})\right)$ and $B = (ca, sb)$ (see equation (7.46) on page 211) and continue to use the same notation. By Pythagoras $\rho^2 = c^2 a^2 + s^2 b^2$. We use $r' = \rho + h$, where $-\rho h$ is the scalar product of OB with OF, thus $\rho h = -(c^4 - s^4)(a^2 - b^2)$. Condition (11.10) becomes for our situation $\rho(\rho + r') = 2\rho^2 + \rho h = a^2 + b^2$, whose verification is now a simple calculation using trigonometric identities.

11. A *complete quadrangle* consists of four points in general position (i.e. no three are collinear). These four points are joined pairwise by six lines that

intersect in *three* additional points. The dual version of Theorem 11.10 can be stated as follows. *The four lines, concurrent in one of these three points and joining the remaining points, are harmonic.*

12. By construction, the columns of S are orthogonal to those of A. The foot of the altitude through A_k is thus given by

$$\begin{bmatrix} f_{1k} \\ f_{2k} \\ f_{3k} \end{bmatrix} = \begin{bmatrix} a_{1k} \\ a_{2k} \\ a_{3k} \end{bmatrix} - \frac{1}{s_{1k}^2 + s_{2k}^2 + s_{3k}^2} \begin{bmatrix} s_{1k} \\ s_{2k} \\ s_{3k} \end{bmatrix}.$$

13. Let Q be an orthogonal matrix, whose columns are a basis of eigenvectors of A. Then $Q^{\mathsf{T}}AQ = D$ is diagonal. We set

$$s_i = \begin{cases} 0 & \text{if } d_{ii} = 0, \\ |d_{ii}|^{-1/2} & \text{else}, \end{cases}$$

$S = \text{diag}(s_1, s_2, s_3)$ and $T = QS$.

14. In homogeneous coordinates, the conics are given by $x^{\mathsf{T}}Ax = 0$ and $x^{\mathsf{T}}Bx = 0$ with some real symmetric 3×3 matrices A and B. Let x_0 be the homogeneous coordinates of the point P. Then its polars have the form $x_0^{\mathsf{T}}Ax = 0$ and $x_0^{\mathsf{T}}Bx = 0$. These two lines coincide, if there exists a real number $\lambda \neq 0$ with $Ax_0 = \lambda Bx_0$. Such a condition is called a *generalised eigenvalue problem*. In geometrically relevant situations, both matrices A and B are invertible and thus x_0 is the solution of the standard eigenvalue problem $B^{-1}Ax_0 = \lambda x_0$.

References

Italic numbers in square brackets indicate the pages on which the corresponding reference is mentioned in the present book.

A. Abdulle and G. Wanner (2002), *200 years of least squares method*, Elem. Math. 57 (2002) 45–60. *[p. 323]*

A.C. Aitken (1964), *Determinants and Matrices*, Oliver and Boyd, Edinburgh and London 1964. *[p. 267]*

Apollonius of Perga ($\sim$ 230 B.C.), *Conics*, see Heath (1896) and Ver Eecke (1923).

Apollonius of Perga ($\sim$ 230 B.C.), *De locis planis*, see Simson (1749).

Archimedes ($\sim$ 250 B.C.), *On the sphere and cylinder*, 2 books, see Heath (1897), p. 1 and Ver Eecke (1921), I, pp. 1–124.

Archimedes ($\sim$ 250 B.C.), *Measurement of a circle*, see Heath (1897), p. 91 and Ver Eecke (1921), I, pp. 127–134.

Archimedes ($\sim$ 250 B.C.), *On conoids and spheroids*, see Heath (1897), p. 99 and Ver Eecke (1921), I, pp. 137–236.

Archimedes ($\sim$ 250 B.C.), *On spirals*, see Heath (1897), p. 151 and Ver Eecke (1921), I, pp. 237–299.

Archimedes ($\sim$ 250 B.C.), *On the equilibrium of planes*, 2 books, see Heath (1897), p. 189 and Ver Eecke (1921), I, pp. 303–350.

Archimedes ($\sim$ 250 B.C.), *Quadrature of the parabola*, see Heath (1897), p. 233 and Ver Eecke (1921), II, pp. 377–404.

J.-R. Argand (1806), *Essai sur une manière de représenter les quantités imaginaires dans les constructions géométriques*, chez Mme Vve Blanc, Paris 1806. *[p. 242]*

J.-L. Ayme (2003), *Sawayama and Thébault's theorem*, Forum Geometricorum 3 (2003) 225–229. *[p. 227]*

O. Baier (1967), *Zur Rytzschen Achsenkonstruktion*, Elem. Math. 22 (1967) 107–108. *[p. 69]*

P. Baptist (1992), *Die Entwicklung der neueren Dreiecksgeometrie*, B.I. Wissenschaftsverlag, Mannheim 1992. *[p. 93, 216, 359]*

I. Barrow (1735), *Geometrical Lectures: Explaining the Generation, Nature and Properties of Curve Lines*, London 1735. *[p. 315]*

A. Ostermann and G. Wanner, *Geometry by Its History*,
Undergraduate Texts in Mathematics, DOI: 10.1007/978-3-642-29163-0,
© Springer-Verlag Berlin Heidelberg 2012

B. Beckman (2006), *Feynman says: "Newton implies Kepler, no calculus needed"*, J. of Symbolic Geometry 1 (2006) 57–72. *[p. 147]*

E. Beltrami (1868), *Saggio di interpretazione della geometria non-euclidea*, Giornale di Matematiche 6 (1868) 284–312; French transl. (traduit de l'italien par J. Hoüel) Ann. Sci. École Norm. Sup. 6 (1869) 251–288. *[p. 214]*

J.L. Berggren (1986), *Episodes in the Mathematics of Medieval Islam*, Springer-Verlag, New York 1986. *[p. 127]*

Jac. Bernoulli (1694), *Constructio Curvæ Accessus & Recessus æquabilis, ope rectificationis Curvæ cujusdam Algebraicæ, addenda numeræ Solutioni mensis Junii*, Acta Erud. (Sept. 1694) 336–338. *[p. 207]*

Jac. Bernoulli (1695), *Explicationes, Annotationes et Additiones ad ea, quæ in Actis superiorum annorum de Curva Elastica, Isochrona Paracentrica, & Velaria, hinc inde memorata, & partim controversa leguntur; ubi de Linea mediarum directionum, aliisque novis*, Acta Erud. (Dec. 1695) 537–553; in *Opera*, vol. 1, pp. 639–663. *[p. 207]*

Jac. Bernoulli (Opera), *Jacobi Bernoulli, Basileensis, Opera*, 2 vols., Genevæ 1744.

Jac. Bernoulli (Werke), *Die Werke von Jakob Bernoulli*, 4 vols., Birkhäuser, Basel 1969–1993. *[p. 243]*

Joh. Bernoulli, *Opera Omnia*, 4 vols., Lausannæ et Genevæ 1742; reprinted Georg Olms 1968. *[p. 88, 380, 390]*

G.D. Birkhoff (1932), *A set of postulates for plane geometry, based on scale and protractor*, Annals of Mathematics 33 (1932) 329–345. *[p. 53]*

J. Bolyai (1832), *Appendix scientiam spatii absolute veram exhibens: a veritate aut falsitate Axiomatis XI Euclidei (a priori haud unquam decidenda) independentem; adjecta ad casum falsitatis quadratura circuli geometrica*, pp. 1–26; Appendix to F. Bolyai: *Tentamen juventutem studiosam in elementa matheseos purae, elementaris ac sublimioris, methodo intuitiva, evidentiaque huic propria, introducendi cum Appendici triplici*, Tomus primus, Maros Vásárhelyini 1832. *[p. 53]*

British Museum (1898), *Facsimile of the Rhind Mathematical Papyrus*, London 1898. *[p. 12]*

L.A. Brown (1949), *The Story of Maps*, Little, Brown and Company, Boston 1949; Dover reprint 1979. *[p. 137]*

R.C. Buck (1980), *Sherlock Holmes in Babylon*, Amer. Math. Monthly 87 (1980) 335–345. *[p. 347]*

M. Cantor (1894), *Vorlesungen über die Geschichte der Mathematik. Erster Band*, zweite Auflage, B.G. Teubner, Leipzig 1894. *[p. 125, 164]*

M. Cantor (1900), *Vorlesungen über die Geschichte der Mathematik. Zweiter Band*, zweite Auflage, B.G. Teubner, Leipzig 1900. *[p. 193]*

L.N.M. Carnot (1803), *Géométrie de position*, Duprat, Paris 1803. *[p. 87, 176]*

J.-C. Carrega (1981), *Théorie des corps. La règle et le compas*, Hermann, Paris 1981. *[p. 122]*

F.B. Cavalieri (1647), *Exercitationes geometricæ sex*, Bononiæ 1647. *[p. 49]*

A. Cayley (1846), *Sur quelques propriétés des déterminants gauches*, J. Reine Angew. Math. (Crelle) 32 (1846) 119–123; reprinted in Cayley (1889), vol. 1, 332–336. *[p. 305]*

A. Cayley (1858), *A memoir on the theory of matrices*, Phil. Trans. R. Soc. Lond. 148 (1858) 17–37; reprinted in Cayley (1889), vol. 2, 475–496. *[p. 291]*

A. Cayley (1889), *The Collected Mathematical Papers of Arthur Cayley*, Cambridge Univ. Press, Cambridge 1889–1897.

M. Chasles (1837), *Aperçu historique sur l'origine et le développement des méthodes en géométrie*, Paris 1837; second ed. 1875; German transl. 1839. *[p. 87, 88]*

G. Chrystal (1886), *Algebra. An Elementary Text-Book*, 2 vols., first ed. 1886; reprint of the sixth ed. Chelsea Publ. Company, New York 1952. *[p. 45]*

T. Clausen (1828), *Geometrische Sätze*, J. Reine Angew. Math. (Crelle) 3 (1828) 196–198. *[p. 109]*

N. Copernicus (1543), *De revolutionibus orbium cœlestium*, Nürnberg 1543; second impression Basel 1566. *[p. vii, viii, 142]*

H.S.M. Coxeter (1961), *Introduction to Geometry*, John Wiley & Sons, New York 1961. *[p. viii, 91, 105, 111, 339]*

H.S.M. Coxeter and S.L. Greitzer (1967), *Geometry Revisited*, Math. Assoc. of America, Washington 1967. *[p. 105, 217, 225, 226]*

G. Cramer (1750), *Introduction à l'analyse des lignes courbes algébriques*, Genève 1750. *[p. 61, 157, 267, 292]*

A.L. Crelle (1821/22), *Sammlung mathematischer Aufsätze und Bemerkungen*, 2 vols., Berlin 1821–22. *[p. 82]*

M. Crouzeix (2004), *Bounds for analytical functions of matrices*, Integr. equ. oper. theory 48 (2004) 461–477. *[p. 193]*

R. Descartes (1637), *La Geometrie*, Appendix to the *Discours de la methode*, Paris 1637; translated from the French and Latin by D.E. Smith and M.L. Latham, The Open Court Publ. Company, Chicago 1925; Dover reprint 1954. *[p. 159, 188]*

H. Dörrie (1933), *Triumph der Mathematik. Hundert berühmte Probleme aus zwei Jahrtausenden mathematischer Kultur*, Verlag von Ferdinand Hirt, Breslau 1933; English translation (by D. Antin) *100 Great Problems of Elementary Mathematics: Their History and Solution*, Dover Publ. 1965. *[p. 193]*

H. Dörrie (1943), *Mathematische Miniaturen*, Wiesbaden 1943; reprinted 1979. *[p. 55, 225, 273, 314]*

U. Dudley (1987), *A Budget of Trisections*, Springer-Verlag, New York 1987. *[p. 253]*

J.D. Dunitz and J. Waser (1972), *The planarity of the equilateral, isogonal pentagon*, Elem. Math. 27 (1972) 25–32. *[p. 280, 299]*

A. Dürer (1525), *Underweysung der messung*, Nürnberg 1525. *[p. 26, 69, 112, 290]*

A. Eisenlohr (1877), *Ein mathematisches Handbuch der alten Ägypter, übersetzt und erklärt*, Leipzig 1877; second ed. (without plates) 1891. *[p. 12]*

J.-P. Ehrmann (2004), *Steiner's theorems on the complete quadrilateral*, Forum Geometricorum 4 (2004) 35–52. *[p. 220]*

D. Elliott (1968), *M.L. Urquhart*, J. Austr. Math. Soc. 8 (1968) 129–133. *[p. 232]*

P. Erdős, L.J. Mordell and D.F. Barrow (1937), *Problem 3740 and solution*, Amer. Math. Monthly 44 (1937) 252–254. *[p. 222]*

Euclid (∼ 300 B.C.), *The Elements*, hundreds of editions, commentaries, translations during the centuries, first scientific book printed (in 1482), Clavius' edition 1574, definitive Greek text by Heiberg 1883–1888, English transl. see Heath (1926).

L. Euler (1735), *Solutio problematis astronomici ex datis tribus stellae fixae altitudinibus et temporum differentiis invenire elevationem poli et declinationem stellae* [E14], Commentarii Academiae Scientiarum Imperialis Petropolitanae 4, ad annum 1729 (1735) 98–101; reprinted in *Opera Omnia*, series secunda, vol. 30, 1–4, Orell Füssli, Zürich 1964. *[p. 116]*

L. Euler (1748), *Introductio in Analysin Infinitorum* [E101, E102], Lausannae 1748; reprinted in *Opera Omnia*, series prima, vol. 8, Teubner, Leipzig and Berlin 1922; vol. 9, Orell Füssli, Zürich and Leipzig, Teubner, Leipzig and Berlin 1945. *[p. 73, 183, 185, 189, 192, 208, 246, 309, 363, 375]*

L. Euler (1750), *Variae demonstrationes geometriae* [E135], Novi Commentarii Academiae Scientiarum Petropolitanae 1, 1747/8 (1750) 49–66; reprinted in *Opera Omnia*, series prima, vol. 26, 15–32, Orell Füssli, Zürich 1953. *[p. 170, 173, 174, 233]*

L. Euler (1751), *Recherches sur les racines imaginaires des équations* [E170], Histoire de l'Académie Royale des Sciences et Belles-Lettres 5, Berlin 1749 (1751) 222–288; reprinted in *Opera Omnia*, series prima, vol. 6, 78–147, Teubner, Leipzig and Berlin 1921. *[p. 385]*

L. Euler (1753), *Solutio problematis geometrici* [E192], Novi Commentarii Academiae Scientiarum Petropolitanae 3, 1750/1 (1753) 224–234; reprinted in *Opera Omnia*, series prima, vol. 26, 60–70, Orell Füssli, Zürich 1953. *[p. 69, 78]*

L. Euler (1755), *Principes de la trigonométrie sphérique tirés de la méthode des plus grands et plus petits* [E214], Histoire de l'Académie Royale des Sciences et Belles-Lettres 9, Berlin 1753 (1755) 233–257; reprinted in *Opera Omnia*, series prima, vol. 27, 277–308, Orell Füssli, Zürich 1954. *[p. 116]*

L. Euler (1758), *Elementa doctrinae solidorum* [E230], Novi Commentarii Academiae Scientiarum Petropolitanae 4, 1752/3 (1758) 109–140; reprinted in *Opera Omnia*, series prima, vol. 26, 71–93, Orell Füssli, Zürich 1953. *[p. 59]*

L. Euler (1760), *Demonstratio theorematis Fermatiani omnem numerum sive integrum sive fractum esse summam quatuor pauciorumve quadratorum* [E242], Novi Commentarii Academiae Scientiarum Petropolitanae 5, 1754/55 (1760) 13–58; reprinted in *Opera Omnia*, series prima, vol. 2, 338–372, Teubner, Leipzig and Berlin 1915. *[p. 261]*

L. Euler (1761), *Theoremata circa residua ex divisione potestatum relicta* [E262], Novi Commentarii Academiae Scientiarum Petropolitanae 7, 1758/1759 (1761) 49–82; reprinted in *Opera Omnia*, series prima, vol. 2, 493–518, Teubner, Leipzig and Berlin 1915. *[p. 248]*

L. Euler (1765), *Du mouvement de rotation des corps solides autour d'un axe variable* [E292], Histoire de l'Académie Royale des Sciences et Belles-Lettres 14, Berlin 1758 (1765) 154–193; reprinted in *Opera Omnia*, series secunda, vol. 8, 200–235, Orell Füssli, Zürich 1965. *[p. 300]*

L. Euler (1767a), *Solutio facilis problematum quorundam geometricorum difficillimorum* [E325], Novi Commentarii Academiae Scientiarum Petropolitanae 11, 1765 (1767) 103–123; reprinted in *Opera Omnia*, series prima, vol. 26, 139–157, Orell Füssli, Zürich 1953. *[p. 91, 214]*

L. Euler (1767b), *Recherche sur la courbure des surfaces* [E333], Histoire de l'Académie Royale des Sciences et Belles-Lettres 16, Berlin 1760 (1767) 119–143; reprinted in *Opera Omnia*, series prima, vol. 28, 1–22, Orell Füssli, Zürich 1955. *[p. 213]*

L. Euler (1774), *Demonstrationes circa residua ex divisione potestatum per numeros primos resultantia* [E449], Novi Commentarii Academiae Scientiarum Petropolitanae 18, 1773 (1774) 85–135; reprinted in *Opera Omnia*, series prima, vol. 3, 240–281, Teubner, Leipzig and Berlin 1917. *[p. 248]*

L. Euler (1781), *De mesura angulorum solidorum* [E514], Acta Academiae Scientiarum Imperialis Petropolitanae 1778, pars posterior (1781) 31–54; reprinted in *Opera Omnia*, series prima, vol. 26, 204–223, Orell Füssli, Zürich 1953. *[p. 130, 139]*

L. Euler (1782), *Trigonometria sphaerica universa, ex primis principiis breviter et dilucide derivata* [E524], Acta Academiae Scientiarum Imperialis Petropolitanae 1779, pars prior (1782) 72–86; reprinted in *Opera Omnia*, series prima, vol. 26, 224–236, Orell Füssli, Zürich 1953. *[p. 127, 133]*

L. Euler (1783), *Quomodo sinus et cosinus angulorum multiplorum per producta exprimi queant* [E562], Opuscula Analytica 1 (1783) 353–363; reprinted in *Opera Omnia*, series prima, vol. 15, 509–521, Teubner, Leipzig and Berlin 1927. *[p. 152, 363]*

L. Euler (1786), *De symptomatibus quatuor punctorum, in eodem plano sitorum* [E601], Acta Academiae Scientiarum Petropolitanae 1782, pars prior (1786) 3–18; reprinted in *Opera Omnia*, series prima, vol. 26, 258–269, Orell Füssli, Zürich 1953. *[p. 298, 313]*

L. Euler (1790), *Solutio facilis problematis, quo quaeritur circulus, qui datos tres circulos tangat* [E648], Nova Acta Academiae Scientiarum Imperialis Petropolitanae 6, 1788 (1790) 95–101; reprinted in *Opera Omnia*, series prima, vol. 26, 270–275, Orell Füssli, Zürich 1953. *[p. 233]*

L. Euler (1815), *Geometrica et sphaerica quaedam* [E749], Mémoires de l'Académie Impériale des Sciences de St.-Pétersbourg 5, 1812 (1815) 96–114; reprinted in *Opera Omnia*, series prima, vol. 26, 344–358, Orell Füssli, Zürich 1953. *[p. 180, 370]*

L. Euler (Opera Omnia), *Leonhardi Euleri Opera Omnia*, series I–IV, 76 volumes published, several volumes in preparation.

Giul. C. Fagnano (Giulio Carlo Fagnano dei Toschi) (1750), *Produzioni Matematiche*, 2 vols., Pesaro 1750; reprinted 1912 in *Opere Matematiche*, Volterra, Loria, Gambioli ed. *[p. 82, 90, 239, 382]*

Giov. F. Fagnano (Giovanni Francesco Fagnano dei Toschi, son of Giulio) (1770), *Sopra le proprietà antiche e nuove dei triangoli*, inedito. *[p. 86]*

Giov. F. Fagnano (Giovanni Francesco Fagnano dei Toschi, son of Giulio) (1779), *Problemata quaedam ad methodum maximorum et minimorum spectantia*, Nova Acta Erud. anni 1775, Lipsiae 1779, pp. 281–303. *[p. 86, 197, 198, 236]*

P. Fermat (1629a), *Apollonii Pergæi libri duo de locis planis restituti*, manuscript from 1629, published posthumously in Fermat, *Opera*, pp. 12–27. *[p. 6]*

P. Fermat (1629b), *Methodus ad disquirendam maximam & minimam*, first manuscript from 1629, sent 1638 to Mersenne, published in *Opera*, pp. 63–73. *[p. 200]*

P. Fermat (1629c), *Porismatum Euclidæorum Renovata Doctrina, & sub formâ Isagoges recentioribus Geometris exhibita*, manuscript from 1629, published posthumely in *Opera*, pp. 116–119. *[p. 180]*

P. Fermat (Opera), *Varia opera mathematica D. Petri de Fermat, Senatoris Tolosani*, apud J. Pech, Collegium PP. Societatis JESU, Tolosæ 1679.

P. Fermat (Oeuvres), *Œuvres de Fermat*, 4 vols. + 1 suppl. (vol. 1 contains original Latin texts, vol. 3 French translations, vol. 2 and 4 letters), edited by P. Tannery and C. Henry, Gauthier-Villars et fils, Paris 1891–1912. *[p. 180, 194, 195, 197, 234, 236]*

K.W. Feuerbach (1822), *Eigenschaften einiger merkwürdigen Punkte des geradlinigen Dreiecks und mehrerer durch sie bestimmten Linien und Figuren. Eine analytisch-trigonometrische Abhandlung*, Nürnberg 1822. *[p. 216]*

R.P. Feynman, R.B. Leighton, M. Sands (1964), *The Feynman Lectures on Physics*, vol. 1, Addison-Wesley, Reading, Mass. 1964. *[p. 141]*

R.P. Feynman, D.L. Goodstein and J.R. Goodstein (1996), *The Motion of Planets Around the Sun*, W.W. Norton Comp., New York 1996; French transl. Diderot 1997; including an audio CD. *[p. 147]*

J.V. Field (1996), *Rediscovering the Archimedean polyhedra: Piero della Francesca, Luca Pacioli, Leonardo da Vinci, Albrecht Dürer, Daniele Barbaro, and Johannes Kepler*, Arch. Hist. Exact Sci. 50 (1996) 241–289. *[p. 282]*

P. Finsler (1937/38), *Einige elementargeometrische Näherungskonstruktionen*, Comm. Math. Helvetici, 10 (1937/38) 243–262. *[p. 255, 256]*

W.A. Förstemann (1835), *Umkehrung des Ptolomäischen Satzes*, J. Reine Angew. Math. (Crelle) 13 (1835) 233–236. *[p. 150]*

A.-F. Frézier (1737), *La théorie et la pratique de la coupe des pierres et des bois, pour la construction des voûtes et autre parties des bâtimens civils & militaires, ou traité de stéréotomie à l'usage de l'architecture*, vol. 1, Guerin, Paris 1737. *[p. 69]*

G. Galilei (1638), *Discorsi e dimostrazioni matematiche, intorno à due nuove Scienze, Attenti alla mecanica & i movimenti locali, del Signor Galileo Galilei Linceo*, published "in Leida 1638"; critical edition by Enrico Giusti, Giulio Einaudi Editore, Torino 1990; German translation Arthur von Oettingen, Ostwald's Klassiker, Leipzig 1890/91. *[p. 141, 143]*

C.F. Gauss (1799), *Demonstratio nova theorematis omnem functionem algebraicam rationalem integram unius variabilis in factores reales primi vel secundi gradus resolvi posse*, PhD thesis, C.G Fleckeisen, Helmstedt 1799. *[p. 242]*

C.F. Gauss (1809), *Theoria motus corporum coelestium in sectionibus conicis solem ambientium*, F. Perthes and I.M. Besser, Hamburg 1809; *Werke*, vol. 7, pp. 1–288. *[p. 266]*

C.F. Gauss (1828), *Disquisitiones generales circa superficies curvas*, Comm. Soc. Reg. Sci. Gotting. Rec. 6 (1828), pp. 99–146; *Werke*, vol. 4, pp. 217–258. *[p. 213]*

C.F. Gauss (Werke), *Werke*, 12 vols., Königl. Gesell. der Wiss., Göttingen 1863–1929; reprinted by Georg Olms Verlag, 1973–1981. *[p. 43, 86, 177, 233, 241, 348]*

J.W. Gibbs and E.B. Wilson (1901), *Vector Analysis. A text-book for the use of students of mathematics and physics*, founded upon the lectures of J.W. Gibbs by E.B. Wilson, Yale University Press, New Haven 1901. *[p. 261]*

J. Gray (2007), *Worlds Out of Nothing*, Springer-Verlag, London 2007. *[p. 53]*

P. Griffiths and J. Harris (1978), *On Cayley's explicit solution to Poncelet's porism*, L'Enseignement Mathématique 24 (1978) 31–40. *[p. 328]*

B. Grünbaum (2009) *An enduring error*, Elem. Math. 64 (2009) 89–101. *[p. 287]*

B. Grünbaum and G.C. Shephard (1993), *Pick's theorem*, Amer. Math. Monthly 100 (1993) 150–161. *[p. 277]*

A. Gut, J. Waldvogel (2008), *The feet of the altitudes of a simplex*, Elem. Math. 63 (2008) 25–29 *[p. 336]*

M. de Guzmán (2001), *The envelope of the Wallace–Simson lines of a triangle. A simple proof of the Steiner theorem on the deltoid*, Rev. R. Acad. Cien. Serie A. Mat. 95 (2001), 57–64. *[p. 220]*

E. Hairer, C. Lubich and G. Wanner (2006), *Geometric Numerical Integration*, Springer-Verlag, Berlin 2002; second ed. 2006. *[p. 144]*

E. Hairer and G. Wanner (1997), *Analysis by Its History*, second printing, Springer-Verlag, New York 1997. *[p. 44, 73, 157, 195, 202, 203, 204, 211, 234, 246, 253, 365, 380, 381]*

G.H. Hardy, P.V. Seshu Aiyar, B.M. Wilson (1927), *Collected papers of Srinivasa Ramanujan*, Cambridge Univ. Press, Cambridge 1927; reprinted by Chelsea Publ. 1962.

R. Hartshorne (2000), *Geometry: Euclid and Beyond*, Springer-Verlag, New York 2000. *[p. 24, 53, 54]*

T.L. Heath (1896), *Apollonius of Perga: Treatise on Conic Sections, edited in modern notation with introductions including an essay on the earlier history of the subject*, Cambridge: at the University Press, Cambridge 1896. *[p. 61]*

T.L. Heath (1897), *The Works of Archimedes, edited in modern notation with introductory chapters*, Cambridge: at the University Press, Cambridge 1897; Dover reprint 2002. *[p. 20, 81, 84]*

T.L. Heath (1920), *Euclid in Greek*, Cambridge University Press, Cambridge 1920. *[p. 14]*

T.L. Heath (1921), *A History of Greek Mathematics*, 2 vols., Clarendon Press, Oxford 1921; Dover reprint 1981. *[p. 4, 17, 20, 79, 161, 162, 284, 333, 346, 398]*

T.L. Heath (1926), *The Thirteen Books of Euclid's Elements, translated from the text of Heiberg, with introduction and commentary*, second ed., 3 vols., Cambridge University Press, Cambridge 1926; Dover reprint 1956. *[p. 3, 16, 21, 27, 28, 29, 30, 33, 36, 42, 54, 173, 280, 347, 352]*

P.P.A. Henry (2009), *La solution de François Viète au problème d'Adriaan van Roomen*, manuscript 2009. *[p. 150, 169]*

J.G. Hermes (1895), *Ueber die Teilung des Kreises in 65537 gleiche Teile*, Nachrichten von der Königl. Gesellschaft der Wissenschaften zu Göttingen, Mathematisch-physikalische Klasse aus dem Jahre 1894. Göttingen (1895) 170–186. *[p. 251]*

O. Hesse (1861), *Vorlesungen über analytische Geometrie des Raumes, insbesondere über Oberflächen zweiter Ordnung*, Teubner Leipzig 1861. *[p. 270]*

O. Hesse (1865), *Vorlesungen aus der analytischen Geometrie der geraden Linie, des Punktes und der Kreise in der Ebene*, Teubner Leipzig 1865. *[p. 270]*

D. Hilbert (1899), *Grundlagen der Geometrie*, Teubner, Leipzig 1899. *[p. 52, 53, 397]*

D. Hilbert and S. Cohn-Vossen (1932), *Anschauliche Geometrie*, Springer-Verlag, Berlin 1932; Engl. translation *Geometry and the Imagination* by P. Nemenyi, Chelsea Publ. Company, New York 1952. *[p. 53]*

E.W. Hobson (1891), *A Treatise on Plane Trigonometry*, Cambridge: at the University Press, Cambridge 1891, third enlarged edition 1911, seventh edition 1928; Dover reprint *(A Treatise on Plane and Advanced Trigonometry)*, New York 1957. *[p. 117, 118, 119, 382]*

J.E. Hofmann (1956), *Über Jakob Bernoullis Beiträge zur Infinitesimalmathematik*, L'Enseignement Mathématique 2 (1956) 61–171. *[p. 207, 237]*

K. Hofstetter (2005), *Division of a segment in the golden section with ruler and rusty compass*, Forum Geometricorum 5 (2005) 135–136. *[p. 54]*

H. Hunziker (2001), *Albert Einstein, Maturitätsprüfung in Mathematik 1896*, Elem. Math. 56 (2001) 45–54. *[p. 179, 234]*

A. Hurwitz and R. Courant (1922), *Funktionentheorie*, Grundlehren der Math. Wiss. 3, Springer, Berlin 1922. *[p. 342]*

C. Huygens (1673), *Horologium oscillatorium sive de motu pendulorum ad horologia aptato demonstrationes geometricæ*, F. Muguet, Paris 1673; reprinted in *Oeuvres*, vol. 18, pp. 69–368. *[p. 202]*

C. Huygens (1691), *Appendice IV à la pars tertia de l'horologium oscillatorium*, in *Oeuvres*, vol. 18, pp. 406–409. *[p. 237]*

C. Huygens (1692), *Correspondance N° 2794*, in *Oeuvres*, vol. 10, pp. 418–422. *[p. 211, 212, 213]*

C. Huygens (1692/93), *Appendice VI à la pars quarta de l'horologium oscillatorium*, in *Oeuvres*, vol. 18, pp. 433–436. *[p. 203]*

C. Huygens (1724), (Christiani Hugenii Zulichemii) *Opera varia*, 2 vols., Lugduni Batavorum (= Leiden) MDCCXXIV. *[p. 254]*

C. Huygens (1833), (Christiani Hugenii) *Exercitationes mathematicæ et philosophicæ*, ex manuscriptis in Bibliotheca Lugduno-Batavæ (= Leiden) MDCCCXXXIII. *[p. 380]*

C. Huygens (Oeuvres), *Œuvres complètes de Christiaan Huygens*, publ. par la Société Hollandaise des Sciences, 22 vols., Den Haag 1888–1950.

H. Irminger (1970), *Zu einem Satz über räumliche Fünfecke*, Elem. Math. 25 (1970) 135–136. *[p. 280, 282]*

J. Kepler (1604), *Ad Vitellionem paralipomena, quibus Astronomiae pars optica traditur, potissimum de artificiosa observatione et aestimatione diametrorum deliquiorumque Solis & Lunae, cum exemplis insignium eclipsium*. Francofurti 1604; reprinted in *Gesammelte Werke*, vol. 2. *[p. 61]*

J. Kepler (1609), *Astronomia Nova ΑΙΤΙΟΛΟΓΗΤΟΣ, seu Physica Coelestis, tradita commentariis De Motibus Stellæ Martis, Ex observationibus G. V. Tychonis Brahe*, Jussu & sumptibus Rudolphi II. Romanorum Imperatoris &c; Plurium annorum pertinaci studio elaborata Pragæ, A S^æ. C^æ. M^{tis}. S^æ. Mathematico Joanne Keplero, Cum ejusdem C^æ. M^{tis}. privilegio speciali Anno æræ Dionysianæ MDCIX; reprinted in *Gesammelte Werke*, vol. 3; French transl. by Jean Peyroux "chez le traducteur" 1979; English transl. by W. Donahue, Cambridge 1989. *[p. 61, 141]*

J. Kepler (1619), *Harmonices Mundi*, Lincii Austriæ 1619; reprinted in *Gesammelte Werke*, vol. 6. *[p. 70, 142, 162, 168, 282]*

J. Kepler (Gesammelte Werke), *Johannes Kepler Gesammelte Werke*, 21 vols., Beck'sche Verlagsbuchhandlung, München 1938–2002.

C. Kimberling (1994), *Central points and central lines in the plane of a triangle*, Math. Mag. 67 (1994) 163–187. *[p. 94]*

C. Kimberling (1998), *Triangle Centers and Central Triangles*, vol. 129 of *Congressus Numerantium*, Utilitas Mathematica Publ. Inc., Winnipeg 1998 *[p. 94]*

F. Klein (1872), *Vergleichende Betrachtungen über neuere geometrische Forschungen* (so-called Erlanger Programm), Verlag von A. Deichert Erlangen, 1872; *Werke I*, p. 460; Engl. transl. M.W. Haskell, Bull. New York Math. Soc. 2, (1892–1893), 215–249. *[p. 319, 338]*

F. Klein (1926), *Vorlesungen über die Entwicklung der Mathematik im 19. Jahrhundert*, 2 vols., Springer-Verlag, Berlin 1926–1927. *[p. 53, 176, 291]*

F. Klein (1928), *Vorlesungen über Nicht-Euklidische Geometrie*, Springer-Verlag, Berlin 1928. *[p. 330, 338]*

M. Kline (1972), *Mathematical Thought from Ancient to Modern Times*, Oxford Univ. Press, New York 1972. *[p. 61, 291]*

A.A. Kochański (1685), *Observationes Cyclometricæ ad facilitandam Praxin accomodatæ*, Acta Erud. 4 (1685) 394–398. *[p. 256]*

N. Kritikos (1961), *Vektorieller Beweis eines elementargeometrischen Satzes*, Elem. Math. 16 (1961) 132–134. *[p. 254]*

E.D. Kulanin and O. Faynshteyn (2007), *Victor Michel Jean-Marie Thébault zum 125. Geburtstag am 6. März 2007*, Elem. Math. 62 (2007) 45–58. *[p. 227]*

J.D. Lawrence (1972), *A Catalog of Special Plane Curves*, Dover Publications, New York 1972. *[p. 200]*

A.-M. Legendre (1794), *Éléments de Géométrie*, first ed. 1794; 43th ed., Firmin-Didot, Paris 1925. *[p. 11, 52]*

G.W. Leibniz (1693), *Supplementum geometriæ dimensoriæ, seu generalissima omnium tetragonismorum effectio per motum: similiterque multiplex constructio linea ex data tangentium conditione*, Acta Erud. (1693) 385–392; Corrigenda p. 527. *[p. 211]*

S.A.J. Lhuilier (1810/11), *Théorèmes sur les triangles, relatifs à la page 64 de ces Annales*, Annales de Mathématiques (Gergonne) 1 (1810/11) 149–159. *[p. 93, 173, 215]*

N.I. Lobachevsky (1829/30), О началахъ геометріи *[On the foundations of geometry]*, Kazan Messenger 25 (1829) 178–187, 228–241; 27 (1829) 227–243; 28 (1830) 251–283, 571–636; first publications in Germany: *Géométrie imaginaire*, J. Reine Angew. Math. (Crelle) 17 (1837) 295–320 and *Geometrische Untersuchungen zur Theorie der Parallel-Linien von Nicolaus Lobatschefsky*, Berlin 1840. *[p. 53]*

E.S. Loomis (1940), *The Pythagorean Proposition*, second ed. 1940, reprinted by The National Council of Teachers of Mathematics, Washington 1968, second printing 1972. *[p. 18]*

G. Loria (1910/11), *Spezielle algebraische und transzendente ebene Kurven. Theorie und Geschichte*, 2 vols., second ed., B. G. Teubner Verlag, Leipzig und Berlin, 1910/11. *[p. 200, 207]*

G. Loria (1939), *Triangles équilatéraux dérivés d'un triangle quelconque*, Math. Gazette 23 (1939) 364–372. *[p. 105]*

Z. Lu (2008), *Erdős–Mordell-type inequalities*, Elem. Math. 63 (2008) 23–24. *[p. 224]*

C. Maclaurin (1748), *A Treatise of Algebra in three parts. Containing I. The Fundamental Rules and Operations. II. The Composition and Resolution of Equations of all Degrees; and the different Affections of their Roots. III. The Application of Algebra and Geometry to each other*, London MDCCXLVIII. *[p. 267]*

J.B.M. Meusnier (1785), *Mémoire sur la courbure des surfaces*, Mémoires de Mathématique et de Physique, Acad. Royale des Sciences, Paris, vol. 10, M.DCC.LXXXV, pp. 477–510, Pl. IX and X. *[p. 213]*

G. Miel (1983), *On calculations past and present; the Archimedean algorithm*, Amer. Math. Monthly 90 (1983) 17–35. *[p. 353]*

J.W. Milnor (1982), *Hyperbolic geometry: the first 150 years*, Bull. Amer. Math. Soc. 6 (1982) 9–24. *[p. 53]*

F. Minding (1839), *Wie sich entscheiden läßt, ob zwei gegebene krumme Flächen auf einander abwickelbar sind oder nicht; nebst Bemerkungen über die Flächen von unveränderlichem Krümmungsmaaße*, J. Reine Angew. Math. (Crelle) 19 (1839) 370–387. *[p. 214]*

A. Miquel (1838a), *Théorèmes de géométrie*, J. math. pures et appl. 3 (1838) 485–487. *[p. 8, 95]*

A. Miquel (1838b), *Théorèmes sur les intersections des cercles et des sphères*, J. math. pures et appl. 3 (1838) 517–522. *[p. 95, 126, 352]*

R. Müller (1905), *Über die Dreiecke, deren Umkreis den Kreis der 9 Punkte ortho-
    gonal schneidet*, Zeitschrift für mathematischen und naturwissenschaftlichen
    Unterricht, ein Organ für Methodik, Bildungsgehalt und Organisation der
    exakten Unterrichtsfächer an den höheren Schulen, Lehrerseminaren und
    gehobenen Bürgerschulen, 36 (1905) 182–184. *[p. 238]*

R.B. Nelsen (2004), *Proof without words: four squares with constant area*, Math.
    Mag. 77 (2004) 135. *[p. 26]*

I. Newton (1668), *Analysis of the properties of cubic curves and their classification
    by species*, manuscript (1667 or 1668), in *Mathematical Papers*, vol. II, pp. 10–
    89. *[p. 185]*

I. Newton (1671), *A Treatise of the Methods of Series and Fluxions*, manuscript
    of 1671, in *Mathematical Papers*, vol. III, p. 32; first published by J. Colson
    as *The Method of Fluxions and Infinite Series with its Application to the
    Geometry of Curve-lines*, "translated from the author's Latin original not
    yet made publick", London 1736; French translation "par M. de Buffon",
    Paris 1740. *[p. vii, 210]*

I. Newton (1680), *The geometry of curved lines*, manuscript, 1680, in *Mathematical
    Papers*, vol. IV, pp. 420–505. *[p. 108]*

I. Newton (1684), *De motu corporum in gyrum (On the motion of bodies in an
    orbit)*, manuscript (autumn 1684), in *Mathematical Papers*, vol. VI, pp. 30–
    91. *[p. 368]*

I. Newton (1687), *Philosophiae Naturalis Principia Mathematica*, Londini anno MD-
    CLXXXVII; 2nd edition 1713; first English edition by A. Motte, London
    1729. *[p. 141]*

I. Newton (Mathematical Papers), *The Mathematical Papers of Isaac Newton*, 8
    vols., edited by D.T. Whiteside, Cambridge Univ. Press, Cambridge 1967–
    1981. *[p. 86, 127]*

C.O. Oakley and J.C. Baker (1978), *The Morley trisector theorem*, Amer. Math.
    Monthly 85 (1978) 737–745. *[p. 105]*

B. Odehnal (2006), *Three points related to the incenter and excenters of a triangle*,
    Elem. Math. 61 (2006) 74–80. *[p. 110]*

A. Oppenheim (1961), *The Erdős inequality and other inequalities for a triangle*,
    Amer. Math. Monthly 68 (1961) 226–230 and 349. *[p. 222]*

A. Ostermann and G. Wanner (2010), *A dynamic proof of Thébault's theorem*, Elem.
    Math. 65 (2010) 12–16. *[p. 227]*

Pappus of Alexandria (∼ 300 A.D.), *Collection* (or συναγωγή, *Synagoge*), first Latin
    translation by F. Commandino (published in 1588 and 1660); definite Greek-
    Latin edition by F. Hultsch, Berolini 1876–1878; French translation: Ver Eecke
    (1933). *[p. 58, 62, 69, 78, 79, 80, 90, 102, 106, 121, 161, 176, 178, 188, 223,
    232, 325, 332, 342, 398]*

T.E. Peet (1923), *The Rhind Mathematical Papyrus*, British Museum 10057 and
    10058, Liverpool 1923. *[p. 1, 12, 19]*

R. Penrose (2005), *The Road to Reality; a Complete Guide to the Laws of the Uni-
    verse*, Alfred A. Knopf, New York 2005. *[p. 16]*

G. Pick (1899), *Geometrisches zur Zahlenlehre*, Sitzungsberichte des deutschen na-
    turwissenschaftlich-medicinischen Vereins für Böhmen 'Lotos' in Prag, Neue
    Folge 19 (1899), 311–319. *[p. 277]*

L. Pisano (Leonardo da Pisa; Fibonacci) (1202), *Liber Abaci*, Codice Magliabechi-
    ano, Badia Fiorentina, publ. Roma 1857; Engl. translation *Fibonacci's Liber
    Abaci* by L.E. Sigler, Springer-Verlag, New York 2002. *[p. 180]*

L. Pisano (Leonardo da Pisa; Fibonacci) (1220), *Practica Geometriae*, Codice Urbinate, Bibl. Vaticana, publ. Roma 1862. *[p. 17, 171]*

J. Plücker (1830a), *Über ein neues Coordinatensystem*, J. Reine Angew. Math. (Crelle) 5 (1830) 1–36. *[p. 331]*

J. Plücker (1830b), *Über eine neue Art, in der analytischen Geometrie Puncte und Curven durch Gleichungen darzustellen*, J. Reine Angew. Math. (Crelle) 6 (1830) 107–146. *[p. 331, 336, 339]*

J.-V. Poncelet (1817/18), *Géométrie des courbes. Théorèmes nouveaux sur les lignes du second ordre*, Annales de mathématiques (Gergonne) 8 (1817/18) 1–13. *[p. 78]*

J.-V. Poncelet (1822), *Traité des propriétés projectives des figures*, Bachelier, Paris 1822. *[p. 3, 215, 329]*

J.-V. Poncelet (1862), *Applications d'analyse et de géométrie, qui ont servi, en 1822, de principal fondement au traité des propriétés projectives des figures*, Mallet-Bachelier (vol. 1) and Gauthier-Villars (vol. 2), Paris 1862–1864.

J.-C. Pont (1974), *La topologie algébrique, des origines à Poincaré*, Presses Univ. de France, Paris 1974. *[p. 353]*

J.-C. Pont (1986), *L'aventure des parallèles, histoire de la géométrie non euclidienne, précurseurs et attardés*, Peter Lang, Bern 1986. *[p. 52, 53]*

Ptolemy (∼ 150), see Regiomontanus.

L. Puissant (1801), *Recueil de diverses propositions de géométrie, résolues ou démontrées par l'analyse algébrique, suivant les principes de Monge et de Lacroix*, Paris 1801. *[p. 183]*

S. Ramanujan (1913), *Squaring the circle*, J. Indian Math. Soc. 5 (1913) 132; Collected Papers p. 22. *[p. 257]*

S. Ramanujan (1914), *Modular equations and approximations to π*, Quart. J. Math. 45 (1914) 350–372; Collected Papers pp. 23–39. *[p. 257]*

S. Ramanujan (1957), *Notebooks of Srinivasa Ramanujan*, facsimile edition in 2 vols., Bombay 1957; critical edition in 5 vols. by B.C. Berndt, Springer 1985 (see vol. IV, p. 8). *[p. 179]*

S. Ramanujan (Collected Papers), see Hardy, G.H.

J. Ratcliff (1994), *Foundations of Hyperbolic Manifolds*, Springer-Verlag 1994, second ed. 2006. *[p. 53]*

J. Regiomontanus = Johannes Müller from Königsberg (1464), *De triangulis omnimodis libri quinque*, written 1464, printed 1533. *[p. 116]*

J. Regiomontanus (1496), *Epitoma in Almagestum Ptolemaei*, Latin (commented) translation by G. Peu[e]rbach & J. Regiomontanus, Venice 1496. *[p. 113]*

F.J. Richelot (1832) *De resolutione algebraica aequationis $X^{257} = 1$, sive de divisione circuli per bisectionem anguli septies repetitam in partes 257 inter se aequales commentatio coronata*, J. Reine Angew. Math. (Crelle) 9 (1832) 1–26, 146–161, 209–230, 337–358. *[p. 251]*

E. Sandifer (2006), *How Euler did it; 19th century triangle geometry*, MAA Online, www.maa.org/news/howeulerdidit.html, May 2006. *[p. 370]*

R.A. Satnoianu (2003), *Erdős–Mordell-type inequalities in a triangle*, Amer. Math. Monthly 110 (2003) 727–729. *[p. 222]*

K. Schellbach(1853) *Eine Lösung der Malfattischen Aufgabe*, J. Reine Angew. Math. (Crelle) 45 (1853) 91–92. *[p. 122]*

F. van Schooten (1646), *Francisci Vietæ Opera Mathematica*, Leiden 1646; reprinted with a preface by J.E. Hofmann, Georg Olms Verlag, Hildesheim 2001.

F. van Schooten (1657), *Exercitationum mathematicorum liber primus. Continens propositionum arithmeticarum et geometricarum centuriam*, Academia Lugduno-Batava 1657. *[p. 75]*

F. van Schooten (1683), *Trigonometria triangulorum planorum cum sinuum, tangetium et secantium canone accuratissimo*, Bruxellis 1683. *[p. 152]*

P. Schreiber, G. Fischer and M.L. Sternath (2008), *New light on the rediscovery of the Archimedean solids during the Renaissance*, Arch. Hist. Exact Sci. 62 (2008), 457–467. *[p. 282, 290]*

F.-J. Servois (1813/14), *Géométrie pratique*, Annales de mathématiques (Gergonne) 4 (1813/14) 250–253. *[p. 217]*

R. Shail (2001), *A proof of Thébault's theorem*, Amer. Math. Monthly 108 (2001) 319–325. *[p. 227]*

R. Simson (1749), *Apollonii Pergaei Locorum planorum Libri II. Restituti a Roberto Simson M.D.*, Glasguae MDCCXLIX. *[p. 6, 90]*

S. Šmakal (1972), *Eine Bemerkung zu einem Satz über räumliche Fünfecke*, Elem. Math. 27 (1972) 62–63. *[p. 280]*

D.E. Smith and M.L. Latham (1925), *The Geometry of René Descartes* with a facsimile of the first edition, The Open Court Publ. Company, Chicago 1925; Dover reprint 1954. *[p. 159]*

R. Stärk (1989), *Eine weitere Lösung der Thébault'schen Aufgabe*, Elem. Math. 44 (1989) 130–133. *[p. 227]*

G.K.C. von Staudt (1847), *Geometrie der Lage*, Nürnberg 1847. *[p. 353]*

A. Steiner and G. Arrigo (2008), *Passeggiate matematiche; A proposito di ...*, Bollettino dei docenti di matematica (Ticino) 57 (2008) 93–98. *[p. 256]*

A. Steiner and G. Arrigo (2010), *Passeggiate matematiche; Pane e ... trigonometria*, Bollettino dei docenti di matematica (Ticino) 60 (2010) 107–109. *[p. 352]*

J. Steiner (1826a), *Allgemeine Theorie über das Berühren und Schneiden der Kreise und Kugeln*, manuscript from 1823–1826, published posthumely by R. Fueter and F. Gonseth, Zürich and Leipzig 1931. *[p. 6, 41, 98, 233]*

J. Steiner (1826b), *Einige geometrische Sätze*, J. Reine Angew. Math. (Crelle) 1, (1826) 38–52; see Weierstrass (1881/82), vol. 1, pp. 1–16. *[p. 221]*

J. Steiner (1826c), *Einige geometrische Betrachtungen*, J. Reine Angew. Math. (Crelle) 1 (1826) 161–184, *Fortsetzung* 252–288; see Weierstrass (1881/82), vol. 1, pp. 17–76. *[p. 8, 98, 102, 104, 122, 183]*

J. Steiner (1826d), *Leichter Beweis eines stereometrischen Satzes von Euler, nebst einem Zusatze zu Satz X auf Seite 12*, J. Reine Angew. Math. (Crelle) 1 (1826) 364-367; see Weierstrass (1881/82), vol. 1, pp. 95–100. *[p. 353]*

J. Steiner (1827), *Aufgaben und Lehrsätze, erstere aufzulösen, letztere zu beweisen*, J. Reine Angew. Math. (Crelle) 2 (1827) 286–292; see Weierstrass (1881/82), vol. 1, pp. 155–162. *[p. 377]*

J. Steiner (1827/1828), *Questions proposées. Théorème sur le quadrilatère complet*, Annales de math. (Gergonne) 18 (1827/1828) 302–304; see Weierstrass (1881/82), vol. 1, pp. 221–224 (with another title). *[p. 96, 220]*

J. Steiner (1828a), *Bemerkungen zu der zweiten Aufgabe in der Abhandlung No. 17. in diesem Hefte*, J. Reine Angew. Math. (Crelle) 3 (1828) 201-204; see Weierstrass (1881/82), vol. 1, pp. 163–168 (with another title). *[p. 109]*

J. Steiner (1828b), *Vorgelegte Aufgaben und Lehrsätze*, J. Reine Angew. Math. (Crelle) 3 (1828) 207–212; see Weierstrass (1881/82), vol. 1, pp. 173–180. *[p. 109, 378]*

J. Steiner (1832), *Systematische Entwickelung der Abhängigkeit geometrischer Gestalten von einander, mit Berücksichtigung der Arbeiten alter und neuer Geometer über Porismen, Projections-Methoden, Geometrie der Lage, Transversalen, Dualität und Reciprocität, etc.*, G. Fincke, Berlin 1832; see Weierstrass (1881/82), vol. 1, pp. 229–460. *[p. 332]*

J. Steiner (1835a), *Aufgaben und Lehrsätze, erstere aufzulösen, letztere zu beweisen*, J. Reine Angew. Math. (Crelle) 13 (1835) 361–363; see Weierstrass (1881/82), vol. 2, pp. 13–18. *[p. 197, 198]*

J. Steiner (1835b), *Einfache Construction der Tangente an die allgemeine Lemniscate*, J. Reine Angew. Math. (Crelle) 14 (1835) 80–82; see Weierstrass (1881/82), vol. 2, pp. 19–23. *[p. 209]*

J. Steiner (1842), *Sur le maximum et le minimum des figures dans le plan, sur la sphère et dans l'espace en général*, J. Reine Angew. Math. (Crelle) 24 (1842) 93–162 (this first part also published in *Liouville's* Journal 6 (1841) 105–170) and 189–250; German original text: see Weierstrass (1881/82), vol. 2, pp. 177–308. *[p. 193]*

J. Steiner (1844), *Elementare Lösung einer Aufgabe über das ebene und sphärische Dreieck*, J. Reine Angew. Math. (Crelle) 28 (1844) 375–379; see Weierstrass (1881/82), vol. 2, pp. 321–326. *[p. 224]*

J. Steiner (1846a), *Über eine Eigenschaft des Krümmungshalbmessers der Kegelschnitte*, J. Reine Angew. Math. (Crelle) 30 (1846) 271–272; see Weierstrass (1881/82), vol. 2, pp. 339–342. *[p. 343]*

J. Steiner (1846b), *Sätze über Curven zweiter und dritter Ordnung*, J. Reine Angew. Math. (Crelle) 32 (1846) 300–304; see Weierstrass (1881/82), vol. 2, pp. 375–380. *[p. 317]*

J. Steiner (1857), *Ueber eine besondere Curve dritter Classe (und vierten Grades)*, J. Reine Angew. Math. (Crelle, Borchardt) 53 (1857) 231-237; see Weierstrass (1881/82), vol. 2, pp. 639–647. *[p. 204, 218]*

H. Steinhaus (1958), *One Hundred Problems in Elementary Mathematics*, Orig. in Polish *Sto zadań* Wrozław 1958, Engl. transl. Basic Books 1964, Dover reprint 1979. *[p. 129, 315]*

M. Stewart (1746), *Some General Theorems of Considerable Use in the Higher Parts of Mathematics*, Sands, Murray, and Cochran, Edinburgh 1746. *[p. 90]*

I. Stewart (2008), *Professor Stewart's Cabinet of Mathematical Curiosities*, Profile Books LTD, London 2008, French. transl. Flammarion 2009. *[p. 342]*

C.-F. Sturm (1823/24), *Autre démonstration du même théorème (de géométrie)*, Annales de mathématiques (Gergonne) 14 (1823/24) 286–293; Addition à l'article: pp. 390–391. *[p. 120, 221]*

S. Tabachnikov (1993), *Poncelet's theorem and dual billiards*, L'Enseignement Mathématique 39 (1993) 189–194. *[p. 328]*

S. Tabachnikov (1995), *Billiards*, Panoramas et Synthèses 1, Soc. Math. France 1995. *[p. 232]*

N. Tartaglia (1560), *General trattato vol. 3 (La quarta parte del general trattato de numeri et misure)*, Venetia 1560. *[p. 171, 298]*

F.G. Taylor and W.L. Marr (1913), *The six trisectors of each of the angles of a triangle*, Proc. Edinburgh Math. Soc. 32 (1913) 119–131. *[p. 365]*

F.G. Teixeira (1905), *Tratado de las curvas especiales notables*, Madrid, 1905. *[p. 200]*

V. Thébault (1930), *Sur le triangle isoscèle*, Mathesis XLIV (1930) 97. *[p. 225]*

416   References

V. Thébault (1931), *Sur l'expression de l'aire du triangle en fonction des côtés*, Mathesis XLV (1931) 27–28. *[p. 181]*

V. Thébault (1938), *Problem 3887, Three circles with collinear centers*, Amer. Math. Monthly 45 (1938) 482–483. *[p. 227]*

V. Thébault (1945), *The area of a triangle as a function of the sides*, Amer. Math. Monthly 52 (1945) 508–509. *[p. 181]*

S. Thorvaldsen (2010), *Early numerical analysis in Kepler's new astronomy*, Sci. Context 23 (2010) 39–63. *[p. 143]*

D. Tournès (2009), *La construction tractionnelle des équations différentielles*, Collection sciences dans l'histoire, Blanchard, Paris 2009. *[p. 211]*

M. Troyanov (2009), *Cours de géométrie*, Presses polytechniques et universitaires romandes, Lausanne 2009. *[p. 52]*

G. Turnwald (1986), *Über eine Vermutung von Thébault*, Elem. Math. 41 (1986) 11–13. *[p. 227]*

B.L. van der Waerden (1970), *Ein Satz über räumliche Fünfecke*, Elem. Math. 25 (1970) 73–78; *Nachtrag*, Elem. Math. 27 (1972), p. 63. *[p. 280, 304]*

B.L. van der Waerden (1983), *Geometry and Algebra in Ancient Civilizations*, Springer-Verlag, Berlin 1983. *[p. 14, 17]*

D.E. Varberg (1985), *Pick's theorem revisited*, Amer. Math. Monthly 92 (1985) 584–587. *[p. 278]*

P. Ver Eecke (1921), *Les Œuvres complètes d'Archimède*, 2 vols., first ed. 1921, second ed. Vaillant-Carmanne 1960. *[p. 353, 379]*

P. Ver Eecke (1923), *Les Coniques d'Appolonius de Perge*, first ed. 1923, reprinted by Albert Blanchard, Paris 1963. *[p. 61, 72]*

P. Ver Eecke (1933), *Pappus d'Alexandrie, La collection mathématique*, 2 vols., Paris-Bruges 1933.

F. Viète (1593a), *Supplementum Geometriæ*, Tours 1593; see van Schooten (1646), *Opera Mathematica*, pp. 240–257. *[p. 162, 165]*

F. Viète (1593b), *Variorum de rebus Mathematicis Responsorum Liber VIII*, Tours 1593; see van Schooten (1646), *Opera Mathematica*, pp. 347–436. *[p. 62, 151, 232]*

F. Viète (1595), *Responsum ad Problema, quod omnibus Mathematicis totius Orbis construendum proposuit Adrianus Romanus*, Paris 1595; see van Schooten (1646), *Opera Mathematica*, pp. 305–324. *[p. 164, 165, 170]*

F. Viète (1600), *Apollonius Gallus*, Paris 1600; see van Schooten (1646), *Opera Mathematica*, pp. 325–346. *[p. 6, 45, 233]*

J. Wallis (1685), *A Treatise of Algebra, both Historical and Practical. Shewing, The Original, Progress, and Advancement thereof, from time to time; and by what Steps it hath attained to the Heighth at which now it is*, London: Printed by John Playford, for Richard Davis, Bookseller, in the University of Oxford, M.DC.LXXXV. *[p. 158]*

G. Wanner (2004), *Elementare Beweise des Satzes von Morley*, Elem. Math. 59 (2004) 144–150. *[p. 105]*

G. Wanner (2006), *The Cramer–Castillon problem and Urquhart's 'most elementary' theorem*, Elem. Math. 61 (2006) 58–64. *[p. 176]*

G. Wanner (2010), *Kepler, Newton and numerical analysis*, Acta Numerica 19 (2010) 561–598. *[p. 143]*

K. Weierstrass (1881/82), *Jacob Steiner's Gesammelte Werke*, 2 vols., herausgegeben auf Veranlassung der Königlich Preussischen Akademie der Wissenschaften, G. Reimer, Berlin 1881–82; second ed., AMS Chelsea Publishing, New York 1971.

C. Wessel (1799), *Om Directionens analytiske Betegning, et Forsøg, anvendt fornem-melig til plane og sphæriske Polygoners Opløsning*, Nye Samling af det Kongelige Danske Videnskabernes Selskabs Skrifter 5 (1799) 469–518. *[p. 242]*

C. Wilson (1968), *Kepler's derivation of the elliptic path*, Isis 59 (1968) 4–25. *[p. 143]*

Figure Sources and Copyright

Institutions having granted permissions for reproduction:

BGE   Bibliothèque de Genève
BibMa   Bibliothèque Georges de Rham, Section de mathématiques, Genève
CaLib   Cambridge University Library, Cambridge
MFO   Mathematisches Forschungsinstitut Oberwolfach, Germany

page	Figure	Instit.	Catalogue	page	Figure	Instit.	Catalogue
4	Fig. 1.2	Photo S. Favre, Dept. of Anthropology, University of Geneva					
6	Fig. 1.5	BGE	Ka 75, Ka 467, Ka 475, Ka 434, Ka 457				
12	Fig. 1.12	BGE	Y 293				
13	Fig. 1.13	Yale Babylonian Collection YBC 7289					
15	Fig. 1.16	Edisud éditions, Aix-en-Provence, Math. en Méd., 1988					
17	Fig. 1.19	Edisud éditions, Aix-en-Provence, Math. en Méd., 1988					
20	Fig. 1.23	BGE	Y 293	24	Fig. 1.30	BGE	Ka 461
25	Fig. 1.32	Umschau Verlag 1965, Frankfurt					
47	Fig. 2.27	BGE	Kb 32	50	Fig. 2.34	BGE	Ka 648
59	Fig. 2.46	BGE	Ka 461	59	Fig. 2.47	BGE	Ka 461
66	Fig. 3.5	© Richard Feynman 1965		66	Fig. 3.5	BGE	Ka 477
67	Fig. 3.6	BGE	Ka 461	68	Fig. 3.8	BGE	Ka 477
69	Fig. 3.9	BibMa	01.40/DUE(u),	BGE	Kb 32		
73	Fig. 3.13	BGE	Ka 461	74	Fig. 3.16	BGE	Ka 477
82	Fig. 4.5	BGE	Ka 461	97	Fig. 4.18	BGE	Ka 433
99	Fig. 4.19	BGE	T 1047	99	Fig. 4.19	Photo B. Kummer	
101	Fig. 4.23	BGE	Ka 434	107	Fig. 4.29	BGE	Va 900
110	Fig. 4.33	BGE	Ka 434				
112	Fig. 4.36	BibMa	01.40/DUE(u)				
114	Fig. 5.1	BGE	Kb 12, Kb 31				
115	Fig. 5.2	BGE	Ta 2291/1	115	Fig. 5.3	BGE	Kb 12
115	Fig. 5.4	BGE	Kb 467(1)	116	Fig. 5.5	BGE	Kb 12
116	Fig. 5.6	BibMa	P 43/3	124	Fig. 5.12	BGE	Fa 260
125	Fig. 5.14	Hölder-Pichler-Tempsky, Wien 1956					
126	Fig. 5.15	Klaudia Wanner, Wien	132	Fig. 5.19	BGE	Ka 473
138	Fig. 5.24	Jean-Paul Kauthen, Bern	140	Fig. 5.26	BGE	Kb 157
142	Fig. 5.28	BGE	Ka 123	143	Fig. 5.29	BGE	Kb 127
145	Fig. 5.30	CaLib	Add. 3965[6]	150	Fig. 5.34	BGE	Ka 467

A. Ostermann and G. Wanner, *Geometry by Its History*,
Undergraduate Texts in Mathematics, DOI: 10.1007/978-3-642-29163-0,
© Springer-Verlag Berlin Heidelberg 2012

page	Figure	Instit.	Catalogue	page	Figure	Instit.	Catalogue
152	Fig. 5.35	BGE	Ka 272	152	Fig. 5.35	Photo M. Borello	
154	Fig. 5.36	Kathrin Galehr-Nadler, Hohenems					
158	Fig. II.1	BGE	Ka 465, Ka 218, T 1198				
158	Fig. II.2	BGE	Ka 219	160	Fig. 6.2	BGE	Ka 143
161	Fig. 6.3	BGE	Ka 143	162	Fig. 6.4	BGE	Ka 467
163	Fig. 6.6	BGE	Ka 467	165	Fig. 6.9	BGE	Ka 467
167	Fig. 6.10	BGE	Ka 467	167	Fig. 6.11	BGE	Kb 12
168	Fig. 6.12	BGE	Kb 32	169	Fig. 6.13	BGE	Ka 467
169	Fig. 6.14	BGE	Ka467	171	Fig. 6.16	BGE	Ka 105
172	Eq. (6.24)	BGE	Ka 105	181	Fig. 6.23	BGE	Ka 190
183	Fig. 6.25	BGE	Ka 385	186	Fig. 7.1	BGE	Ka 335
197	Fig. 7.9	BGE	Rb 1***	199	Fig. 7.11	BGE	Rb 1***
203	Fig. 7.15	MFO	W 04034	207	Fig. 7.18	BGE	Rb 1
210	Fig. 7.22	BibMa	01.40/NEM(m)	213	Fig. 7.24	BGE	Rb 1
213	Fig. 7.25	BGE	Ra 247, Ka 434	233	Fig. 7.40	BGE	Ka 467
237	Fig. 7.47	MFO	W 04034	241	Drawing by Klaudia Wanner		
242	Fig. 8.1	Georg Olms Verlag, Hildesheim (Gauss, *Werke*, vol. 10)					
255	Fig. 8.10	BGE	Ka 481	255	Fig. 8.12	BGE	Rb 1
256	Fig. 8.13	BibMa	01.40/RAM				
261	Fig. 9.1	BGE	Ra 471*	262	Fig. 9.2	BGE	Zv 385/1
268	Fig. 9.10	BGE	Ka 296, Ka 305	283	Fig. 9.21	BGE	Kb 32
285	Fig. 9.25	Stift Stams, Tirol		298	Fig. 10.6	BGE	Ka 105
300	Fig. 10.7	BGE	Ra 3	308	Fig. 10.12	BGE	Kc 110
315	Fig. 10.20	CaLib	7350.d.56				
319	Fig. 11.1	BibMa	01.40/DUE(u)	320	Fig. 11.2	Stift Stams, Tirol	
322	Fig. 11.5	Bernard Gisin, Genève					
328	Fig. 11.13	BGE	Ka 316	337	Fig. 11.21	BibMa	50/121

All other figures and photos are owned and copyrighted by the authors.

Index

"The author has tried to provide as complete an index as possible."
(D.E. Knuth, *The TeXbook*, 1986, p. 457)

A. Ostermann and G. Wanner, *Geometry by Its History,*
Undergraduate Texts in Mathematics, DOI: 10.1007/978-3-642-29163-0,
© Springer-Verlag Berlin Heidelberg 2012